LINKING INDUSTRIES ACROSS THE WORLD

Linking Industries Across the World

Processes of Global Networking

Edited by

CLAES G. ALVSTAM
School of Economics and Commercial Law,
Göteborg University, Sweden

EIKE W. SCHAMP
Johann Wolfgang-Goethe University,
Frankfurt/Main, Germany

Routledge
Taylor & Francis Group
LONDON AND NEW YORK

First published 2005 by Ashgate Publishing

Reissued 2019 by Routledge
2 Park Square, Milton Park, Abingdon, Oxon, OX14 4RN
52 Vanderbilt Avenue, New York, NY 10017

Routledge is an imprint of the Taylor & Francis Group, an informa business

A Library of Congress record exists under LC control number:

ISBN 13: 978-0-8153-9024-4 (hbk)
ISBN 13: 978-1-351-15392-8 (ebk)

Contents

List of Figures

List of Tables

List of Contributors

Alvstam, Claes, G is Professor in Economic Geography at the Department of Human and Economic Geography, School of Economics and Commercial Law, Gothenburg, Sweden, Claes.alvstam@geography.gu.se.

Braun, Boris is Professor at the Department of Geography, Otto-Friedrich-University Bamberg, Bamberg, Germany, boris.braun@ggeo.uni-bamberg.de.

Buzás, Norbert is Associate Professor of Innovation Management at the Department of Regional and Applied Economics, University of Szeged, Hungary, buzas@eco.u-szeged.hu.

Domański, Boleslaw is Professor at the Department of Regional Development, Jagiellonian University, Cracow, Poland, b.domanski@geo.uj.edu.pl.

Eraydin, Ayda is Professor of Urban and Regional Planning at the Department of Urban and Regional Planning, Faculty of Architecture, Middle East Technical University, Ankara, Turkey, ayda@metu.edu.tr.

Fuchs, Martina is Professor of Geography at the Department of Social and Economic Geography, University of Cologne, Cologne, Germany, fuchs@wiso.uni-koeln.de.

Hsu, Jinn-yuh is Associate Professor at the Department of Geography, National Taiwan University, Taipei, Taiwan, jinnyuh@ntu.edu.tw.

Mather, Charles is Senior Lecturer at the School of Geography, Archaeology and Environmental Studies, University of the Witwatersrand, Johannesburg, South Africa, matherc@geoarc.wits.ac.za.

Park, Sam Ock is Professor at the Department of Geography and Dean of College of Social Sciences, Seoul National University, Seoul, Korea, parkso@snu.ac.kr.

Rogerson, Christian M. is Professor of Human Geography at the School of Geography, Archaeology and Environmental Studies, University of the Witwatersrand, Johannesburg, South Africa, rogersonc@geoarc.wits.ac.za.

Schamp, Eike, W. is Professor in Economic Geography at the Department of Economic and Social Geography, Johann Wolfgang-Goethe University, Frankfurt am Main, Germany, schamp@em.uni-frankfurt.de

Stryjakiewicz, Tadeusz is Professor of Economic Geography at the Institute of Socio-Economic Geography and Spatial Management, Adam Mickiewicz University, Poznań, Poland, tadek@amu.edu.pl.

Tong, Xin is Assistant Professor at the Department of Urban and regional Planning, College of Environmental Sciences, Peking University, tongxin@urban.pku.edu.cn.

Wang, Jici is Professor of Economic Geography at the Department of Urban and Regional Planning, Peking University, Beijing, China, wjc@pku.edu.cn.

Wong, Shuang Yann is an Associate Professor in Humanities and Social StudiesEducation, National Institute of Education, Nanyang Technological University, Singapore, sywong@nie.edu.sg.

Introduction

Linking Industries Across the World – New Challenges to the World Economy

Claes G. Alvstam and Eike W. Schamp

Globalisation is inside everything – globalisation seems to work at all times and in all places, to penetrate all spheres of economic life and to determine both the worlds of practice and scientific discourse. For many it is obviously *the* challenge to economic development, coming for the individual enterprise as the need for restructuring under increasing competitive squeeze, for sectors as acceleration of ageing in their life cycle and/or growth perspective in new technologies, and for regions – and their population – as a threat or promise to their welfare. There is, of course, a broad scientific discourse on the various aspects of current globalisation. Nothing seems to be really new, however, when globalisation is merely seen as a process of increasing economic exchange between nations and regions. This perspective would open queries about the kind of globalisation in earlier historical periods, in particular the 19th century (Bordo et al., 2003). Most scholars would now agree, however, that something was different in the 20th century. We would argue in this book that globalisation always differs from time to time and from place to place, that it is continuously changing its face and that new ways are emerging for industries to link to each other across the world and across the regions.

More than two decades ago there was a lively debate about the emerging New International Division of Labour (NIDL) (Fröbel et al., 1980). One of the driving forces, it was realised, was the new ability of firms to separate labour intensive from capital intensive processes in sectors like the textile and clothing industries. As a consequence, firms were able to relocate labour intensive production to locations that were characterised by both low wages and a sufficient supply of qualified labour. The emerging 'globalisation' seemed to be driven by cost minimising strategies of firms. In fact, it required a host of accompanying innovations, e.g. in transport technology, resulting in a continuing decline of transportation costs and in process organisation, resulting in new routines. It was based on a diversity of changes in the regulatory environment of the firms, such as the implementation of new standards, regulations for re-imports to the country of origin, foreign investment protection laws in host countries and a new philosophy in the macro-economic policies of developing countries, i.e. export-promoting strategies. In an *ex-post* perspective it could be argued that this laid the base for learning processes among all kinds of economic and political actors, propelling what some scholars preferred to term the roller-coaster of current globalisation (cf. Dicken, 2003).

In the tradition of Fröbel, Heinrich and Kreye globalisation is often looked upon from a macro-perspective as a coherent aggregate process, either as a prerequisite for increased global welfare (Gilpin, 2001; Bhagwati, 2004; Wolf, 2004), or as a threat and a restraint towards increasing equality between and within countries (Bello, 2002; Abrahamsson, 2003). We argue that the basic determinants of the degree of globalisation are to be found at the level of the decision-maker(s) within an individual company, who are struggling to understand, to grasp and to digest the vast flow of external information that wells forth and that requires an ever more rapid pace of decision-making with the emergence of new information and communication technology. In this sense globalisation becomes an 'umbrella' concept, covering and containing a multitude of contradictory and among themselves unstructured events.

While the NIDL in terms of physical manufacturing has been extensively described, analysed and debated since the study by Fröbel et al., an entirely new reality has been created by the emergence of new information and communication technology. The revelation that services are an independent element in the value-added process of manufacturing has altered the scope and contents of the globalisation discourse. From having been looked upon as a uniform entity comprising all kinds of non-physical production, services have finally become subject to the same detailed and multi-purpose oriented classification and systematisation attempts that had hitherto been the privileged realm of the physical products. In particular, the necessity of creating a common taxonomical platform for service production within the framework of the General Agreement of Trade in Services (GATS) played a crucial role in the 'visibilisation' of the role of services in the understanding of the spatial dynamics of global industries.

One specific element that has attracted particular attention in recent years has been the phenomenon of 'outsourcing', i.e. the sub-contracting of manufacturing or services from one company to another, thus normally expanding the corporate space. Misleadingly, it has also been applied to the more specific trend towards importing services from low-cost providers located offshore (*Financial Times* April 15, 2004). In this context it is important to make a clear distinction between the initial relatively straightforward and technologically easily accessible forms of outsourcing, and the more sophisticated varieties emerging today (ibid.). The new forms of outsourcing that link industries across the world are much more complex and closely interactive processes, in which the principal company brings together and works 'side-by-side' with various service providers scattered all over the world within a common 'e-room'. Further technological improvement and increased efficiency of human interaction, regardless of geographical distance, imply a huge potential for an entirely new logic in the spatial organisation of manufacturing, and, thus, they contribute to forming entirely new patterns of production, trade and investment. One growing field of outsourcing, complementing the traditional areas of call centres and back-office processing centres, is purchasing, in which, according to a recent survey by the Aberdeen Group (cit. in *Financial Times* April 15, 2004), the per centage of US companies outsourcing procurement services has grown to nearly 50 per cent in travel services, 40 per cent in printing services, 35 per cent in contract labour services and 30 per cent in IT equipment and office equipment services. The latest area of outsourcing comprises human resource functions. About three-quarters of large European and US companies, according to the US-based research group Conference Board (cit. in *Financial Times*, ibid.), report that they outsource at least one important HR function, such as pensions management, health benefit schemes, stock and options administration, payrolls and employee hotlines. It is important to keep in mind that although the cost factor is a significant aspect in the spatial dispersal of value-added production, there is a wide variety of other factors that matter in the decision to commence outsourcing, e.g. smoothing out the administrative peaks on a daily or a seasonal basis and thus freeing up resources for more productive work within the company. Apart from the creation of large-scale companies specialised in manufacturing outsourcing, like the Singapore-based Flextronics, or outsourcing of business processes, like Indian Infosys, there is also a growing trend towards the creation of 'hyper-footloose' micro companies, often comprising only one person, offering a wide range of professional business services to a network of customers in the old industrial core at a long distance. Not infrequently they are located close to agreeable tourist spots along the Mediterranean, in the West Indies or in Southeast Asia, but also in 'creative' and dynamic environments in 'cool' cities, e.g. New York, Barcelona and Singapore (Alvstam and Jönsson, 2000). The paradoxical tendency towards the clustering of this type of firms, despite the obvious advantages of having the liberty to choose any location, and accordingly to contribute to a dispersed pattern of service industries, may be explained by the occurrence of locally produced 'tacit knowledge', and what Meric Gertler (1995; 2003) has coined 'the undefinable tacitness of being (there)'. The vivid 'death of distance', or 'death of geography' discourse of the 1990s, stimulated by contributions from a wide variety of academic scholars, journalists and politicians (see e.g. Ohmae, 1990; Martin, 1996; Cairncross, 1997), has more recently been given a far more nuanced treatment that elucidates the complexity of the concept of geographical proximity (Oinas, 2000), reveals the relationship between geographical space and cyberspace (Dodge & Kitchin, 2001) and points out the previous over-estimation of the distance-destroying capacity of ICT due to the conflation of *spatial* reach with *social* depth (Morgan, 2004). Morgan argues, drawing on evolutionary theories of innovation and recent corporate practice that firms continue to set a high premium on physical proximity in the organisational design of complex activities (ibid. p. 4).

The issue of how the constant transformation of the exogenous prerequisites behind the development of the division of labour changes the spatial pattern of production and consumption continues to attract other disciplines within social science as well. Two labour economists, Frank Levy and Richard Murnane, recently published *The New Division of Labour* (Levy & Murnane, 2004), in which they argue that it is computerisation, not outsourcing and off-shoring, that is wiping out what they describe as 'moderately skilled' labour in the developed world. In this context, all work tasks that rely on rules-based logic and that can be computerised are subject to outsourcing and off-shoring, be they based in manufacturing or in services. Computers substitute for labour in some occupations but complement human work in others. Accordingly, Levy & Murnane argue that while computers are good at rules-based logic, they lack the mental models and experiences that help humans navigate through an unpredictable world. 'The result is a picture in which the number of menial jobs is growing but the general shift of occupations is towards higher-end jobs' (ibid.).

The institutional factors that regulate the process of linking industries across the world should not be overlooked either. The continuous liberalisation and deregulation of tariff- and non-tariff barriers to trade within the multilateral GATT/WTO framework, as well as in its regional forms, such as the EU single internal market, have contributed to speeding up the spatial dispersal of industrial activities, just as the demand from the business world to enhance productivity and profits has pushed the liberalisation process across national boundaries. At the same time, specific national regulations and requirements have constrained industrial enterprises in their internal strategic decision-making processes involving exports, foreign direct investment and the anatomy of their global production network at large. The institutionalist turn in economic geography (Barnes, 1999; Martin, 2000) has opened up new opportunities to describe and explain both constraints and incentives provided by institutional actors at different geographical levels to curb or endorse technological innovations, e.g. within the field of information and communication technology. Roger Hayter (2004) has pleaded for a dissented institutionalism, which recognises the importance of political and social processes without eliminating the economic from consideration (ibid. p. 96). Dissenting institutionalism, in Hayter's terms, evolves around the three interlocking principles of embeddedness, evolution and difference, in which embeddedness should be viewed as a symbiosis of economic and non-economic factors. The evolutionary principle argues that changes in this symbiosis over time are path dependent, crisis ridden over business cycles, and transformative. Finally the difference principle states that the evolution of embedded social economies is geographically (and socially) segmented (ibid. pp. 97-98).

In this discourse, geographers have been discussing globalisation as a powerful driver of regional development for nearly two decades. Without any doubt, their work has increasingly contributed to a better understanding of what globalisation really means and what chances and challenges are emerging from globalisation for regions – in a global economy that is perceived as constantly being driven by increasing 'global' competition. We claim, however, that the political, institutional and technical environment that once gave rise to the concept of the New International Division of Labour has changed fundamentally since then. At first instance, there was the fall of the wall – a metaphor for the radical change in the global political system. It is now common understanding that the collapse of the socialist system in the Soviet Union caused the breakdown of the bi-polar world system that had prevailed for over 50 years and paved the way for a new political-economic world order, which is currently emerging. Many countries, regions and (new) firms were newly integrated into the competitive structures of the world economy. As a result, they were forced to undertake basic internal restructuring – a process often called transformation or transition – while simultaneously firms and regions in the 'old' capitalist world faced a changing environment and had to adapt and restructure too.

Another field of radical change in the near past is that of basic information and communication technologies, which provided new means for a new kind and reach of exchanges in information, knowledge and capital. The digital revolution did not arrive in large parts of the world until, at the end of the 1980s and the beginning of the 1990s, tremendous investments went into a new, fibreglass based telecommunication infrastructure that encompassed the world. This process was triggered by deregulation and liberalisation tendencies in the US first, and later in a host of other countries. Often, it was not realised that the new fibreglass technology, the construction of worldwide communication networks and satellite technology were necessary preconditions for the time-space compression that we now understand as a basis of globalisation (see, Hudson, 1997). It was only during the 1990s that the Internet, Intranets, e-commerce and further services of the electronic age gained in global importance (Roche and Blaine, 2000, p. 68).

Furthermore, scholars of macroeconomics and other disciplines increasingly identified a new kind of organisation of production processes in global production patterns, which they called 'fragmentation' (Arndt and Kierzkowski, 2001). Fragmentation, as the further deepening of the division of labour between firms and locations, is a proverbial geographical process. It is closely tied to efficient new service links that are subject to increasing returns to scale, such as telecommunication or transportation services. This has been mentioned above. It is, however, similarly related to changing strategies and structures within the 'classical' multinational enterprise and between firms. Curzon Price, for example, hypothesises that in the new fragmentation processes 'the giant, unfocussed MNCs' will weaken in favour of more numerous, smaller and highly specialised firms (Curzon Price, 2001, p. 106). In any case, the dominating organisational form in the international context, the multinational or transnational firm, faced new challenges to cope with increasing competition, a new volatility of markets, new requirements for flexibility, and changing conditions for the control of resources. This is extensively debated in the organisational and management science literature (e.g., Bartlett and Ghoshal, 1989, Morgan et al., 2001), which discusses the emergence of less hierarchical and more decentralised forms of management in and between firms across different countries. Neither the 'traditional' hierarchical organisation in multinational firms nor arms-length exchange, the simple coordination by the 'market', seem to grasp the widening linkages between value chains that extend worldwide, both in the global branch plant systems within a firm and between individual firms. Rather, they imply new forms of global networking and new ways of 'governing' value chains. Scholars from different disciplines have developed new systemic concepts. For instance, Gereffi and Korzeniewicz (1994) introduced the idea of different forms of global commodity chains – the producer driven versus the buyer driven commodity chain – though this was later criticised for its dualistic simplicity. The concept stimulated a lively debate; however, on related concepts such as the global value chain (Gereffi et al., 2003) and the global production network (Henderson et al., 2002). Other concepts for the industrial organisation of value chains are emerging, such as 'global network flagships' – some multinationals that mutate into foci that govern global production networks (e.g., Ernst and Kim, 2002) or newly emerging 'modular production networks', which are formed around lead firms of product development, marketing and distribution and 'turn-key suppliers' in large-scale manufacturing (Sturgeon, 2003). It seems obvious that some fundamental changes in the global interrelationship of industries are taking place, which are still rather unclear. More information about the new 'reality' is required, on the one hand, but more so new interpretations and new ways of thinking, on the other.

In sum, we believe that global linkages between industries and regions still are – if not more than ever before – an important issue in social and geographical research. In recent decades, scholars in different sciences have extensively discussed the internationalisation of the enterprise, while geographers have analysed the impact of transnational firms on regional development in different local environments. If we consider the new conditions in the global economy, as mentioned above, we conclude that understanding the global system has not become easier. We claim, instead, that its complexity has increased; we further claim that the linkages that have emerged are not yet really understood, either in their diversity or in their effects. Increasingly, new forms of networking make up new forms of 'embedding' of firms and regions into the global system. Networks have a major and increasing impact on regional development. Global competition and spatial (global) division of labour result ever less in converging regional development; in contrast, very probably, they intensify regional differentiation and regional disparities. We see this aspect of increasing interrelationship at different regional levels as the main element of globalisation. This is what we want to highlight in this book. For various reasons, we do not pursue a systematic textbook approach to regional development that is embedded in global networks. Though some may call it eclectic, we want to discuss new aspects of global networking in a geographical context, where they seem to be important and where the 'new' forms of networking are more visible. This applies more to nations and regions that were integrated into the global economy during the 1990s, in a new manner compared to the well-known core countries of the global triad (e.g., Dicken, 2003). In any case, the new forms of global networking deserve further attention among geographers, wherever they do their empirical work.

The Themes of the Book

The book is divided into four parts that attempt to analyse the interplay of global networks and regional structures from different perspectives. The *first* part deals with two issues that foster the growth of global networks, describe a new quality of global networking and, finally, have an important impact on regional structures and regional development. Everybody will agree that the time and costs of global transportation were drastically reduced during the 19th century and that this decline continued all throughout the 20th century (Kaukiainen, 2001). Transportation costs nowadays are only a minor part of average costs. As a result, other costs such as labour costs gain in importance for investment, procurement or locational decisions. Transaction costs have become crucial in decisions on the range and manner of global networks. Without exaggeration, we may claim that the radical decline of transaction costs in recent times has created new opportunities in the global networking of industries and regions. This is partly due to the new communication technologies mentioned above. Moreover, it has to be stressed that new institutions and new forms of organisation have considerably reduced the transaction costs in global exchange. Since they often create the opportunities for the global exchange of information and commodities, they represent a crucial factor for the increasing competition and increasing global division of labour. Both political and private standards are among those institutions that reduce transaction costs. We consider the formation of new standards important, in accordance with the literature on long-term technological development, where standards are part of the institutionalisation of a new techno-economic paradigm (Freeman and Perez, 1988; Tate, 2001). In chapter 1, *Boris Braun* introduces the reader to a 'new generation of comprehensive management standards', i.e. the 'management system metastandards' which, he claims, form a new base for the global networking of firms. Though these standards are voluntary, they are crucial for the global extension of supply chains and are becoming particularly important since they are generic, not specific (Nadvi and Wältring, 2002). Chapter 1 reminds us, however, that the diffusion of new standards is a fragmented, heterogeneous process that reproduces the uneven development of the global economy. Braun's findings intersect with conclusions by other scholars, who still expect the dominance of national variants of standards in the global economy (Tate, 2001). In any case, standards have a major impact on the geographies of global networks by bringing about new contexts of inclusion and exclusion.

In a different manner, *Charles Mather* picks up the issue of fragmentation of the global economy in chapter 2. He focuses on the changing structures of regional production – of citrus fruits in South Africa – in the course of the reorganisation of the global commodity chain. Currently, this is one of the most challenging concepts for a better understanding of global (vertical) networks. In a sympathetic critique, various authors have further developed the concept in recent times (e.g., Henderson et al., 2000, Raikes et al, 2000; Gereffi, 2002). Its perspective is a systems approach that combines regions of consumption and regions of production through the concept of the value chain. Its main dimensions of 'governance' are the configuration and the co-ordination of the value chain with regard to its territoriality (Gereffi and Korzeniewicz, 1994). Generally speaking, Gereffi differentiates between a traditional 'producer driven' value chain and a more recent 'buyer driven' value chain in the global context. In current empirical research, these types seem to be largely bound to different sectors of production. However, Charles Mather further develops this idealised dichotomy into a sequential model by demonstrating how a producer driven chain shifts into a buyer driven one. He analyses two different but intersecting processes of fragmentation, i.e., one at the organisational level where new agents enter the chain while others lose their position, and one at the regional level where – as a result of organisational fragmentation – some are increasingly excluded from the chain. Although the chains differ fundamentally with respect to their governance, capitalism still shows its well-known face, that of producing uneven regional development.

The *second part* of the book turns its attention to the most powerful driver of contemporary global competition, i.e., innovation, and the knowledge networks that are necessary for the generation of innovation. Geographers as well as scholars from other different social sciences have extensively discussed the local conditions that enable knowledge generation and innovativeness. One of their most renowned concepts is the regional innovation system. However, systemic concepts tend to overemphasise a system's internal processes at the expense of its environment, as *Ayda Eraydin* argues in chapter 3. She claims that the necessity to link the regional innovation system with 'the world' often remains underestimated and unresearched. Consequently, she offers a sophisticated typology of networks and clusters from which a variety of possible regional systems and their global 'embeddedness' emerge. How far we are still in a phase of trial in our attempt to better understand complex linkages in systemic relationships and to differentiate between them at the same time will become clear in the following chapter 4. Here, *Sam Ock Park* starts from a similar perspective and yet chooses another approach in order to develop, first, a typology of regional innovation clusters from a larger set of characteristics. In a subsequent systematic classification thereafter, he stresses different dimensions of relationships that prevail between the types of regional clusters developed before. These are very useful considerations that help to further develop the issue, in particular in times when we do not yet possess a general theory on the subject.

Implementing new technologies in the new world of the so-called knowledge economy imposes new challenges to the stimulation and organisation of co-operation between state and para-state intermediaries and small and medium-sized companies. Starting from the shift from a linear model of technology transfer to a network model, in chapter 5 *Norbert Buzás* offers a closer description of R&D networks with two examples from Finland and Hungary. The author wishes to demonstrate that there is no one best way to organise 'knowledge sharing networks'. Instead, these are contingent on the societal environment, even if both countries faced considerable transformation of their economies during the 1990s.

The *third part* of the book brings to the fore the international enterprise and its diverse ways of adapting to the host country and learning in the host environment. The management literature offers much about the current diversity of strategies for going abroad, as was demonstrated by Bartlett and Ghoshal (1989) long ago. Additionally, the national business systems approach emphasises the basic differences between national host environments (Whitley, 1999; 2000). International enterprises, then, seem to be embedded into social practices and institutional arrangements that combine their general routines – which have mostly been 'learned' in the home country – with the particular 'local' environment of the host country. We are far from knowing how these diversities fit together in a proper way. This holds particularly true when the shifting firm structure in international linkages is recognised. For a long time internationalisation and networking has no longer been the privilege of large transnational companies alone. Small and medium-sized enterprises increasingly form part of the global division of labour, either through their own branch plants or through diverse forms of partnership.

Bridging the distance – whether geographical or 'cultural' – is only one side of the coin. The problem of 'managing knowledge across borders' in more or less decentralised networks has been recently tackled by taking a closer look at the role of the expatriate manager (Beaverstock, 2004) or that of a transnational technical community that evolved from earlier immigration to high technology regions and re-migration to the home country (e.g., Saxenian, 2002). Here, in part three, the other side, i.e. the adapting and learning 'behaviour' of the managers of the branch plant in their new regional and national environment, is discussed. The following chapters attempt to understand the formation of networks as a process that emerges from a variety of practices of different agents and in which agents need time to learn.

In the more decentralised 'network' organisation of many multinationals of today, local managers become agents in their own right, who play a role in the process of knowledge creation and learning. In chapter 6, *Martina Fuchs* points out the specific understandings of the local managers as a driver of the diversity of internationalisation forms in small multinationals. Where to locate an R&D lab in the branch plant pattern of a small multinational, then, is co-determined by the specific 'shared visions' among the local managers of the firm, who may respond to chances and challenges of the 'local' business environment.

When the decision has been made to invest in a foreign country, the enterprise has to develop various practices in order to establish functioning relations to relevant actors within the host country. This kind of creating 'embeddedness' in a foreign environment is the argument put forward in chapter 7 by *Boleslaw Domański*. He displays a diversity of networks that have to be established: networks such as with the state and different state organisations, trade unions, linked enterprises on the supply side and demand side, and, last but not least, with the representatives of the 'local' society. These are all different fields where different practices have to be learned by the foreign firm. Creating this kind of embeddedness and learning is not an easy task, since both the environment and the foreign firm change their positions over time. Finally, however, Boleslaw Domański sees the successful foreign enterprise as increasingly linked within the host country through learning. The successful establishment of a foreign firm reflects the learned capability to cope with conflicts in the host country rather than a harmonious state of social relations.

The requirements for 'learning the ropes', as Boleslaw Domański puts it, are based on the common understanding that societies are different. The notion of a national business system refers to differences in institutional arrangements. This is indeed the starting point of the next chapter, in which *Jinn-yuh Hsu* sheds some light on the question as to how far foreign enterprises have to adapt to the institutional environment of their host country. In the literature there are two contrasting views, one referring to 'simple' transfers of 'best practice' from the foreign country to the host country, the other referring to hybridisation, i.e. the necessity to adapt foreign practices to the host country. As argued in chapter 8, this dichotomy is far too simple. Jinn-yuh Hsu calls for a context specific analysis on 'how did the organisational reshuffling process exactly happen'. He analyses Taiwanese investments in Mainland China, or more precisely, Taiwanese informatics industrial firms that have located in the Yangtze River Delta. One of the dimensions of this specific context often referred to is *guanxi*, but it will be shown that *guanxi* applies in very different situations of 'embeddedness'. Obviously, the firms made use of *guanxi* in the first phase to transfer the 'regional embeddedness' they had at home, while in a later phase, when they had been allowed to sell in China, it was helpful for local adaptive processes. As Hsu puts it, this is still a fragile situation today that may even end in a return to the transfer model of the early phase.

There is a second aspect put forward in this part of the book. While the preceding chapters introduce time as a crucial dimension into the analysis of networking and emphasise the importance of the changing context for entrepreneurial action as well as of learning processes, it is rather obvious that we could expect simultaneity of different forms of networks. Furthermore, in the current debate in social sciences we are all too readily willing to judge network formation as a positive factor stimulating regional growth and development. The following chapter 9 presents – from the territorial perspective of Poland – a diversity of networks that exist simultaneously. In chapter 9 *Tadeusz Stryjakiewicz* points out the lack of network structures during the socialist command economy. His main concern is whether the networks that emerged during the simultaneous process of transition and integration into a global spatial division of labour could be seen as 'disastrous' or 'beneficial' to development. Chapter 9 offers five types of current networks in Poland, along the dimensions of old vs. new and policy-driven vs. market driven. In assessing their impact, a confusing puzzle arises where neither the old nor the policy-driven networks per se are 'bad'. In fact, Stryjakiewicz comes to the conclusion that it is the context that determines the effects of networks, and there are always ambiguities in political processes that decide which networks will make their way.

The *final part* of the book is devoted to the formation of 'local' business systems in the global economy through multifarious ways of networking. The chapters go beyond the micro-perspective on entrepreneurial interaction in the formation of networks. They re-integrate the sector's and network's development into a broader societal context. If not explicitly, the contributions implicitly take a systemic perspective that has recently been widely discussed in the social sciences, e.g., by the national business systems approach (Whitley, 2000) or in the ongoing debate on convergence or divergence of different forms of national capitalisms (e.g. Hall and Soskice, 2001). This makes clear that the nation-state still matters in globalisation, not passively as a territorial platform on which different actors construct their networks according to their goals, but actively in still setting more or less rigid rules of the game according to the dominant philosophies of the political system. We consider this a major aspect that improves our understanding of the apparent contradiction between increasing globalisation and the simultaneous new regionalisation of the world economy.

In chapter 10, *Jici Wang* and *Xin Tong* claim that recent industrial clustering in China has taken a unique development path. Given the youth of the clusters the 'government's' policies have an elsewhere unknown impact on the structure of these clusters. Moreover, the clusters seem to be frequently embedded in external rather than internal networks. The authors' analysis of four cases reveals a 'dual structure' on two dimensions, i.e. export-orientation against domestic production and state-owned against privately owned firms. In their short history, clusters have become subject to changing network relations that sometimes foster and at other times erode local networking. In pointing to local 'embedding' and 'disembodying' processes, the authors make a clear statement in favour of local networks as the basis for local development.

In chapter 11, *Shuang Yann Wong* takes an evolutionary perspective on the formation of a Chinese banking sector in Singapore, based on the emergence of ethnic networks in a given political territory – the state of Singapore. Shuang-Yann Wong offers a broader argument on the emergence of a particular business sector. In fact, chapter 11 analyses in considerable depth how far the emergence of ethnic networks (that gave rise to a new business community) was embedded in the broader political-economic history of the territory. In a clear evolutionary perspective, her argument refers to the meaning of past events that enabled the formation of (ethnic) networks and laid the ground for the current role of these networks in the banking sector. The basic context was obviously the shift from a liberal state philosophy during colonial times to the well-known developmental state philosophy after independence.

The final chapter, chapter 12, takes up the concept of global value chains that were introduced in the first part of the book. *Christian M. Rogerson* wants to show that the emergence of regional networks has become one of the major responses to 'global changes'. He applies a political economic approach to the analysis of the clothing industry, an exemplary and 'classical' research subject in social research on the global economy. This chapter highlights the complex intersection of different kinds of 'changes' in the global economy as a driving force for 'local' response. On the one hand, the emergence of large and powerful retailers in the major import countries has been changing the structure and governance of the sector and thus affects the export regions. On the other, a more radical change of the territorial political system is re-integrating the Republic of South Africa into the global economic system. This change consists of a shift from the Apartheid State to a democratic government and, as a consequence, from an autarkic to a liberal economic policy. In coping with the increasing competitive squeeze on the sector and the resulting segmentation of the sector's subgroups and regions, both firms and governments struggle to establish regional production networks in order to improve the sector's global competitiveness.

To conclude, networking in various forms, between a host of agents and with different geographical reach is the major issue of *Linking Industries Across the World*, and both authors and editors are fully aware that networking matters more than ever before in the 21st century. The four parts of the book offer different perspectives of the following dimensions, which determine global networking processes and their specific context in place and time: institution, innovation, organisation and the mix of regional outcomes. These form an entangled reality of the global world, which the chapters in the book try to uncover. Some might conclude from the diversity of aspects presented in the chapters that the world is a kaleidoscope of how industries are linking together. Indeed, we strictly believe that context matters for understanding the global economy, and that we could much learn from contexts, which are not so familiar in the generally Europe- and US-centred world. The traditional perspective of geographers of place and time specificity in economic processes still applies even in a world of global value chains, global diffusion of standards and institutions, the global reach of TNCs and the global adoption of management practices.

Acknowledgement

The editors very much appreciate the assistance of Kathrine Dege in substantially improving grammar and style of many chapters, and of Ramsin Yakob in his role as an editorial coordinator of this work.

References

Abrahamsson, H. (2003) *Understanding World Order and Structural Change - Poverty, Conflict and the Global Arena.* Basingstoke: Palgrave.

Alvstam, C.G. and Jönsson. A. (2000), On Promoting Regional Economic Growth in Industrial Peripheries: Hyper-Footloose Business Services - The Case of Swedish teleworkers in Spain. *NETCOM,* Vol. 14, Nos 1-2. Montpellier; GEMS - UMR ESPACE. pp. 25-52.

Arndt, S.W. and Kierzkowski, H. (2001) *Fragmentation. New Production Patterns in the World Economy.* Oxford: Oxford University Press.

Barnes, T.J. (1999) 'Industrial geography, institutional economics and Innis'. In: Barnes, T.J. and Gertler, M. (eds.) *The New Industrial Geography: Regions, Regulations and Institutions.* London: Routledge, pp. 1-22.

Bartlett, Ch. A. and Ghoshal, S. (1989) *Managing Across Borders. The Transnational Solution.* Boston: Harvard Business School Press.

Beaverstock, J. (2004) Managing across borders: knowledge management and expatriation in professional service legal firms. *Journal of Economic Geography,* Vol. 4, pp. 157-179.

Bello, W. (2002) *Deglobalization - ideas for a new world economy.* London: Zed Books.

Bhagwati, J. (2004) *In Defense of Globalisation.* New York: Wiley.

Bordo, M., Taylor, A.M. and Williamson, J.G. (2003) *Globalization in Historical Perspective.* Chicago: NBER, University of Chicago Press.

Braczyk, H.J., Cooke, Ph. and Heidenreich, M. (eds) (1998) *Regional Innovation Systems.* London: UCL Press.

Cairncross, F. (1997) *The Death of Distance.* Boston: HBS Press.

Curzon Price, V. (2001) Some Causes and Consequences of Fragmentation, in S. W. Arndt and H. Kierzkowski (eds), *Fragmentation. New Production Patterns in the World Economy.* Oxford: Oxford University Press, pp. 88-107.

Dicken, P. (2003) *Global Shift: 4th* edition, *Reshaping the Global Economic Map in the 21st Century.* London: SAGE Publications.

Dodge, M and Kitchin, R., (2001) *Mapping Cyberspace.* London: Routledge.

Ernst, D. and Kim, L. (2002) Global production networks, knowledge diffusion, and local capability formation. *Research Policy,* Vol. 31, pp. 1417-1429.

Financial Times (2004) Services on the assembly line: new technology brings the methods of Henry Ford to offices around the globe. 15 April.

Freeman, C. and Perez, C. (1988) in G.Dosi et al. (eds) *Technical Change and Economic Theory.* London: Pinter, pp. 38-66.

Fröbel, F., Heinrichs, J. and Kreye, O. (1980) *The New International Division of Labour.* Cambridge: Cambridge University Press.

Gereffi, G. and Korzeniewicz, M. (1994) *Commodity Chains and Global Capitalism.* Westport, Conn.: Praeger.

Gereffi, G., Humphrey, J. and Sturgeon, Th. (2003) *The Governance of Global Value chains: An Analytic Framework.* Review of International Political Economy. (forthcoming).

Gertler, M.S. (1995) 'Being there': proximity, organization and culture in the development and adoption of advanced manufacturing technologies. *Economic Geography,* Vol. 71, No. 1, pp. 1-26.

Gertler, M.S. (2003) Tacit knowledge and the economic geography of context, or The undefinable tacitness of being (there). *Journal of Economic Geography*, Vol. 3, No. 1, pp. 75-99.

Gibbons, M. et al. (1994) *The new production of knowledge. The dynamics of science and research in contemporary societies.* London: Sage.

Gilpin, R. (2001) *Global Political Economy.* Princeton N.J.: Princeton University Press.

Hall, P.A. and D. Soskice (eds) (2001) *Varieties of Capitalism. The Institutional Foundations of Comparative Advantage.* Oxford: Oxford University Press.

Hayter, R. (2004) Economic geography as dissenting institutionalism: The embeddedness, evolution and differentiation of regions. *Geografiska Annaler*, 86B, No. 2, pp. 95-115.

Henderson, J. et al. (2002) Global production networks and the analysis of economic development, in *Review of International Political Economy,* Vol. 9, pp. 436-464.

Hudson, H.E., (1997) Global Connections. *International Telecommunications Infrastructure and Policy.* New York: Wiley John & Sons.

Kaukiainen, Y. (2001) Shrinking the world: Improvements in the speed of information transmission, c. 1820-1870, in *European Review of Economic History* X, pp. 1-28.

Levy, F. and R. Murnane (2004) *The New Division of Labour - How computers are creating the next job market.* Princeton N.J.: Princeton University Press.

Martin, P. (1996) The Death of Geography. *Financial Times*, 22 February.

Martin, R. (2000) Institutional approaches in economic geography. In: Sheppard, E. and Barnes, T.J., (eds.), *A Companion to Economic Geography.* Oxford: Blackwell, pp. 77-94.

Morgan, G., Kristensen, P. H. and Whitley, R. (eds) (2001) *The Multinational Firm: Organizing Across Institutional and National Divides.* Oxford: Oxford University Press.

Morgan, K. (2004) The exaggerated death of geography; learning, proximity and territorial innovation systems. *Journal of Economic Geography,* Vol. 4, No. 1, pp. 3-21.

Nadvi K. and Wältring F. (2002) Making Sense of Global Standards. INEF Report 58, http://inef.uni-duisburg.de/page/englisch/documents/report58.pdf.

Nelson, R. (ed) (1993) *National Innovation Systems. A Comparative Analysis.* Oxford: Oxford University Press.

Ohmae, K. (1990) *The Borderless World.* New York: Harper.

Oinas, P. (2000) Distance and learning: Does proximity matter? In: F. Boekema et al. (eds.) *Knowledge, Innovation and Economic Growth.* Cheltenham: Edward Elgar.

Raikes, Ph., Jensen, M. F. and Ponte, St. (2000) Global commodity chain analysis and the French filière approach: comparison and critique. *Economy and Society,* Vol. 29, pp. 390-417.

Roche, E.M. and Blaine, M.J. (2000) Telecommunications and Governance in Multinational Enterprises, in *Information Tectonics. Space, Place and Technology in an Electronic Age*, (eds) M.I. Wilson and K.E. Corey, Chichester, John Wiley, pp. 65-88.

Saxenian, A. (2002) Transnational Communities and the Evolution of Global Production Networks: The Cases of Taiwan, China and India. *Industry and Innovation, Vol.* 9, pp. 183-202.

Smith, A., Rainnie, A., Dunford, M., Hardy, J., Hudson, R. and Sadler, D. (2002) Network of values, commodities and regions: reworking divisions of labour in macro-regional economies. *Progress in Human Geography*, Vol. 26, pp. 41-63.

Sturgeon, Th. (2003) What really goes on in Silicon Valley? Spatial clustering and dispersal in modular production networks. *Journal of Economic Geography, Vol.* 3, pp. 199-225.

Tate, J. (2001) National Varieties of Standardization, in P. A. Hall and D. Soskice (eds) *Varieties of Capitalism. The Institutional Foundations of Comparative Advantage*, Oxford: Oxford University Press. pp. 442-473.

Whitley, R. (1999) *Divergent Capitalisms: The Social Structuring and Change of Business Systems*. Oxford: Oxford University Press.

Whitley, R. (2000) *Divergent Capitalisms. The Social Structuring and Change of Business systems*. Oxford, Oxford University Press (paperback).

Wolf, M. (2004) *Why Globalisation Works*. New Haven CT: Yale University Press.

PART I
MAKING SENSE OF GLOBALISATION: THE WORLD OF VALUE CHAINS AND STANDARDS

Chapter 1

Building Global Institutions: The Diffusion of Management Standards in the World Economy – An Institutional Perspective

Boris Braun

Introduction

It is by now widely accepted in modern economic geography that in the operation of the economy institutions matter (Peck 2000). Still, there is no consensus about how institutions matter – neither in economic geography nor in related academic fields. This situation is partly due to the fact that at least three academic disciplines, i.e. economics, political science, and sociology, have had long traditions in institutional and, since the 1960s, in neo-institutional ideas (see Hodgson 1998, Jessop 2001). Economic geography has taken up many of the institutionalists' claims quite openly in recent years (Amin 1999 & 2001, Martin 2000, Peck 2000, Schamp 2000). This can be explained by the genuine interdisciplinarity of institutionalism, its explicit theoretical openness and the deeply rooted mistrust of many economic geographers in the abstract and – in Granovetter's (1985) words – 'undersocialised' modelling of neoclassical economics. Equally important is the set of explanations institutionalism has to offer with regard to spatial or territorial (mostly local and regional) variation in economic activity. In fact, institutional thought has given new meaning to some of the older concepts of economic geography, such as agglomeration, localisation and urbanisation economies (see Amin 1999). Probably for this reason, most geographical research in this field has concentrated on ideas such as 'institutional' embeddedness and 'institutional thickness' on a local or regional scale. In contrast, despite their growing importance, very little attention has been paid to the emerging new types of global institutions and their rapid diffusion around the globe.

Since international standards first emerged in the early 20th century with the formation of the International Electrotechnical Congress (IEC) in 1906 and the

International Organisation for Standardisation (ISO) immediately after World War II, they have been typical examples of institutional norms.[1] Processes of standardisation and their diffusion can be regarded as perfect examples of institutional evolution (Meeus/Oerlemans 2002). More recently, consensus-based management standards are emerging rapidly as a new form of industrial co-ordination. They can be considered as a distinct social form of co-ordination especially suited to domains in which markets and hierarchies are weak and fragmented (Brunsson 2000). Extending global supply chains, the still growing importance of Transnational Corporations (TNC) and the emergence of more and more Newly Industrialised Countries (NIC) in Asia have contributed considerably to the creation of private standards and the diffusion of these standards across countries and industries.

Even though the important role of standardisation in the world economy has been casually mentioned by geographers favouring an institutional view of the economy (e.g. Amin 2001, Schamp 2000), the details of the standardisation process and the diffusion of standards have so far been largely ignored by economic geography. Against this background, my paper aims to analyse the process of standard creation as well as the rapid diffusion of international management standards in the global economy. It seeks to provide a closer understanding of the rapidly evolving worldwide standardisation of management practices from an institutional perspective. In this sense, management standards can be seen as an important part of the institutional setting which enhances global transactions. I will concentrate my analysis on the quality management and environmental management standards of the International Organisation for Standardisation (ISO), because they have been responsible for some of the most dramatic changes in the management of industrial companies in recent years.

Institutional Theories of Organisational Change

The core ideas of institutionalism concern conventions, norms, values, habits, rules, and their evolution. However, there is no single, general model or theory on the basis of those ideas. According to Hodgson (1998: p. 168), 'institutionalism does not attempt to build an all-embracing, general theory. Instead complex phenomena are approached with a limited number of common concepts and specific theoretical tools'. Some of the most central of these common concepts are the notion of embeddedness of economic actors in specific social and cultural contexts and, closely related to the former, the idea of bounded rationality of economic actors and bounded optimisation in economic transactions. Therefore, human and organisational behaviour is not simply modelled as rational and utility-maximising, but rather as bound by rules, conventions and common values as well as oriented towards legitimacy in an environment of uncertainty. In the institutionalist's world

1 ISO was reconstituted as the successor organisation to the short-lived International Standards Association which had already been established in the 1920s.

individuals and organisations take at least some things for granted without questioning them or constantly looking for alternatives in their search for efficiency.

Many problems of institutionalism stem from the fact that the underlying definitions are often vague or even mutually inconsistent. Especially the meaning of and the distinction between institutions and organisations are not always clear. As Jessop (2001: p. 1221) has pointed out, the well-known geographical concept of 'institutional thickness' introduced by Amin and Thrift rests on a far-reaching conflation of 'institution' and 'organisation'. Other researchers, however, draw a clearer distinction between the two terms (see e.g. Jessop 2001, MacLeod 2001). Along this line, institutions are here defined as persistent and connected sets of formal and informal rules, norms and conventions. Organisations, in contrast, are seen as particular subsets of institutions that involve strategic co-ordination (e.g. private firms, standardisation bodies, families). Particular organisations in this sense are results of institutions, but they are also agents of institutional change that put forward, (re-)formulate and shape (new) institutional arrangements.

Both institutions and organisations change over time in an evolutionary process. Organisations such as large industrial firms change through the adoption of innovations. Institutional theories of innovation diffusion argue that this process is not only driven by the expected technical benefits of adopting an innovation, but mostly by institutional factors. In particular, the new institutionalism in organisational analysis argues that adoption and implementation of organisational ideas and practices takes place in an institutionalised social and cultural context, which is distinctively different from the technical context (DiMaggio/Powell 1991, Scott 1995). According to a central postulate of institutionalism, institutional demands for legitimacy push business organisations to become increasingly similar. Even powerful firms tend to copy or imitate each other in many aspects.

Conformity to institutional norms creates structural similarities or isomorphism across organisations. As a result, management practices, for example, can become more and more alike or even 'standardised'. Isomorphism can stem from the interconnectedness of organisations (e.g. in network relations or in supply chains) as well as from the broader socio-cultural environment. In other words, it can be derived from relational embeddedness as well as from structural embeddedness of economic action. Organisations facing the same set of environmental constraints will tend to be isomorphic to one another and to their environment. Thus, organisations will come to resemble their environment and each other over time (Dacin 1997).

It is a well accepted idea among researchers, and especially institutional sociologists, that organisations become more like each other through three specific mechanisms: coercive, normative and mimetic forces (Scott 1995). These three mechanisms coincide with regulative, normative and cognitive processes (Table 1.1).

Table 1.1 The three pillars of institutions (after Scott)

	Regulative	Normative	Cognitive
Basis of compliance	Expedience	Social obligation	Taken for granted
Mechanisms	Coercive	Normative	Mimetic
Logic	Instrumentality	Appropriateness	Orthodoxy
Indicators	Rules, laws, sanctions	Certification, accreditation	Prevalence, isomorphism
Basis of legitimacy	Legally sanctioned	Morally governed	Culture, knowledge

Source: Scott (1995), modifications by the author.

Coercive isomorphism refers to the homogeneity pressures stemming from political influence and the need to achieve legitimacy within a specific context. Pressures originating from the state or from other powerful organisations are the most direct mechanism of institutional diffusion. The state's role in imposing adoption of organisational practices has been described in many empirical studies (Guler/Guillén/MacPherson 2000: p. 11). Regulatory systems, tort laws, antitrust laws or intellectual property regimes constitute important regulatory aspects of the institutional environment. They influence the set of organisation structures that are possible within that context and the agent's ability to efficiently contract with other agents (Delmas 2002: p. 96). In addition to the state, large multinationals are a second influential type of organisation that may cause coercive isomorphism. Like the state, large enterprises determine certain procedures and standards that suppliers must meet. Some global players such as Microsoft or IBM have the power to shape the institutional environment of the respective industries.

Normative isomorphism occurs when actors follow social and moral obligations. According to Scott (1995: pp. 38-39) the normative pillar of institutions refers to sets of expectations within particular organisational contexts of what constitutes appropriate, and thus legitimate, behaviour. Even rational action and rational choice are grounded in a social context. Therefore, choices and decisions are expected to be pre-structured by socially mediated values and normative frameworks. Normative isomorphism often originates in professional circles. This is how normative factors also account for the co-operative stance of many standard organisations.

Cognitive isomorphism in institutional theory refers to the fundamental effects of definitions of the world and frameworks of meaning that shape the roles and perceptions of social actors (Scott 1995). Mimetic isomorphism (also called appropriation isomorphism) happens when organisations copy others located in their own social system. It includes processes of imitation and copying of organisational models in order to cope with uncertainty. However, it is possible to distinguish between competitive and institutional mimicry, between competitive and institutional bandwagon effects. Competitive bandwagon effects arise from fear of losing competitive advantage. Firms may adopt the same practices because not doing so would disadvantage them relative to their competitors and erode their edge in the marketplace. The second bandwagon effect follows a logic of appropriateness. Here, institutional mimicry is the result of the threat of lost legitimacy and stakeholder support. In both cases, institutional mimicry is most likely under conditions of uncertainty, i.e. when an innovation is poorly understood, and the benefits of adoption are not clear.

There are, however, also limitations to a purely institutional view of organisational change. Institutional pressures always operate in concert with other forces, especially with competitive or market pressures. Moreover, a company is always dependent on both technical and institutional environments. Echoing Granovetter's (1985) concerns about over-socialised theory, it was suggested by Powell (1991) that 'much of the imagery of institutional theory portrays organizations too passively and depicts environments as overly constraining'. Institutional forces are not always primary. There is, for example, a notable tendency for early innovation adoption to be driven by technical, as opposed to legitimacy considerations (Roberts/Greenwood 1997). However, the new institutionalism in economics has also incorporated ideas of individualistic economic rationality into the general concept as it attempts to explain the emergence of institutions by referring to a model of rational individual behaviour (Hodgson 1998, North 1990, Williamson 1975). One of the most central concepts of individualistic rationality in new institutional economics is the transactions cost approach (see especially Williamson 1985). According to this concept, the rationale for the existence of any given organisational design is its cost-efficiency compared to the set of available alternatives (e.g. markets, hierarchies, networks). More recently, Roberts and Greenwood (1997) attempted to integrate transaction cost and orthodox institutional theories in order to provide a better understanding of the process by which new organisational designs are adopted.

International Standardisation

Standards in the broadest sense can be generated in three ways: first, by a formal regulatory process (such as most safety requirements); second, as a result of de facto market processes (such as a product specification that gains a predominant market share; e.g. the IBM compatibility standard for personal computers); third, through a voluntary process of consensus formation. A consensus standard always arises from co-ordination efforts of key parties in a specific sphere (e.g. technical experts, but also delegates of consumers or regulatory authorities) which are brought together in designated organisations for standard development, such as the ISO or the IEC. Even though the ISO and the IEC are powerful players, international standard setting does not necessarily take place under monopolistic conditions. In some areas, e.g. in telecommunications, even standard setting is a highly fragmented process with a large number of competing organisations and standards (Genschel 1997).

ISO is the most important international standard organisation. When ISO began operations in 1947 it had 26 member countries. Today its total membership comprises the principal national standard institutes of 138 countries, on a one member per country basis. ISO is a non-governmental body. Its general goal is to promote the development of standardisation and related activities in order to facilitate the international exchange of goods and services and to develop co-operation in the spheres of intellectual, scientific, technological and economic activity. All 13,000 ISO standards are purely voluntary, although a certain percentage of ISO's standards has been adopted by governments as part of their legislation. The ISO secretariat is located in Geneva, Switzerland, but the overall structure is extremely decentralised. The actual work of developing ISO standards is carried out by 187 Technical Committees (TCs), each with its own technological specialisation. The TCs are, in turn, made up of Subcommittees (SCs) and Working Groups (WGs). In total, there are 2,858 of these technical bodies. It is estimated that every working day some 15 ISO meetings take place around the world (Eicher 1997, ISO 2001a).

The outcome of standardisation processes are standards, which normally take the form of codified technical or procedural rules that are prepared for the common benefits of the parties involved. Standards are laid down in official documents. These documents are then sold by national or international standardisation organisations (e.g. the IEC for electrical and the ISO for non-electrical standards) at a certain price and are available for every organisation interested.[2] Even though the adoption of most standards remains voluntary, non-adoption can have severe negative economic consequences. The compliance to standards may make transactions considerably easier and cheaper. Moreover, standards can reduce uncertainties and foster the legitimacy of a firm's activities.

From a macroeconomic viewpoint standardisation serves the simplification and efficiency of social and economic exchange, as well as the improvement of the quality of goods and services. International standards also play a key role in levelling out trade barriers (Meeus/Oerlemans 2002: p. 2). For this reason, standardisation has also gained attention among such organisations as the World Trade Organisation (WTO) and especially the European Union (EU). Whereas the WTO is interested in international standardisation as a means for dismantling non-tariff trade barriers, the prominence of standardisation in the EU has arisen to a great extent from its usefulness in promoting European integration. As mutual recognition of national policies or direct negotiations of common regulations have proven difficult, this important task has largely been delegated to standard-making organisations. The standards resulting from the deliberations of these bodies of experts

---

2 National organisations for standardisation are, for example, the American National Standards Institute (ANSI), the British Standards Institution (BSI) and the Deutsches Institut für Normung (DIN).

are then incorporated or referred to in EU directives. According to Mendel (2000), this approach has in actual fact led to a unique contribution of the EU to global governance – explicitly fusing the professional authority of non-governmental organisations with the EU's gradually evolving regulatory powers. One of the most important outcomes of this 'new approach' is the European Eco-Management and Audit Scheme (EMAS).

As a result of these developments, Europeans are particularly well represented in the global standards sector, both in number of organisations and involvement in international standardisation bodies. Regional standardisation also began in Europe with the founding of the Comite Européen de Normalisation (European Committee for Standardisation – CEN) and CENELEC, the European version of the IEC. Comparable bodies in other parts of the world have followed, including the Pan American Standards Commission (COPANT), the Arab Organisation for Standardisation and Metrology (AOSM), and the Pacific Area Standards Congress (PASC). These organisations, however, tend to concentrate on promoting regional trade and standardisation, rather than on developing completely new standards themselves.

Basic Types of International Standards

Two of the most important basic types of international industry standards are compatibility and quality standards. Compatibility standards define interface requirements that allow different core products from different manufacturers to use the same complementary goods and services, or to be connected together in networks. Examples are telephone and internet protocols, videocassettes or many auto parts such as sparking-plugs or exhaust pipes. Positive network externalities become apparent when the value of using a particular technology increases with the number of agents who are also using the same technology standard (Cowan/Miller 1998, Katz/Shapiro 1994). The very existence of positive network externalities can, however, prevent the development of more advanced and efficient product alternatives. In these situations (de facto) standardisation and path dependency may even result in irreversible lock-ins of technological trajectories (Boschma/Lambooy 1999). The cost of switching may deter users from adopting a new and better standard, even when the old standard has objectively outlived its usefulness. A well-documented though highly disputed example is the still dominant, but – as some say – technologically obsolete QWERTY-keyboard design for typewriters and computers (see David 1985 & 2000; for a contrasting view see Gomes 1988, Lewin 2001, Liebowitz/Margolis 1990 & 1999).

This danger of inefficient standardisation presents itself to a lesser degree with quality standards. Quality standards are less affected by specific network externalities, as they are broader in scope and cover only basic product requirements for measurement and minimum quality. Examples are safety standards, packaging standards, weights and measures. However, after a period of time quality standards also run into the danger of defining lower technological solutions than the best available technology.

A major disadvantage of the traditional types of standards – compatibility and quality standards – is that they tend to fix the outcome of innovation processes in relation to past experiences instead of shaping the innovation management processes themselves. More generally speaking, traditional compatibility or quality standards are poor predictors of future outputs. For future outputs, it is the management process that really matters. This emphasis on management processes instead of results has been central to the development of new types of management standards since the mid-1980s. Management standards for relatively simple management tasks such as specific test procedures or job descriptions are nothing new. However, as tasks are assembled into larger management systems and interactions become more complex, an exact codification of all system specifications is almost impossible. Therefore, the new generation of comprehensive management standards concentrates on the codification of design rules to guide the creation of an open system of management systems. Referring to a concept in systems theory, Uzumeri (1997) coined the term 'metastandards' to characterise this new approach to standardisation. Figure 1.1 provides a classification of standards in two dimensions: (a) the stage of organisational process that is being standardised, and (b) the conceptual level of the standardisation with respect to system complexity. The central role of management metastandards in both dimensions becomes obvious.

Stage

Level	**Inputs** Suppliers	**Management Process**	**Outputs** → Customers Stakeholders
Metastandard *(rules for designing systems of items)*	performance criteria for inputs *(e.g. technical building codes)*	**Management system metastandard** *(e.g. ISO 9000, ISO 14000)*	performance criteria for outputs *(e.g. goals for customer satisfaction)*
System standard *(description of systems of items)*	required system of inputs *(e.g. complex technical specifications)*	design of a specific management system *(e.g. systems for responding to customer complaints)*	required system of outputs *(e.g. complex technical product specifications)*
Standard *(description of an item)*	specific input requirement *(e.g. specifications for supply parts)*	detailed task description *(e.g. codified routine task procedures)*	specific output requirement *(e.g. finishing tolerances)*

Figure 1.1 Categories of standardisation in organisations
Source: Uzumeri (1997), modifications and additions by the author.

The ISO 9000 and ISO 14000 Management Standards

Management metastandards fulfil two important needs. First, they synthesise information on best management practices and thus facilitate industry improvement. Second, they help to differentiate well-managed, well-performing firms from poor performers. The significance of management metastandards became evident when the first important example emerged that affected thousands and thousands of companies around the world. ISO 9001, the core of the ISO 9000 family of standards on quality management, is arguably the most influential single metastandard to date. ISO created the first version of the ISO 9000 family of quality management standards in 1987. The standards were developed by ISO/TC 176, which comprises experts from business and other organisations around the world. The most important forerunner of ISO 9000 was the BS 5750 series of quality assurance system standards, developed by the British Standards Institution in 1979.

Europeans initially were seen as lagging behind both Japan and the USA in the development of the contemporary quality movement, now usually labelled 'Total Quality Management' (TQM).[3] TQM encompasses a variety of techniques that share a vision of results-oriented productive activity, including employee participation, continuous improvement, integrated process control, competitive benchmarking, and innovation (Cole 1999). Given their predominance in international standardisation, however, European countries, and especially the United Kingdom, quickly took the leading role in popularising the integrated quality approach through the ISO 9000 family of standards, which is based on TQM ideas.

Unlike traditional quality standards, the ISO 9000 standards do not attempt to evaluate a finished product, a delivered service, an organisation's operational effectiveness or competitive performance in meeting market requirements. ISO 9000 ignores inputs and outputs. It assumes that customers and suppliers will negotiate these in the sales contracts. ISO 9000 standardises procedures, rights, and roles, rather than goals or outcomes. Hence, a company's certification to the ISO 9000 family of standards is not a sign of product quality in the conventional sense. ISO 14000, the ISO family of standards for good environmental management, which was first published in 1996, shares this procedural spirit and does not call for specific emission reductions or environmental performance levels. However, ISO 14001 requires an environmental policy, a delineation of responsibilities, a training and documentation system, and a facility to monitor and correct deficiencies.

The ISO 9000 standards are actually a series of five documents, separated into two 'guidance standards' intended as interpretative references and three

3 In the late 1970s and early 1980s Japanese companies were the first to establish product quality as a competitive weapon as manufactures recognized that more and more complex products and supply chains are vulnerable to bad components and poor supplier practices. In order to minimize their vulnerabilities, they demanded more information on their supplier's quality management systems.

'contractual standards' (ISO 9001, 9002 and 9003) according to which organisations may choose to become certified, depending on the scope of system elements to be considered (Mendel 2000). ISO 14000 was designed for compatibility with the ISO 9000 family and adheres to the same basic structure, although currently only one contractual standard is offered (ISO 14001). Interestingly, the ISO 14000 family of standards was also developed according to a British model – the BS 7750 on environmental management. The BS 7750 was also the point of reference for the European EMAS initiative (Braun/Grotz 2002: p. 275).

Figure 1.2 shows how ISO 9001, ISO 14001 and other metastandards are typically applied. The ISO 9001 standard, for example, lists 20 essential quality management subsystems. To comply to the standard, a organisation must effectively implement every subsystem. The 20 clauses of ISO 9001 specify which subsystems are required, rather than how they should operate. The standard implicitly acknowledges that the required subsystems may operate differently from company to company. Once the company has designed the management system and all subsystems have been implemented, an auditor is engaged to perform a quality audit.

ISO does not itself issue certificates of conformity to the ISO management standards. This is carried out independently by certification bodies in different countries. Successful certification requires implementing a given set of rules and management procedures. These rules address a wide range of functions, such as contract reviews, production control, internal audits, and training. Third-party certification ensures that certified companies have implemented the necessary management elements and comply with the letter (although not necessarily the spirit) of the guidelines (Terlaak/King 2001). Certificates are awarded for a period of three years, after which they are either extended or withdrawn as a result of a second audit. Organisations must engage in a continuous effort to remain ISO certified. However, the certification process has been subject to substantial criticism because of its commercialisation and because of the considerable discretionary powers of the certification bodies, which are not always completely independent (Homburg/Becker 1996).

Certificates which can be referred to in advertisements and promotions signal to customers and other external stakeholders, such as governments, consumers or environmental organisations, that a quality and/or environmental management system is in place. While external stakeholders may not know exactly how a company has implemented the management subsystems, they can be confident that all necessary subsystems exist and that they function to some acceptable degree. According to Uzumeri (1997: p. 26) it is exactly this logic of completeness that gives metastandards their meaning.

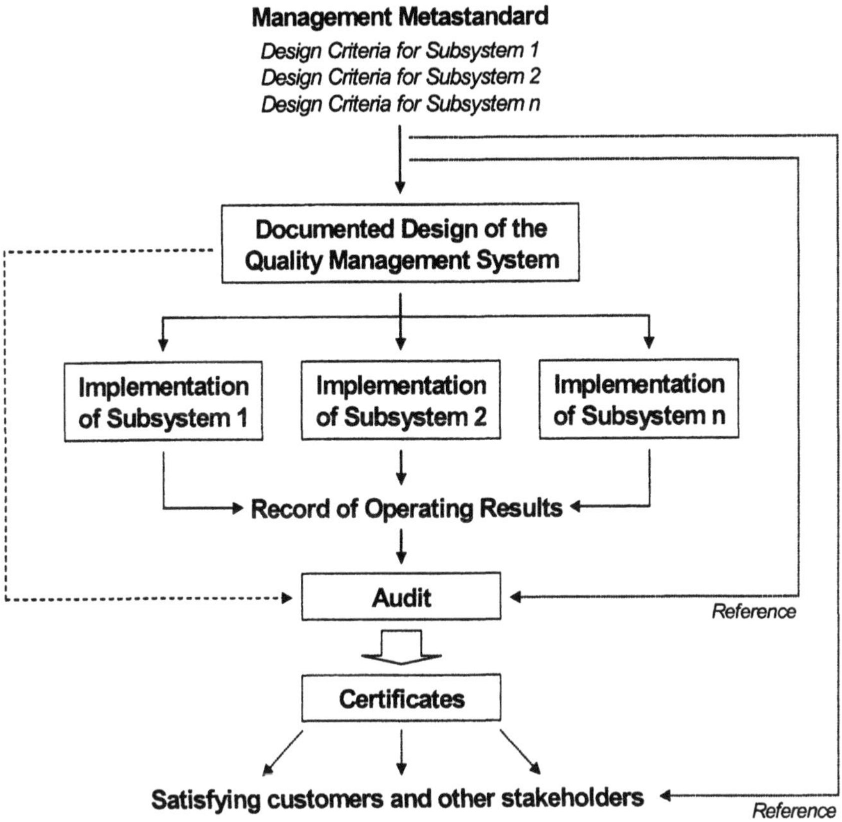

Figure 1.2 Structure of a metastandard compliance system
Source: Uzumeri (1997), modifications and additions by the author.

Global Diffusion of ISO 9000 and ISO 14000

Since the early 1990s, the number of business enterprises and other organisations complying with industry management metastandards has increased dramatically. Especially the worldwide success of the ISO 9000 family of standards is very impressive in quantitative terms. Up to the end of 2000, almost 409,000 certificates had been awarded in 158 countries all over the world. Europe, with already more than 220,000 certificates awarded, is still increasing rapidly with about 54,000 (+ 32 per cent) more certificates than two years before. Especially the United Kingdom, in many respects the centre of management innovation, is still in a leading position, albeit with decreasing growth rates in recent years due to a beginning saturation of the market (Table 1.2). The same is true, though to a lesser degree, for some other Western European countries such as the Netherlands and Germany.

Table 1.2 Countries with highest numbers of ISO 9000 and ISO 14000 certificates 2000

ISO 9000			ISO 14000		
Country	Certificates 2000 (no.)	Growth 1998-2000 (%)	Country	Certificates 2000 (no.)	Growth 1998-2000 (%)
UK	63,725	8	Japan	5,556	260
USA	35,018	40	UK	2,534	175
Germany	32,500	35	Sweden	1,370	351
Italy	30,367	68	Germany	1,260	94
Australia	24,772	75	Australia	1,049	198
China	25,657	215	USA	1,042	258
Japan	21,329	148	Netherlands	784	130
France	17,170	21	France	710	141
South Korea	15,424	100	Switzerland	690	92
Spain	12,576	96	Spain	600	266
Canada	11,435	51	Denmark	580	85
Netherlands	11,036	4	Korea	544	107
Switzerland	8,660	35	Italy	521	324
Brazil	6,719	81	China	510	443
Israel	6,140	66	Finland	508	147

Source: International Organisation for Standardisation (2001b), calculations by the author.

In East Asian countries, with 82,000 certificates already awarded, the number of certificates is still growing quickly. The 44,000 increase (+ 116 per cent) between 1998 and 2000 confirms the growing interest of the region's industries in the ISO 9000 certification process. China, Japan and South Korea are among the leading countries in this respect.

The number of certified companies in North America is quite surprising, if one takes into account that initially many U.S business leaders had reservations about the European-dominated standardisation process and some even called it a 'bureaucratic approach' to quality management. On a country basis the USA is now second in the number of ISO 9000 certificates issued. The number of ISO 9000 certificates in Australia is remarkable given its relatively small economic base. This may be a reflection of two factors: first, the institutional and cultural proximity to Britain and Europe and, second, the increasing integration of Australia's economy into global supply chains.

While the absolute figures for the ISO 14001 standard are significantly lower, they, too, have shown a steady increase since 1996. Up to the end of 2000 just fewer than 22,900 certificates were held in 98 countries. One reason for the lesser popularity of the ISO 14001 certification may be its later publication date and the much shorter time of diffusion. Yet, this is also a reflection of the fact that quality is still a more important issue for most industries than environmental concerns.

Although the list of leading countries does not differ greatly between ISO 14000 and ISO 9000, some countries like Japan or most Scandinavian states rank considerably higher on the environmental management list. These national differences are at least partly due to the regulative pressure from the respective environmental policies. Moreover, there are seriously competing voluntary standards in the environmental management field. This is true for industry standards like Responsible Care in the chemical industry (King/Lenox 2000), some independent American (de facto) standards and first of all for EMAS, which has been available for voluntary participation of companies in EU and EFTA member states since April 1995. EMAS is comparable in its major objectives, but it has a direct legal status and is somewhat stricter in its requirements than ISO 14001. Even though most companies perceive the two systems as complementary rather than as exclusive, EMAS has been a big competitor for ISO 14001 in some European countries, especially in Germany and Austria (Braun 2003).

The growth in numbers of certificates by world region suggests a roughly similar spatial pattern of diffusion for ISO 9000 and ISO 14000 (ISO 2001b). Looking at different regions of the world, Europe leads the way in adoption of both standards, with East Asia second and North America third. Many countries in Western and Southern Asia and especially in Africa – with the notable exception of South Africa –still lag behind in certification numbers. This pattern is clearly visible in Figures 1.3 and 1.4, where the number of ISO certificates awarded is related to the labour force of the countries in order to control for size effects.

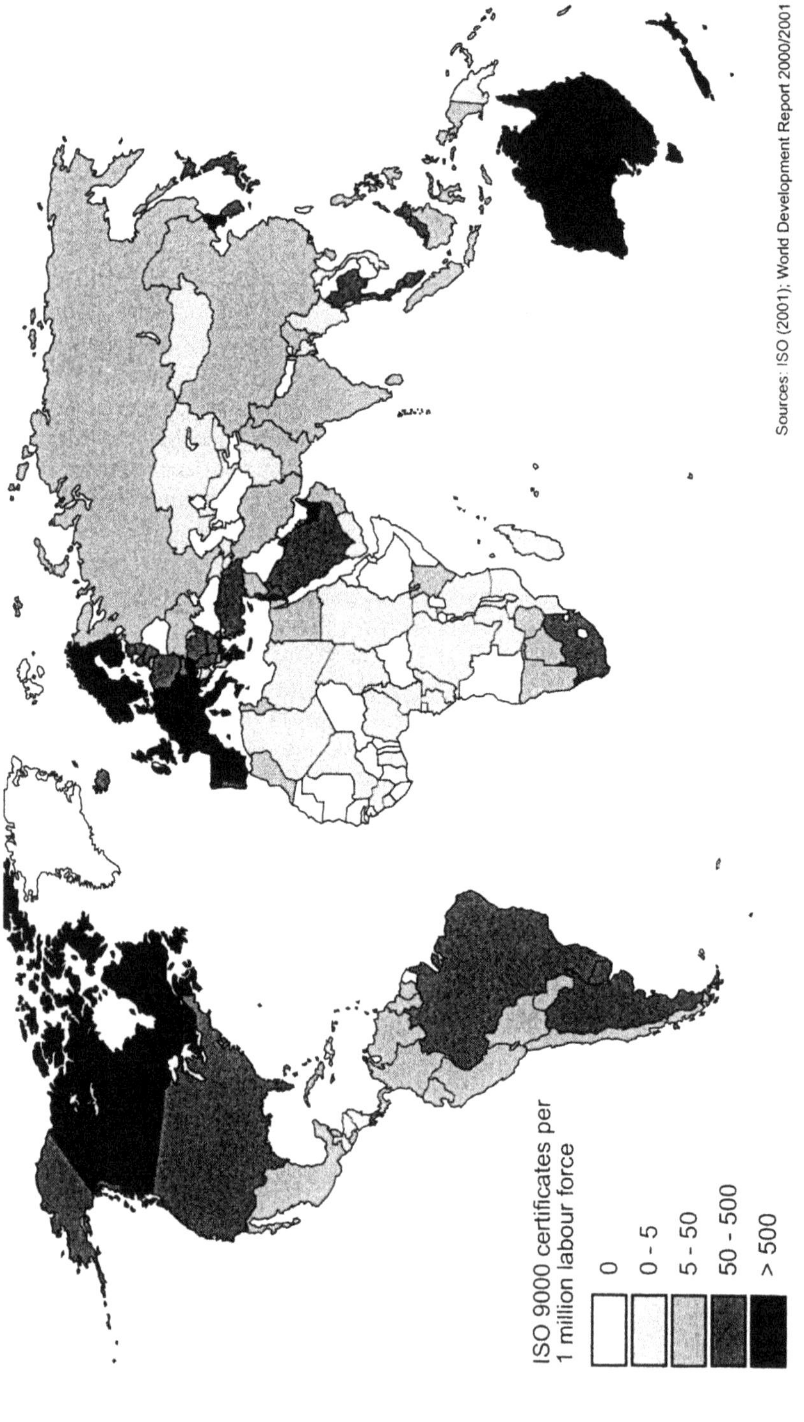

Figure 1.3 Global diffusion of ISO 9000 certifications, end of 2000

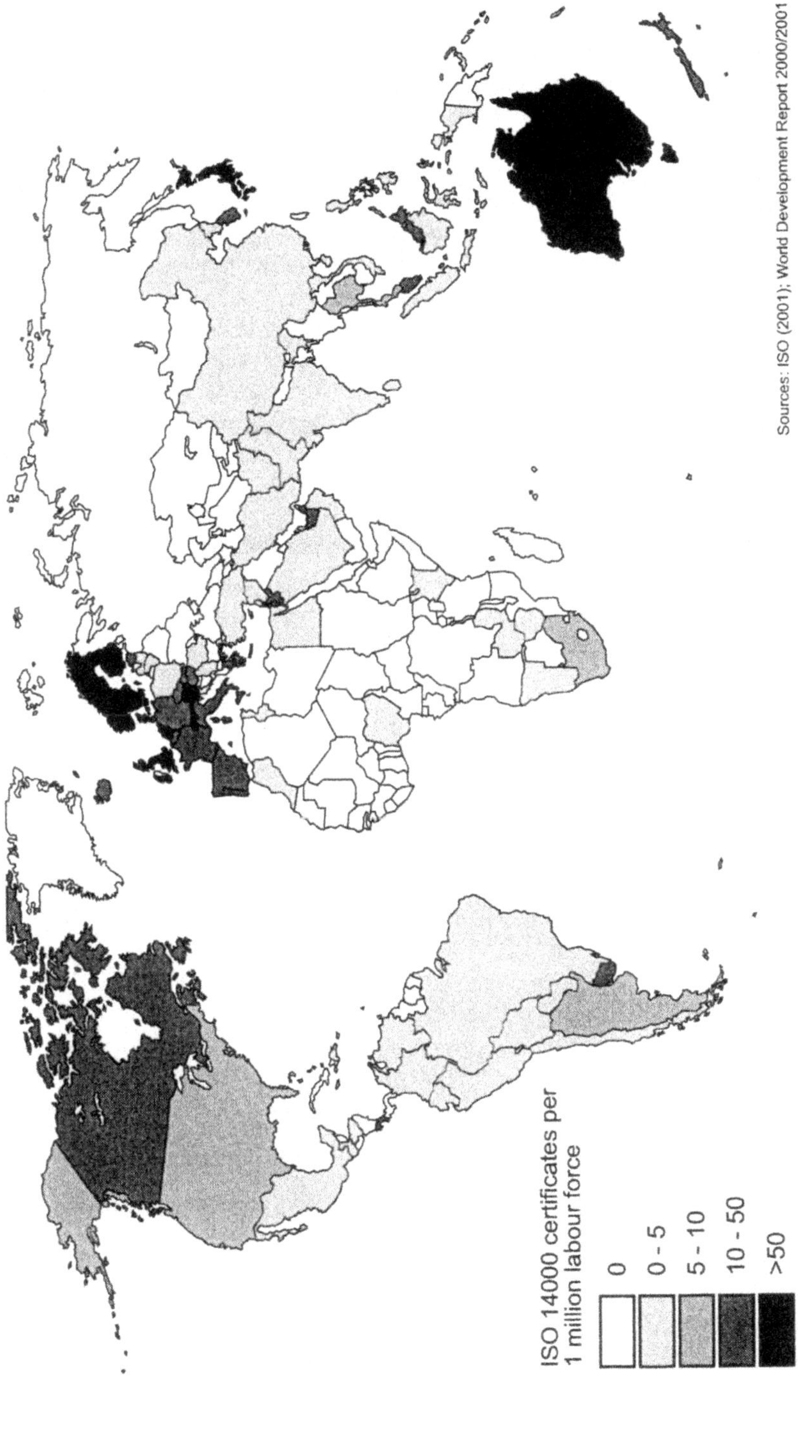

Sources: ISO (2001); World Development Report 2000/2001

Figure 1.4 Global diffusion of ISO 14000 certifications, end of 2000

The slow growth of ISO certifications in many African countries basically reflects their limited integration into world trade and foreign direct investment flows (see e.g. Dicken 1998), as foreign multinationals are very often the first companies in a country to seek certification according to ISO standards.

A breakdown of certificates by industry yields some differences between the adoption patterns of ISO 9000 and ISO 14000. Table 3 displays the percentages of ISO 9000 and ISO 14000 certificates by economic sector. In both cases, the manufacturing sector dominates the distribution, although this picture is more pronounced for ISO 14000 than for ISO 9000. While this may partly be because the two standards are being compared at different points of their development, it also underscores the political sensitivity of environmental issues as the major contrast between the two standards (Mendel 2000). The dominance of the goods producing sector with the ISO 14000 certificates can be regarded as an indicator of growing environmental responsibilities and increasing regulative and market pressure in many manufacturing industries. The same is true for public utilities, such as energy and water supply, recycling and waste disposal.

Table 1.3 ISO 9000 and ISO 14000 certificates by economic sector 2000, worldwide (in %)

Sector	ISO 9000 certificates	ISO 14000 certificates
Agriculture, fishing	0.6	1.2
Mining	0.6	1.0
Manufacturing	61.0	68.1
Construction	10.2	5.9
Energy and water supply, recycling	1.2	6.9
Private services	23.0	13.9
Public administration	0.4	1.1
Education	1.4	0.4
Health and social services	1.7	1.4
Total	100	100

Source: International Organisation for Standardisation (2001b), calculations by the author.

The analysis of international certification data reveals that the ISO management standards are clearly global phenomena by now. However, the diffusion pattern in many respects still reflects the European and Anglo-Saxon roots of the management standardisation movement. Management standards are not only (global) institutions. Rather, their global diffusion pattern is also strongly determined by institutions. This leads to the question which mechanisms determine the rapid diffusion of metastandards from an institutional perspective. Which factors explain the adoption of management metastandards across countries and industries?

Management Metastandards as Global Institutions: Mechanisms Responsible for Diffusion

Preparing a company for quality and environmental management standards costs money. Recent studies revealed that the combined labour and cash expense for preparing a medium-sized plant for ISO 9001 is in the range of US$ 50,000 to US$ 1 million, with a typical cost of about US$ 250,000. Costs are somewhat lower for successful ISO 14001 (and EMAS) certifications (US$ 40,000 to US$ 150,000 on average), basically because in most cases a quality management system is already in place when firms decide to implement an environmental management system (for an overview see, e.g., Braun 2003 and Uzumeri 1997). In addition to these implementation costs firms also have to bear the costs of running the management system.

If private companies opt for metastandard certification despite these expenses, they must have powerful incentives to do so (instead of opting for alternative management designs). Theoretically these incentives can stem from rational and comparative cost-benefit considerations as well as from institutional pressures. In view of recent research on institutional theories of organisational change it seems appropriate to interpret the adoption of management metastandards as driven by a combination of rational choice approaches which aim at a comparative cost-benefit assessment of all possible alternatives of organisational design and institutional mechanisms leading to isomorphism (Roberts/Greenwood 1997).

Generally speaking, firms adopt management metastandards to improve their internal practices and to acquire legitimacy with buyers and suppliers on international or even global markets. Standardisation is always driven by some kind of positive externality as well as by transaction cost considerations. When ISO introduced ISO 9000, it was hoped that the compliance certificates would reduce transaction costs in world trade. Likewise today the adopters of ISO 9000 hope that they will be able to reduce their transaction costs in writing contracts because some kind of 'guarantee' exists in the field of product quality between the two sides due to a third-party certification. However, while management systems according to ISO 9000 or ISO 14000 standards may definitely reduce transaction costs and improve the management efficiency of an organisation, it is unlikely that the certification is the result of a completely independent decision-making process based on pure economic rationality. Rational efficiency or cost-benefit considerations are certainly valuable explaining factors for the success of management standards. Yet, they alone are not sufficient to explain the recent dramatic increase in metastandard certificates around the globe:

1. An ISO metastandard is only one of many theoretical alternatives to achieve management goals. Therefore, the predominance of ISO management standards can hardly be attributed to rational efficiency considerations alone.
2. Many cost-benefit assumptions are vague and are driven by hopes and beliefs rather than by 'hard' economic facts (see also Delmas 2002). Most companies are simply not able to compare the real costs and benefits of an ISO 9000 or ISO 14000 certification because they lack detailed and comprehensive cost accounting. Moreover, it may even prove impossible to evaluate the exact monetary impact of the ISO certifications, as this impact can not be isolated from other factors determining sales figures and cost burdens.
3. Companies work in an environment that is increasingly characterised by uncertainty. Most management decisions have to be taken on an ex ante, rather than on an ex post basis. In this situation social orientation and the imitation of successful or socially accepted models may be more useful than a seemingly rational and time-consuming evaluation of all possible alternatives.
4. In reality, metastandards do not lead to more cost-efficiency in the management process by replacing a large number of 'old' standards and requirements. In most cases, ISO management standards have added to, rather than reduced the multitude of documents, controls, paper work and internal regulations.

All in all, efficiency-based arguments do not pay sufficient attention to social processes of diffusion. Collective agents and other influential actors help to define issues, pre-structure acceptable solutions and lend external support to organisational activities. As a result, the spread of management models, including ISO 9000 and ISO 14000, is strongly conditioned by the institutional and cultural contexts in which organisations find themselves with respect to other actors. These processes can be divided into the three broad institutional categories already mentioned – coercive, normative, and cognitive mechanisms leading to organisational isomorphism (see Guler/Guillén/MacPherson 2000, Mendel 2000, Scott 1995, Werle 1999).

Coercive Mechanisms

Coercive mechanisms refer to the effects of actors having the power to establish rules and manipulate sanctions. Organisations feel pressure to conform to these demands, because they are compelled to do so through regulation or exchange. As mentioned previously, the ISO 9000 and ISO 14000 standards owe much of their success to inclusion in legal rules related to the formation of a Single Market in Europe. For example, the EU published 'Global Approach to Conformity Assessments' in 1989, which advised companies to adopt a certified quality management based on ISO 9000. Another prominent example of a de facto legal status of ISO standards is the recent amendment of EMAS (EMAS II) which now includes parts of ISO 14001.

Legal rules in turn become marketing requirements for suppliers, which generate a pyramiding diffusion down the supply chain. Market-induced requirements have had particularly strong effects. Soon after the introduction of ISO 9000, adoption appeared to be driven more by market and customer pressure than by government regulation. Especially important was the power of large firms from the developed world within global commodity chains. In this sense, the success of management metastandards around the world is also the result of highly asymmetrical power relations. ISO 9000 certification is now frequently used as an easily accessible and visible tool to distinguish among suppliers. As supply chains become increasingly global and spatially extended, firms need some means for differentiating their reliability and quality. Participation in international management standards promises to be an easily accessible way to keep or to gain a place on a customer's bid list. In an increasingly competitive global economy, where customers and other stakeholders have a lot of choices and time constraints, this is certainly an important explanation for the rapid diffusion of quality management standards. The same explanation applies to the diffusion of environmental management standards. However, there are also some remarkable differences between quality management and environmental management standards. On the one hand, there is less pressure within the supply chain, because for most customers the environmental performance is less important than the quality of products. On the other hand, there are some additional incentives for the implementation of an environmental management system, most importantly the need to cope with increasingly stringent government regulations or the wish to reduce expenses for resource inputs and waste disposal.

Normative Mechanisms

The normative strength of ISO management standards is reflected in their perception as a 'best practice' method of demonstrating careful and responsible management. Although this appeal is probably strongest in Europe, ISO 9000 has become a premier standard for optimal customer-supplier relationships all over the world. Prominent carriers of these norms are regulatory agents, professional communities, and multinational enterprises (Mendel 2000).

Regulatory agents influence not only through coercive mechanisms, but also by educational activities and incentive programs promoting 'best practices'. In many developed countries such as Japan or Britain, government authorities have mounted national campaigns for ISO 9000 and ISO 14000 registrations. In many developing countries in Asia ISO management standards are central to industry efforts to become competitive and are backed by promotional campaigns by governments and business leaders.

Professional and scientific communities have catalysed the diffusion by their acceptance and circulation of ISO management standards. Professionals have played an instrumental role in developing and institutionalising the ISO standards through their positions in standardisation and other co-ordinating committees. Moreover, the growth of ISO certifications has opened additional professional services and occupations directly related to certification activities, such as registrars, notifying bodies, and registration consultants. These processes, among others, lead to further professionalisation (and commercialisation) of the standardisation and certification businesses. As a result, the diffusion of management standards becomes increasingly self-supporting.

Lastly, multinational companies have to be recognised as models of organisational forms and influential channels for the transfer of management knowledge across national borders. To the extent that the organisational culture of a multinational firm cuts across national borders, it may supersede local routines and facilitate the spread of international organisational standards. Foreign subsidiaries import management styles and standards into their host countries and export ideas and practices back to other parts of the parent organisation. Hence, one can expect that diffusion rates will be higher in countries and industries with a substantial share of multinationals. Indeed, there is strong empirical support for the thesis that foreign multinationals as well as the global trade network are important facilitators of the international diffusion of management standards (Guler/Guillén/MacPherson 2000).

Cognitive Mechanisms

Regulative and normative mechanisms rely heavily on the relational links between (network) actors. Cognitive pressures operate more diffusely through shared cultural frames and cognitive orders that are taken for granted. Specific cognitive mechanisms relevant to ISO management standards include the prevalence of the models, predispositions to formal organisational solutions, and social roles and positions in relation to others who have already adopted.

Both institutional and ecological theorists have associated the taken-for-grantedness of social form with its dominance or density within a relevant population (Mendel 2000, Scott 1995). Consequently, the proportion of ISO-certified organisations within a country or an industry is relevant for future rates of adoption by creating typical bandwagon effects.

Some organisations, sectors and even countries may exhibit deeply rooted preferences for formal managerial solutions such as ISO metastandards. On the organisational level, these predispositions would be reflected in greater degrees of general 'bureaucratisation' or a dominance of codified procedures. At the country level the 'mentalities' of professional, state, and other institutional elites can shape the diffusion of management innovations. For example, countries with corporatist orientations and/or a more formal and functionalistic style of rationality, e.g. Japan, Germany or France, tend to show relatively high rates of adoption. However, a more decentralised, less rigid shop-floor oriented management style like in the United Kingdom can also create affinities for formalistic solutions (Braun 2003: pp. 261-286). Thus, it is not surprising that Britain was the first country in the world to introduce national management standards (BS 5750, BS 7750) and is still in a leading position with regard to metastandard adoption.

Organisations are much more likely to model the behaviour of those considered similar and those recognised as successful and advanced. Diffusion research has frequently emphasised the impact of initial adoption by 'opinion leaders' or 'high prestige organisations' (see e.g. Terlaak/King 2001). Peripheral organisations may be among the first to try a new, unproven innovation, but adoption by core organisations will increase the rate of diffusion throughout a network or industry. A similar logic applies to countries and economies. Developed countries located in political and economic core positions very often tend to act as reference societies, supplying and demonstrating organisational models for more peripheral states. However, simply using economic development as a predictor of adoption rates may be misleading. Both core and peripheral nation-states engage in organisational mimicry, the former in order to catch up and the latter to maintain their leading position.

Concluding Remarks

This chapter intended to show that standardisation as a process of explicitly formulating general rules defining and regulating economic activity can be interpreted as a process of global 'institution building'. The role of standards as global institutions becomes particularly obvious for management metastandards like ISO 9000 and ISO 14000. Management metastandards constitute an international consensus on good management practice. Firms adopt management metastandards to improve their internal practices and to acquire legitimacy with buyers and suppliers on international or even global markets.

The discussion of coercive, normative and cognitive mechanisms demonstrates that institutional theory with its core idea of organisational isomorphism provides useful concepts for explaining the global diffusion of management innovations such as ISO 9000 and ISO 14000. However, systematic empirical research in this field is just beginning. The available findings lend support to the view that cross-national isomorphism operates primarily through the agency of powerful organisations such as the state, foreign multinationals and 'high prestige organisations'. The process of diffusion is very much an evolutionary one in which history matters and the nature of early adopters plays an important role in determining future adoption patterns. Adoption of new management practices is often path dependent. Initial conditions can critically determine the eventual path.

However, one should keep in mind that globalisation is not a uniform and homogeneous process, but rather a fragmented and heterogeneous one. Along this line, empirical research may well indicate that standardised management practices diffuse across the world in contingent ways, depending on the extent to which firms in each country are exposed to coercive, normative and mimetic forces. The process of adopting standards still seems to be shaped by the characteristics of individual countries, rather than simply by the 'supposedly sweeping effects of globalisation' (Guler/Guillén/MacPherson 2000: p. 26). Moreover, the idea of organisational isomorphism has its limits. Empirical research shows quite clearly that the simple adoption of a management standard does not automatically change and standardise organisational practices (Braun 2003, De Vasconcelos/De Vasconcelos 2000). Implementation dynamics can differ remarkably according to the structural characteristics of a given organisation. Thus, a complete convergence of process, product and work organisations on a global management model can hardly be expected. As Storz (2003: p. 15) puts it with regard to her empirical study of ISO 9000 implementation by Japanese and German firms: '*one* standard is definitely not *one* standard'. In order to make it effective receivers have to interpret it – and this interpretation can differ considerably from firm to firm and from country to country as new standards always have to be integrated into existing structures.

Meanwhile management metastandards regulate management practices in a broad range of industries around the globe. They have emerged as important drivers for the globalisation of supply chains and industry networks and, thus, tend to reduce the organisational and spatial fragmentation of markets. While there is not much hope that the standards can significantly reduce power asymmetries in economic transactions, the risks of technological lock-ins and inefficient path dependencies seem to be limited, because management metastandards are open to interpretation and change and do not prescribe certain organisational and technological solutions. Studying the role and the diffusion of international standardisation can provide closer insights into the question how institutions matter in a global economy. Economic geographers, therefore, should recognise this increasingly important field of theoretical and empirical research more thoroughly than in the past. Especially an institution-oriented economic geography could provide valuable contributions for a better understanding of the role of standardisation in developed and developing economies.

References

Amin, Ash (2001) Moving on: institutionalism in economic geography. *Environment and Planning A* 33: pp. 1237-1241.

Amin, Ash (1998) An institutionalist perspective on regional economic development. *International Journal of Urban and Regional Research* 23, pp. 364-378.

Boschma, Ron A. and Jan G. Lambooy (1999): Evolutionary economics and economic geography. *Journal of Evolutionary Economics* 9, pp. 411-429.

Braun, Boris (2003) *Unternehmen zwischen ökologischen und ökonomischen Zielen*. Münster: Lit.

Braun, Boris and Reinhold Grotz (2002) Environmental management in manufacturing industry: a comparison between British and German firms. In: Schätzl, Ludwig and Javier Revilla Diez (eds.): *Technological change and regional development in Europe*. Heidelberg, New York: Physica, pp. 273-292.

Brunsson, Nils (2000) Standardization as a social form. In: Brunsson, Nils and Bengt Jacobsson (eds.): *A world of standards*. Oxford: Oxford University Press, pp. 52-70.

Cole, Robert E. (1999) *Managing quality fads*. New York: Oxford University Press.

Cowan, Robin and John H. Miller (1998): Technological standards with local externalities and decentralized behaviour. *Journal of Evolutionary Economics* 8, pp. 285-296.

Dacin, M. Tina (1997) Isomorphism in context: The power and prescription of institutional norms. *Academy of Management Journal* 40, pp. 46-81.

David, Paul A. (1985) Clio and the economics of QWERTY. *The American Economic Review* 75: pp. 332-337.

David, Paul A. (2000): *Path dependence, its critics and the quest for 'historical economics'*. Working paper based on the keynote address to the European Association of Evolutionary Political Economy held in Athens, 7-9 November 1997 (http://www.econ.iastate.edu/tesfatsi/pathdep.pdavid.pdf).

De Vasconcelos Flávio C. and Isabella F. F. G. de Vasconcelos (2000) *Isomporphism and the limits of normalization. Two case studies focussing the ISO 9000 norms in the French computer industry*. Working paper (http://in3.dem.ist.utl.pt/downloads /cur2000/papers/ S28P03.pdf).

Delmas, Magali A. (2002) The diffusion of environmental management standards in Europe and in the United States: an institutional perspective. *PolicySciences* 35: pp. 91-119.

Dicken, Peter (1998) *Global shift. Transforming the world economy*. London: Sage.

DiMaggio, Paul J. and Walter W. Powell (1991): The iron cage revisited: institutional isomorphism and collective rationality in organizational fields. In: Powell, Walter W. and Paul J. DiMaggio (eds): *The new institutionalism in organizational analysis*. Chicago: University of Chicago Press, pp. 63-82.

Eicher, Lawrence D. (1997) The ISO 9000 family of international standards. *Accreditation and Quality Assurance* 2: pp. 323-326.

Genschel, Philipp (1997) How fragmentation can improve co-ordination: setting standards in international telecommunications. *Organization Studies* 18: pp. 603-622.

Gomes, Lee (1998): QWERTY spells a saga of market economies. *The Wall Street Journal* February 25, 1998: B1.

Guler, Isin; Guillén, Mauro F. and John Muir MacPherson (2000) *Global competition, institutions, and the diffusion of organizational practices: the international spread of ISO 900 quality certificates*. Working Paper (http://www-management.wharton. upenn.edu/guillen/ pdfs/isopapers.pdf).

Granovetter, M. (1985) Economic action and social structure: the problem of embeddedness. *American Journal of Sociology* 91, pp. 481-510.

Hodgson, Geoffrey M. (1998) The approach of institutional economics. *Journal of Economic Literature* 36, pp. 166-192.

Homburg, Christian and Jan Becker (1996) Zertifizierung von Qualitätssicherungssystemen nach den Qualitätssicherungsnormen DIN.

ISO 9000 ff. Eine kritische Beurteilung. *WiSt* 25: 444-450.

ISO – International Organisation for Standardisation (2001a) *ISO in figures, January 2001*. Geneva: ISO Central Secretariat.

ISO – International Organisation for Standardisation (2001b) *The ISO survey of ISO 9000 and ISO 14000 certificates – tenth cycle*. Geneva: ISO Central Secretariat.

Jessop, Bob (2001) Institutional re(turns) and the strategic-relational approach. *Environment and Planning A* 33: pp. 1213-1235.

Katz, Michael L. and Carl Shapiro (1994): Systems competition and network effects. *Journal of Economic Perspectives* 8, pp. 93-115.

King, Andrew and Michael L. Lenox (2000) Industry self-regulation without sanctions: the chemical industry's responsible care program. *Academy of Management Journal 43*, pp. 698-716.

Lewin, Peter (2001) The market process and the economics of QWERTY: two views. *Review of Austrian Economics* 14: pp. 65-96.

Liebowitz, Stan J. and Stephen E. Margolis (1990) The fable of the keys. *Journal of Law and Economics* 33, pp. 1-25.

Liebowitz, Stan J. and Stephen E. Margolis (1999) *Winners, losers & Microsoft: competition and antitrust in high technology*. Oakland: The Independent Institute.

MacLeod, Gordon (2001) Beyond soft institutionalism: accumulation, regulation, and their geographical fixes. *Environment and Planning A* 33: pp. 1145-1167.

Martin, Ron (2000) Institutional approaches in economic geography. In: Barnes T. and E. Sheppard (eds.): *Companion to economic geography*. Oxford: Blackwell, pp. 77-94.

Mendel, Peter J. (2000) *International standardization and global governance: the spread of quality and environmental management standards*. Paper presented to the research forum "Organizations, Policy, and the Natural Environment: Institutional and Strategic Perspectives", April 28-30, 2000, Evanston, IL.

Meeus, Marius T.H. and Loen A.G. Oerlemans (2002) *Why do firms participate in standardization? An empirical exploration of the relation between isomorphism and institutional dynamics in standardization*. University of Utrecht, Department of Innovations Studies, Working Paper (http://econ.geog.uu.nl/unices/standmeeus.doc).

North, Douglass C. (1990) *Institutions, institutional change, and economic performance*. Cambridge: Cambridge University Press.

Peck, Jamie (2000) Doing regulation. In: Clark, Gordon L., Feldman, Maryann P. and Meric S. Gertler (eds.): *The Oxford handbook of economic geography*. Oxford: Oxford University Press, pp. 61-80.

Powell, Walter W. (1991) Expanding the scope of institutional analysis. In: Powell, Walter W. and Paul J. DiMaggio (eds.): *The new institutionalism in organizational analysis*. Chicago: University of Chicago Press, pp. 183-203.

Roberts, Peter W. and Royston Greenwood (1997) Integrating transaction cost and institutional theories: Toward a constrained-efficiency framework for understanding organizational design adoption. *Academy of Management Review* 22: pp. 346-373.

Schamp, Eike W. (2000) *Vernetzte Produktion. Industriegeographie aus institutioneller Perspektive*. Darmstadt: Wissenschaftliche Buchgesellschaft.

Schmidt, Susanne K. and Raymund Werle (1998) *Coordinating technology. Studies in international standardization of telecommunications*. Cambridge, Mass.: MIT.

Scott, W. Richard (1995) *Institutions and organisations. Theories and research*. London: Sage.

Storz, Cornelia (2003) Implementation of standards: what about the possibility of a convergence of production systems by international rules? In: Pascha, Werner (ed): *Systematic chance in the Japanese and German economies: Convergence and differentiation as a dual challenge*. London, New York: Routledge Curzon (forthcoming).

Terlaak, Ann and Andrew A. King (2001) *Contrasting institutional theory and signaling theory to explain the adoption process of an industry standard*. Working Paper (http://www.stern. nyu.edu/~aking/Contrasting_ISO.pdf).

Uzumeri, Mustafa V. (1997) ISO 9000 and other metastandards: Principles for management practice? *Academy of Management Executive* 11: pp. 21-36.

Werle, Raymund (1999) *Conflict or peaceful co-existence? Standards organizations on a global scope*. Working paper (http://www.unibw-hamburg.de/mweb/nif/fum/ alt/wa99/ werle.pdf).

Williamson, Oliver E. (1975) *Markets and hierarchies, analysis and anti-trust implications: A study in the economics of internal organization*. New York: Free Press.

Williamson, Oliver E. (1985) *The economic institutions of capitalism*. New York: Free Press.

Chapter 2

'Fragmenting the Chain': Economic and Spatial Differentiation in the Post-Liberalisation South African Citrus Export Commodity Chain

Charles Mather

Introduction

A key development in the global food system in the last 30 years has been the rapid increase in the trade of fresh fruits and vegetables (Friedland, 1993; Le Heron and Roche, 1995; Raynolds, 1997; Goodman and Watts, 1997; Dolan and Humphrey, 2000). These 'high value foods' (HVF) have arguably displaced traditional export commodities such as cocoa, coffee, tea and sugar (Watts, 1996). During the 1980s the growth in HVFs was 8 per cent per annum, compared to a decline for most traditional agro-exports. As Watts (1996, p. 232) has argued, there are several specific characteristics of these 'non-traditional' exports including 'perishability, heterogeneity, seasonality, long gestation periods, and externalities associated with marketing'. Although the trade in high value foods is dominated by developed countries, there are several developing countries that are active participants in the trade of non-traditional agro-commodities (Barrett *et al*, 1999; Dolan and Humphrey, 2000). The involvement of these nations in fresh fruit exports has earned them the label new agricultural countries (NACs), the agro-industrial equivalent of newly industrialising countries (Friedman, 1993). The rise of NACs like Mexico, Brazil, Chile and Thailand is linked to a much broader process of global agro-food restructuring associated with what has been tentatively called the 'third food regime' (McMichael, 1996; Friedman, 1993; Le Heron and Roche, 1995).

Changes in the global trade of fresh fruit in the last thirty years suggest the need to disaggregate the category of high value, 'non-traditional' fruit exports (cf. Raikes and Gibbon, 2000). Three commodities – bananas, citrus and apples – dominate the global trade of fresh fruit and together they make up 76 per cent of global fruit exports (Rabobank, 1997). It is becoming increasingly difficult to sustain the definition of 'non traditional' high value fruits for all three categories, but especially for citrus and apples. Although these commodities remain vulnerable to decay once harvested, advances in storage technology have meant that it is possible to keep fruit in cold storage for far longer than before. Together with longer growing seasons in both hemispheres, the result of these changes has been significant overlaps between the northern and southern hemisphere fruit production. Indeed, it is no longer unusual to find citrus and apples from Europe, South Africa, New Zealand, and Chile at the same time on the shelves of multiple retailers. The situation for producers in this ultra competitive and frequently over-supplied market environment has become very difficult (McKenna and Murray, 2002). Multiple retailers, especially in the UK, consider these commodities central to their programmes in that they position themselves in relation to their competitors on the quality, price and variety of their fruit (Dolan and Humphrey, 2000; Barrett *et al*, 1999). Retailers have, however, benefited from a situation of oversupply and seasonal overlaps and now demand much higher internal and external quality standards of traditional varieties, while at the same time insisting on new varieties of apples or citrus. Producers are finding it increasingly difficult to sustain themselves in these oversupplied and increasingly competitive global fresh fruit markets. The liberalisation of agriculture in several agro-exporting countries has further exposed producers to the increasing power of multiple retailers and large northern hemisphere importers. Single channel export schemes provided some protection against the demands of retailers; the dismantling of these systems in many countries has removed whatever bargaining power producers had in the past (van der Laan *et al*, 1999; Shepherd and Faofli, 1999).

This chapter explores recent changes in what might be called a 'newly traditional' fruit commodity: fresh citrus exports from South African. The global commodity (value) chain (GCC) framework is used to explore the impact of changes in the governance of the chain over the last 10 years. For several reasons, the South African citrus sector is an interesting case with which to use the GCC framework. First, there have been dramatic changes in the regulation of the chain following the liberalisation of citrus exports after 1997. Despite the criticism levelled against the 'ideal-typical' categories of producer driven and buyer driven chains, this chapter suggests – with some qualifications – that there has been a shift from one form of chain governance to another. The impact of this change in governance, from a largely producer driven to 'buyer-driven' in the period after 1997, plays a central role in the restructuring of the sector with implications for exporters, producers and farm workers. Second, citrus is produced in several distinct ecological and economic spaces in South Africa (Figure 2.1). The impact of changes in governance has been shaped by the distinctiveness of the citrus growing regions. Exploring the spatial impact of liberalisation allows this chapter to examine the relationship between commodity chains and regional development, an issue that has been of interest to geographers working with the global commodity chain framework (e.g. Coe *et al*, 2003; Henderson *et al*, 2001; Lesley and Reimer, 1999).

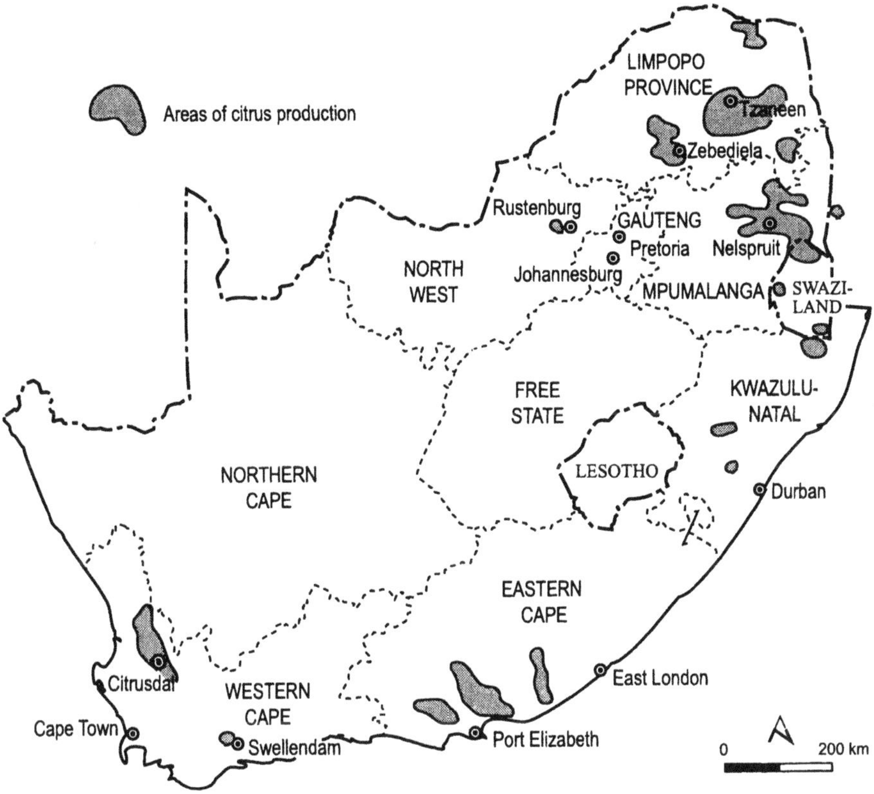

Figure 2.1 Citrus production in South Africa

Producer Driven Chains

In Gereffi's (1994) original formulation of the GCC framework he distinguishes between producer driven and buyer driven chains. The former are characteristically capital or technology intensive industries and are governed by large transnational corporations. These large organisations play a key role in controlling productive activities in the chain. Case studies of buyer driven chains have focused on the automobile, aircraft and semi-conductor sectors, where barriers to entry are very high. Although there are no examples of producer driven chains in agriculture, the organisation of South African citrus exports between the early 1940s and late 1990s comes close, even though it does not fit neatly into the definition of 'producer driven'.

From the 1940s all South African citrus was exported by a cooperative called the South African Citrus Cooperative Exchange (SACCE). This organisation's control of exports was granted by the Citrus Board, which had the statutory power to select a single channel agent for citrus exports. In practice, the relationship between the two organisations was very close and there was never any question that SACCE would not be appointed the official agent of the Citrus Board. SACCE had considerable influence over producers by assisting them in improving their production practices and by suggesting appropriate cultivar choices. The cooperative also assisted cooperative pack-houses in efficient fruit packing and appropriate measures for post harvest storage of citrus. Through the post war period it established a large network of extension officers and laboratories to test soils, leaves, and fruit. The Citrus Exchange also supplied farmers with budwood, the propagation material for citrus trees, to farmers through a network of approved nurseries. By the late 1980s the Citrus Exchange had established a system of production that sourced fruit, first in Zimbabwe and the northern provinces of South Africa, and then from producers in Swaziland and KwaZulu-Natal and finally from citrus growers in the Eastern and Western Cape. Sourcing from several different ecological zones allowed the Citrus Exchange to supply northern hemisphere markets with huge volumes of citrus for 5 or 6 months a year.

From the 1960s the Citrus Exchange used its command of large volumes of citrus – over half the southern hemisphere's fresh crop – to exercise greater control over markets in Europe. In a strikingly 'commodity chain' like statement in 1960, the chairman of the Citrus Exchange called for greater control over distribution: 'it was always our policy to control the distribution of fruit down to the first point of sale...we are now adopting a policy which is designed to enable us to control the flow of our fruit right down to the ultimate consumer' (cited in Cartwright, 1977, p. 84).

To this end the Exchange established a panel of 'master agents' through Europe who were responsible for specific country or regional markets. Using the master agents, and the control over huge volumes of citrus that could be stored for limited periods of time, the Exchange was able to exercise some control over prices, even when the crop itself was of a poor quality.[1] The Exchange was also heavily involved in promoting its Outspan brand, which they did vigorously until they were disrupted by anti-apartheid protestors in the United Kingdom and elsewhere in Europe (Mather and Rowcroft, 2004).

Detailed interviews with UK based importers and retailers and long-term employees of the Citrus Exchange suggest two important qualifications to the single channel organisation's ability to drive the citrus chain. First, although it was possible to shift fruit of poor quality from one market to another and maintain prices by controlling supplies, in practice this was very difficult. In some seasons, the scale of the quality problem was too great; in other cases, their ability to

1 'When we knew what the crop was we decided that we would keep count 56s tight (large sized fruit). We used promotional efforts for the 88s and 105s (less popular small fruit) to get rid of them. We used campaigns like "the small ones are juicier", that people remember to this day. And Outspan used to spend a lot on promotion: as much as 30p a carton based on a £7 carton price.'

manipulate prices was compromised by supplies of citrus from other southern hemisphere producers or overlaps from citrus produced in Spain and Italy. Citrus Board reports suggest that the problem of competition from other southern hemisphere producers, which seriously undermined their ability to protect prices, was a problem as early as the late 1970s (Citrus Board, 1979, p. 2). Second, from the late 1980s, especially in the UK, the Citrus Exchange was facing demands from a retail sector that was growing in power. New demands were being made on internal and external fruit quality, and also for newer varieties of citrus. The Citrus Exchange's inability or reluctance to meet these demands revealed its inflexibility but perhaps also its inability to influence growers who remained 'producer driven'. UK based Citrus Exchange employees found to their great frustration that growers were unwilling to meet retailer's quality standards:

> At the UK office we were really a branch of the head office based in South Africa. By implication we were grower driven. But we couldn't get growers to do what we wanted them to do. There were huge pow-wows over quality standards. We tried to set a common standard for the sugar acid ratio of grapefruit – this was one of Tesco's demands. But the growers couldn't agree so in the end we had to say to Tesco "Do you want the fruit or don't you?"[2]

This extraordinary response to the multiple retailer Tesco reflects the key problem the Citrus Exchange faced in the late 1980s and early 1990s. Citrus growers in South Africa were encouraged to produce large volumes of fruit thanks to a pooling system that rewarded 'pool participation' rather than fruit quality. The incentive to produce higher volumes structured all production practices, usually at the expense of fruit quality.

The Citrus Exchange was aware of the problems it faced in increasingly buyer-driven markets. A consultancy report, commissioned in the late 1980s, revealed that while the single channel exporter had technical expertise that was unrivalled, it had fallen behind its competitors in terms of marketing strategies. In particular the organisation had failed to meet new market demands for a commodity that was increasingly oversupplied and under pressure from retailers pressing for better quality and more variety. The consultants recommended that the organisation restructure with a focus on becoming more 'market driven' through a review of its overseas marketing methods and strategies and by introducing 'performance-linked' incentives to growers (Citrus Board, 1990). The Exchange's own 'in-house' investigation completed two years later confirmed the earlier consultancy report: the Citrus Exchange had failed to adapt to a radically different global

2 This was the comment from Marks and Spencer's fruit buyer who had 20 years experience in the industry: 'Whilst there was generally speaking a controlled marketing of fruit (in the single channel) in a disciplined and fairly sophisticated way, when retailers wanted specific requirements of size, varieties, different standards, it used to be extremely frustrating. We always found it frustrating because we couldn't necessarily get what we wanted. And that would be from a size point of view, an eating quality point of view, and from a varietal point of view.'

trading environment for citrus. In outlining a new strategy the Exchange stressed the need for flexibility in products, in fruit delivery and in pricing. It also confirmed the need for incentives to improve the quality of citrus produced by South African growers. Significantly, the Citrus Exchange recognised that its greatest challenge in adapting to new market environment lay in the 'successful management of the cultural changes within the organisation and the industry' (Citrus Board, 1992, p. 33).

Though the 1990s, the Citrus Exchange embarked on a significant restructuring programme. Based on the earlier consultancy recommendations, it abandoned control of the unprofitable domestic market. The organisation also attempted to respond to the quality demands of retailers by identifying 'good growers' and rewarding them for high quality produce, which was then sent to UK retailers to satisfy their demand for quality and variety. The Exchange also urged cooperative pack-houses to adapt to new packaging requirements and encouraged new cultivar development through its impressive research and extension infrastructure. Finally it embarked on an aggressive campaign to extend its infrastructure both locally and overseas, and it established formal and informal alliances with several domestic and international fruit companies.

The Citrus Exchange appears to have had more success transforming its organisational structure than it did in changing the 'production culture' of its growers and cooperative members. In the early 1990s it transformed itself from a cooperative to a private company called Outspan International. Shares in the new company were distributed to individual growers and cooperatives in line with their most recent supply figures. A few years later Outspan merged with the single channel exporter of deciduous fruit, Unifruco, to create a new company called Capespan. Internationally Capespan restructured and streamlined its overseas offices; it also acquired overseas assets mainly in the form of port and cold storage facilities. In the United Kingdom, the company severed its relationship with its master pannelists, and presented itself to retailers as a preferred supplier of citrus and deciduous fruit. From the mid-1990s the company also began to source fruit from northern hemisphere producers to provide a 'full year around basket of fruit' to its retailer suppliers. The direction of Capespan's restructuring during the 1990s was in line with the strategies of other large multinational fruit traders like Del Monte, Dole and Chiquita. Under pressure due to oversupplied markets and stagnant consumption levels, these global corporations have attempted to improve their competitiveness by diversifying their production range, improving their distribution network and by selling off low profit activities, especially associated with production. The stakes for these companies is extremely high: a recent Rabobank report has suggested that the profit margins for 'market leaders' in the fruit trade, defined according to market share, is much higher than it is for those companies lower down the competitive pecking order: 'market leader in a given industry is generally able to earn above average margins, the number two players earns average margins, whereas the number three player earns below average margins' (Rabobank, 2001, p. 2).

The pressure on exporters to improve quality and variety in the citrus category has been intense. Retailers, especially those based in the United Kingdom, consider citrus to be an important commodity that must be available year around, but also one that is under fierce competition from more exotic fruit varieties. The responsibility of securing a steady supply of high quality citrus while at the same time sourcing new varieties has been passed down the chain to a limited number of preferred suppliers, or what the industry calls 'service providers' in recognition of the additional tasks they perform. From the retailer's perspective service providers now have a greater market share and a more stable relationship, and in return they expect better service. The pressure faced by preferred suppliers in the citrus category is challenging. As one service provider argued, they were faced with the challenge of transforming a 'bloody orange' into something more exciting:

> Our job is to improve turnover and to increase the excitement in the product. It is difficult for them (Marks & Spencer's) to sell their point of difference in citrus. We've got to make it look different even though it's only a bloody orange. That's our job. So we have clear targets in terms of varietal development, quality etc.

Capespan's attempt to communicate a 'buyer driven' culture to its growers before the demise of the single channel was however met with some resistance. The company's efforts to introduce incentives for quality, according to one source, were rejected by cooperatives 'who did not want to incur penalties for fruit that had problems'. And despite its calls for flexibility, it seemed incapable of meeting the demands of retailers and their immediate suppliers. As one UK based importer noted,

> The problem with the South Africans is that they aren't flexible enough. They tell us that they can't pack Clementines and Satsumas at the same time. They don't understand that the customer wants both on the shelves at the same time. They tell us that when they've finished the Clementines they will pack the Satsumas.

There is considerable evidence that Capespan's attempts to communicate changing market demands floundered on a grower community that had always been rewarded for volume rather than quality.[3]

From the mid-1990s Capespan faced the additional problem of losing its control over the single channel. Even though earlier consultant reports had suggested that it shift to a voluntary system of membership and supply, much in the same way the US cooperative Sunkist operates, Capespan spent considerable effort during the 1990s attempting to convince the Department of Agriculture on the necessity of the single channel. Much of this effort was structured by a new marketing act, which was passed in 1996. According to the new legislation a single

3 'Outspan did try to incentivise towards the end of the regulated era. In the early 1990s fruit was assessed from individual growers and if the quality was good then they got different payments. But the system didn't have teeth. It was always blunted by the big coops. They didn't want to incur penalties for fruit that had problems.'

channel could be granted if the majority of growers supported it, and if the single desk agent could demonstrate its commitment to supporting 'previously disadvantaged' growers. Capespan responded to these guidelines by canvassing growers on the single channel in 1996. The results, which were publicised as a press conference, showed that almost 90 per cent of growers were in favour of a single desk for citrus exports. Later the same year the company also highlighted its commitment to 'reconstruction and development' through its assistance to emerging farmers in the Bantustans. It claimed to be a pioneer in a number of 'successful working models of black export citrus ventures' (Outspan Briefing Document, 1995). Capespan pointed out that the establishment of new black citrus growers would be more difficult, if not impossible, in a deregulated market. Here the organisation promised to continue and expand its efforts in assisting existing and new black growers in becoming competitive citrus exporters. Finally, Capespan provided evidence on the higher revenues that research comparing its returns on overseas markets with those of its competitors. Returns for both Australian and Argentinean citrus growers were significantly lower, it claimed, because of the fragmented structure of the industry compared with South Africa (Stanbury, 1996).

Despite Capespan's efforts, it never seemed likely that the single desk would be retained. The Kassier Commission of Inquiry into agricultural marketing was impressed with the industry's achievements during the single channel era, and felt that these provided the basis for 'voluntary association' rather than a statutory system of controlled exports (Kassier, 1992). The survival of the single channel seemed very unlikely in the changing political and economic climate of a democratising South Africa. The largely white controlled financial press warned that the attempt by former marketing boards to retain the single channel was 'hi-jacking' the natural shift to a free market in agriculture and would lead to monopoly private marketing companies (e.g. *Finance Week*, 9 September 1993). The African National Congress was also opposed to private monopolies, but for political reasons. In its pre-election manifesto it expressed its desire to break the 'hidden monopolies' that existed within white commercial agriculture. Not surprisingly, Capespan failed to convince the ANC Minister of Agriculture, Derek Hanekom, who stated on various occasions that he was 'totally unconvinced' by the argument for a single channel marketing system. When the new Marketing of Agricultural Products Act (1996) was passed it came as no surprise that the single channel for citrus, as well as deciduous fruit, was dissolved.

The end of the single channel from the beginning of 1997 marked the end of almost 60 years of controlled citrus exports from South Africa. There is little doubt that the single desk system allowed the Citrus Exchange (and later Capespan) to 'drive' the chain in significant ways. Through its extension and research infrastructure it shaped the production practices of growers. With over half of the southern hemisphere's citrus crop it was also able to influence agents downstream of the chain. In this sense, the single desk exporter, representing the interests of producers, was able to drive the chain in important respects. The South African citrus case raises two important qualifications to the idea of chain governance. First, it was possible for producers to govern chains outside of the ideal typical categories outlined in the GCC framework. Second, the citrus case confirms that there are different degrees of 'driveness': the Citrus Exchange's ability to drive the chain was compromised by poor quality and increasingly by competition from other southern hemisphere producers. From the mid-1980s the pressure on the single channel intensified as overlaps between north and south increasingly led to chronic oversupply, a situation that was exploited by increasingly powerful retailers. The next two sections of this chapter examine the impact of the liberalisation of citrus exports from South Africa. Both sections explore the idea of 'fragmentation', firstly in terms of the chain itself and in the last section in terms of citrus growing regions in South Africa.

Chain Fragmentation

The term 'fragmentation' is appropriate to describe the impact of liberalisation on South Africa's citrus export chain. From 1997 to 1998 the number of agents exporting citrus rose from one (Capespan) to over 200. The new agents that were established ranged from small enterprises exporting a few thousand cases to large multinational fruit companies including Dole and Del Monte. The impact of many new exporters disrupted what was a very stable and secure system of fruit supply from South Africa. During the 60 years of regulated exports, Capespan and its cooperative predecessors were able to guarantee a steady supply of citrus for between 4 and 5 months of the year. Importers and retailers could always count on large volumes fruit, which partially made up for the uneven quality of South African citrus. With the fragmentation of the chain after liberalisation, this carefully managed system broke down. Markets in the UK and on the European continent were oversupplied or importers and retailers found that their fruit had been sold to another agent with a better offer. As the manager of one UK based service provider noted, 'It got crazy. Commitments weren't being met. All of a sudden the 50,000 cartons you were promised weren't there any more'.

The fragmentation of the chain has had an impact on export revenues for South African citrus producers. Since 1998 export earnings have been lower in most markets (Figures 2.2, 2.3). These decreases have occurred despite static or increased volumes of exports; in other words, the unit price for South African citrus has declined (Gibbon, 2003). In 2000, the third year after the dissolution of the single channel, returns to the citrus industry were the lowest in decades and the industry declared itself in crisis. The immediate cause of the crisis in 2000 was a combination of poor weather patterns which resulted in small and poor quality fruit from South Africa, and large seasonal overlaps of production from Spain and other northern hemisphere citrus producers. Capespan and a group of larger exporters blamed the situation on a large number of smaller, and less experienced, exporters which had apparently damaged the reputation of all South African citrus on overseas markets. The cause of *general* declines in export revenues since 1998 is, however, associated with deeper problems related to the impact of liberalisation and the subsequent fragmentation of the citrus export chain.

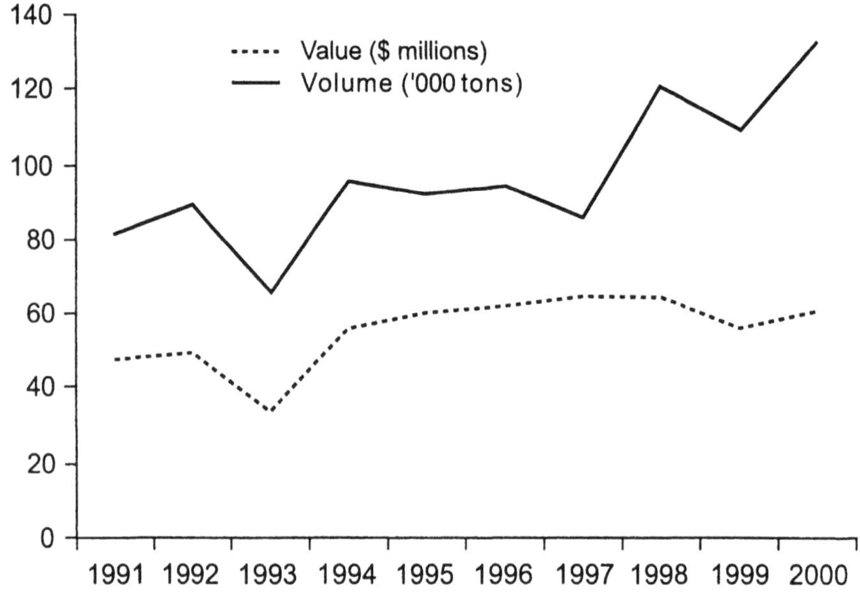

Figure 2.2 Citrus exports to the United Kingdom from South Africa

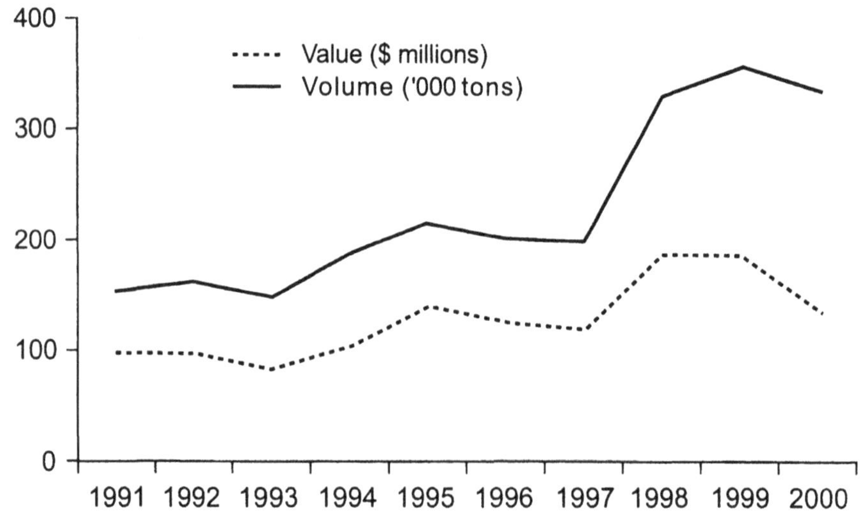

Figure 2.3 Citrus exports to continental Europe from South Africa

There are several inter-related reasons for the decline in export revenues for citrus producers in South Africa since deregulation in 1997. During the regulated era, as noted earlier, Capespan monitored the quality of exports at packhouses and at the ports. The organisation was also able to manage quality problems identified after the six week voyage to Europe by diverting lower quality fruit to less discerning markets, usually to Eastern Europe, or by destroying the fruit. Fruit quality was also monitored by the Perishable Products Export Control Board (PPECB), a state sanctioned entity that certified fruit for export in terms of quality and also in terms of phytosanitary regulations. With the fragmentation of the chain, it was no longer possible to manage quality problems in the same way. Moreover, the PPECB was unprepared for so many new export agents and lacked the capacity to effectively impose quality controls on citrus exports. A consequence of the fragmentation of the chain has been poorer quality control, which has in turn resulted in lower export revenues.

The problem of lower quality citrus was exacerbated by the oversupply of South Africa's traditional markets of the UK and continental Europe. In the last years of the single channel, Capespan remained focus on these two markets, but had also diversified into several other countries including Japan, Hong Kong and Canada. Based on a survey of more than 30 export agents it seems that many new agents focused their export efforts exclusively to continental Europe. The large number of European based importers, who will accept smaller volumes of citrus, made it easier for exporters without established networks to send fruit overseas. However, there are serious disadvantages to South African exporters and growers in this chain. Fruit destined for European importers is sold on consignment and there is no negotiation on prices beforehand. Although domestic exporters consider European importers, especially those based in Belgium and the Netherlands, to be 'excellent traders', the terms of the relationship is heavily weighed against producers especially when markets are oversupplied. When these importers find it difficult to dispose of the crop due to quality or oversupply, they do not hesitate to discount the fruit at the expense of producers. This problem appears to have been particularly severe in the 2000 season.

In the period since 2000 there have been efforts to privately regulate the citrus export chains, with a focus on volume and quality, although these vary according to destination markets. Citrus destined to the UK is now governed by multiple retailers like Tesco and Sainsbury's who play a central role in driving the chain. Their service providers, as noted earlier, have been given the responsibility of sourcing a secure supply of high quality citrus from South Africa. Although Capespan and the large cooperatives continue to supply some fruit to several UK retailers, service providers prefer to use mid-sized export agents or grower/packers. In many cases the export agents are directly involved in primary production. Service providers argue that they are better able to influence the production practices of mid-sized grower/packers, unlike the situation they face with larger cooperatives and Capespan. Establishing closer relationships with producers also means that service providers can discuss quality and the establishment of new varieties of fruit. Finally, mid-sized agents are in a much better position to guarantee traceability, which has been a problem for the larger cooperatives with hundreds of smaller producer members. The result is that the chain to the UK retailers consists of only 3 or 4 players: grower/packers – local export agent – service provider based in the UK – multiple retailers. Fruit exports to this market are usually 'programmed': volumes and varieties of citrus exports are coordinated long before the season starts. As a result of this detailed planning, the relationship between growers with pack-houses and high quality fruit, and retailers is often perceived as long term, even though no formal contracts are signed.

The chain to continental Europe is in flux and retailers appear to be playing a more important role in fresh fruit chains. Imports to Europe are handled by a large number of small, medium and large importers who accept fruit on consignment only, which places South African producers at a decided disadvantage. Yet even in this chain there has been some consolidation: growers are now less likely than before to send fruit on consignment where it is likely to be traded on wholesale markets. They are also more likely to send fruit with local or overseas-based agents who have an established reputation, or a secure market for the product. As a consequence of these changes the number of export agents operating in South Africa has declined from around 200 in 1998 to less than 100 in 2002. There has also been a concentration of market share in that 16 agents now market over 80 per cent of citrus exports from South Africa (Citrus Southern Africa, 2002).

The impact of liberalisation on exporters is reasonably straightforward in that an initial period of change and uncertainty has given way to consolidation and a more stable relationship between growers and exporters. The effect on producers and packers is however more complex and also geographically variable. During the 1990s the pressure on producers for higher quality fruit and new varieties of citrus increased dramatically. New standards were set for the internal quality of citrus fruit based on measurable criteria. Although more difficult to measure, there were also new requirements on the external appearance of fruit including blemishes, the amount of green tingeing and the shape of the fruit. Growers were also under pressure to produce new varieties of fruit including 'easy peelers' or early and late ripening varieties of existing citrus types to exploit 'market windows' in the northern hemisphere. Since the late 1990s growers and packers have been faced with a range of new process criteria formalised in systems such as HACCP or EurepGAP, which require significant capital investments. Failure to meet these requirements can result in being excluded from important markets. Finally, several new 'protocol' markets – which have special phytosanitary demands – were opened for South African exporters in the 1990s. The three most important have been Japan, Taiwan and most recently the United States. Exporting to these markets is extremely difficult due to strict control measures imposed by official representatives of the exporting countries. The sanitary and phytosanitary requirements in most 'non protocol' markets have, at the same time, increased dramatically since the mid-1990s.

A significant number of growers were protected from these rising quality demands during the last years of the single channel. The pool system – which averaged grower returns by variety through the season – ensured that growers with poor quality fruit were not always directly penalised. The better bargaining position of Capespan by virtue of its status as single desk exporter may also have offset demands for quality and various process requirements. Indeed, interviews with UK based fruit buyers suggest that it was difficult to impose quality standards on the South African citrus given that the single channel could supply huge volumes citrus through the northern hemisphere summer. In the period after 1998 individual growers lost the protection of the pool and the industry as a whole lost the bargaining position afforded by the single channel. The new demands for quality, variety and the higher phytosanitary standards for both traditional and new 'protocol' markets has led to sharp differentiation between citrus producers in South Africa. Many citrus growers have been excluded from export chains due to their inability, or reluctance, to meet the quality standards now being demanded by retailers and importers. Those farmers who have not gone bankrupt are shifting their emphasis to tourism opportunities or to less risky agricultural crops (Mather and Greenberg, 2003). Black citrus farmers in agricultural projects in the former homelands have found it especially difficult to meet the new standards of citrus export markets. The assistance they had from the single channel exporter has been withdrawn, as has the financial and other support provided by former homeland governments. Citrus farming in these regions has all but collapsed since liberalisation (Greenberg, 2002; Mann, 2001).

While a category of grower is being excluded from the global citrus chain, the ostensible winners in this process are those farming operations that are more attractive to overseas buyers. These farms normally have larger volumes of fruit, they often have attractive varieties including 'easy peelers', navels and other seedless citrus cultivars, and they usually have their own packing infrastructure. Several of these growers have moved up the chain, by establishing their own marketing companies with varying degrees of success. Indeed, most of these growers talk of the need to get 'closer to the customer' and express the desire to 'shorten the chain' to cut costs, much in the same way that retailers talk about establishing closer ties to producers. Although this category of farmer appears to have succeeded in the post liberalisation era, they operate in what remains a highly buyer driven chain. Improving the quality and variety of the fruit produced is not so much a way of 'upgrading' in the GCC sense; instead, upgrading is a way of staying on global chains, rather than moving up the chain with a view to capturing more value. Even the most highly sought after growers, who are known by name to UK retailers, complain of buying practices and prices which they say makes it difficult for them to farm sustainably. One very prominent Citrusdal farmer, with a reputation for the high quality fruit, remained vulnerable to overseas 'agents': 'There is always the loophole of quality. The agents play with quality. Two weeks ago I was told there was too much fruit (Clementines) in the UK. We know that in a few days we will get a call saying that there is a quality problem with the fruit'. For this Citrusdal farmer, this was a predictable pattern: problems with oversupply soon became a quality issue, which was then used to justify lower than expected returns to citrus growers. Even though these farmers continue to participate in the global citrus chain, they do so in what remains a buyer driven chain.

Cooperative pack-houses have also faced very difficult circumstances in the period since liberalisation. Those that exist outside the Western and Eastern Cape citrus growing regions have tended to be small, with a limited membership of growers and relatively small volumes of fruit. After the end of the single channel in 1998, many of these cooperative have been dissolved and the packing facility, usually outdated and requiring significant investment to meet HACCP or ISO regulations, has been sold or simply abandoned. The situation in the Western and Eastern Cape is considerably more complex. The cooperatives in these two regions are large with memberships of between 100 and 300 growers. During the single channel these cooperatives established a large pack-house infrastructure geared to sorting and packing millions of cartons of fruit during the 4 or 5 months of the citrus harvest season. Many cooperatives in these two regions also invested in retail stores, packing factories and other facilities that allowed it to decrease input costs for its grower members. By the mid-1990s most had a complex asset base that went far beyond the pack-house.

These large cooperatives have faced enormous challenges adapting to liberalisation of the citrus export chain after 1998. Before and after 1998 a considerable number of growers with a reputation for good quality citrus – and often larger volumes of fruit – broke away from the cooperative and established private packing facilities. The loss of volume has severely compromised their profitability, which was always based on using economies of scale to pack as cheaply as possible. Losing volume is not the only challenge facing pack-houses: demands are now being made for traceability, a difficult task for cooperatives with many grower members. Traceability has been a requirement for the UK market since 1990 through the Food Safety Act, which has required retailers to apply 'due diligence' in their sourcing practices (Marsden *et al*, 2001). In 'protocol markets' including the United States, Japan, Korea and Taiwan traceability is a basic requirement for exporters. Indeed, in Taiwan traceability is required to the orchard and not only to the farm. Although the ability to trace an orange back to the farm has not been a demand for exports to continental Europe, it is a key requirement for EurepGap accreditation, which comes into force at the end of 2003. Without EurepGap certification producers and packers face the prospect of being excluded from selling to EurepGap members and will be forced to sell their fruit in less lucrative markets or on wholesale markets where they are more vulnerable to discounting. Achieving traceability in a pack-house designed to pack millions of cartons of fruit for 200 or more growers has proved to be a significant challenge for cooperative pack-houses.

The inflexibility of pack-houses and the difficulties they face in achieving traceability has excluded them, at least until they can guarantee traceability, from some UK retailer programmes. Several service providers in the UK interviewed in 2001 expressed their reluctance to source from former cooperatives, preferring instead to establish relationships with grower packers with large volumes of fruit:

> In terms of cooperatives, our technologist and myself aren't that keen on cooperatives. It is almost impossible to guarantee traceability; our preference is to work with larger players that can offer traceability; they have due diligence; all the

> things you would expect. In fact I haven't visited a citrus grower in South Africa who doesn't have his own pack-house.

Large cooperatives have responded differently to the challenges they face in a fragmented and increasingly demanding citrus export chain. In Citrusdal, the Goodehoop cooperative decided to meet the new market demands 'head on' by selling its various assets and focusing on its 'core business' of fruit packing. In order to meet the demands for traceability it has forced to restructure its packing operations. It has also successfully achieved HACCP accreditation in one pack-house and is in the process of meeting this standard for the other two pack-houses it owns. Not surprisingly, the HACCP accredited pack-house is the newest of the three; achieving accreditation with the other two is a major obstacle for the company. The restructuring process has been difficult with significant job losses, the impact of which is evident in the small town of Citrusdal where unemployment levels have increased dramatically. In the Eastern Cape, Patensie Citrus decided on a more conservative route: in order to shore up losses in volume to other privately owned facilities it has enforced a 'pack-right' clause on its members. This regulation forces growers to deliver fruit to the company's pack-houses or face huge charges for their contribution to previous capital investments. This strategy appears to have failed in that growers have taken Patensie's claim to the competition commission for arbitration. After long deliberation, the Commission found in the favour of growers and rejected the pack-right as 'non-competitive behaviour'. The different responses of these two cooperatives appears to have been a consequence of management culture, but also perhaps differences in the financial situation of the two cooperatives. Due to various large investments in the late 1980s and early 1990s, Patensie's debt situation was considerably worse than Goodehoop's. Despite their different responses, both face considerable difficulties in meeting the new quality and process requirements (e.g. HACCP) of a liberalised market environment.

The fragmentation of the citrus commodity chain since deregulation has a dramatic impact on the South African citrus industry. For growers, the new quality requirements associated with ultra competitive global markets has excluded some growers who are unable or unwilling to restructure and invest to meet the demands of northern hemisphere importers and retailers. These growers are either moving away from agriculture or they are shifting to less input intensive agro-commodities. Growers who continue to participate in the global export chain tend to have larger volumes of fruit, they have an attractive range of citrus varieties and they often have their own packing facilities. These growers have also had to invest and upgrade to meet the escalating demands associated with global citrus markets. Although this category of grower seems to have been 'favoured', they are participating in an ultra competitive and buyer-driven food system where investment is not always sustainable (cf. McKenna and Murray, 2003). There has been an allied process of differentiation in the pack-house node of the chain with cooperatives in all parts of the country facing competition from new and modern privately owned packing facilities. The impact of this competition has depended on the size and asset base of the cooperative pack-house, which in South Africa has a spatial pattern. Outside of the Western and Eastern Cape, cooperative pack-houses have been shut down or they pack fruit on a contract basis. In these two provinces large cooperative pack-houses are facing a significant challenge in meeting the new demands associated with a liberalised market environment. Significantly for South African debates on deregulation in agriculture, they have faced problems using the resources established during single channel and apartheid eras in a new democratic and liberal economic environment. The chapter now shifts to consider a second form of differentiation associated with regional fragmentation of the South African citrus industry.

Spatial Fragmentation

During the single channel era citrus exports were presented to overseas markets as a South African product branded under the name Outspan. In the period since liberalisation this 'national product' has fragmented with many new brand names for South African citrus. Citrus growing regions within the country have also become more prominent than they were before: different regions are now better known for particular varieties or quality of citrus. However, a more powerful driver of this spatial differentiation has been the phytosanitary regulations that increasingly dominate the trade in citrus fruit. These regulations are re-shaping the geography of South African citrus production in complex ways.

Phytosanitary regulations have become central to market access or market exclusion especially between citrus producing countries and regions. Restrictions on citrus imports based on phytosanitary measures have also been used in tit-for-tat trade disputes, especially between the United States and the European Union. The events of September 11th 2001 appear to have increased the concern over unsafe food. Indeed, one Japanese multiple supermarket recently refused to sell Florida grapefruit over the concern that it might carry the anthrax virus. The United States has also responded to September 11 by proposing 'bio-terrorism' measures to control possible contamination of imported food products. In South Africa phytosanitary measures have played a key role in market access in the last decade and exporters now supply several 'protocol' markets including Japan, South Korea, Taiwan, and the United States. Meeting the phytosanitary demands requires considerable effort and significant risk, although if the fruit passes inspection the rewards are high. Fruit destined to Japan, for instance, must be 'cold sterilised', which means the fruit must be maintained at a temperature of -0.6°C for at least 12 days. Inspections occur at South African ports by Japanese plant quarantine authorities and in Japan. Fruit that fails to meet the protocol after arrival in Japan must be incinerated at the expense of the exporter.

Although the protocols vary between different markets, most citrus regulations now restrict the trade of fruit with a disease called 'citrus black spot' (CBS). This particular phytosanitary problem is both controversial and difficult to control. CBS affects the fruit itself, which results in decay, but there are no recorded instances where the disease has been transferred from one citrus producing country to another via exports. Despite this lack of evidence, United States and Japan both restrict citrus infected with black spot. In the early 1990s the European Union declared CBS a 'quarantine organism' and now also restricts fruit exports with this particular disease. The way in which the United States, Japan and the European Union regulate CBS is slightly different. Citrus exports to the United States are restricted to fruit produced in the Western Cape, a region considered to be 'black spot free' due to its Mediterranean climate. While Japan will accept citrus from areas affected by black spot, this particular protocol also specifies various specific districts known for grapefruit or lemon production. The European Union does not specify districts, but has become increasingly strict on fruit that may be affected by black spot.

The nature of the regulations associated with citrus black spot has had important implications for citrus growing regions in South Africa. Since gaining access to the United States market, Western Cape growers have focused much of their efforts in improving the volume and quality of citrus exported to the US. Exports have, as a consequence, risen from 30,000 cartons in 1997 to over one and a half million in 2002. Although this volume is only 3 per cent of total exports from South Africa, in income terms it represents 17 per cent of total income from citrus exports. There have also been very active local efforts to decrease the amount of fruit that does not pass USDA inspections at pack-houses and at the port. Rejections were high in the first years of citrus exports, primarily due to a pest called mealybug. Efforts by industry representatives have since led to dramatic improvements in the amount of fruit approved by the USDA inspectors. Western Cape citrus growers have also been active in improving the quality of fruit, especially the colour of navel oranges and the quality of easy peelers, which initially resulted in a poor reputation of South African producers. Finally, Western Cape growers have established a formal link with citrus growers in Australia to regulate volumes to prevent oversupply. Not surprisingly, there are several other districts, mainly in the Northern Cape, now claiming to be 'black spot free'. Growers in these regions are actively involved in attempting to have their orchards acknowledged to be CBS free.

The extent to which Western Cape citrus producers now see their future tied to the US market was underlined with George Bush's recent visit to South Africa. While his visit was marked by various protests against the US led invasion of Iraq in cities like Pretoria, Johannesburg and Cape Town, farmers and farm workers in Citrusdal, the heart of the Western Cape's citrus industry, 'paraded in decorated tractors and trucks' to celebrate their access to this lucrative market.

The Western Cape is not the only region that is focused on a high value export chain. In the Nkwaleni region of KwaZulu Natal, grapefruit producers see their future in Japan, where consumers are prepared to pay very high price for good quality fruit. For grapefruit producers this is an extremely important development as the market for grapefruit, particularly in the United Kingdom and continental Europe has been depressed for most of the 1990s. Declining consumption is blamed on demographic changes – young people do not like grapefruit – and the lack of innovation in the commodity. In Japan, however, grapefruit consumption remains strong and the fruit is also used in food decoration. As noted earlier there are very strict phytosanitary regulations for this market. Japanese fruit buyers are also extremely quality conscious and have since the late 1990s played an important role in driving quality in regions like Nkwaleni. Fruit buyers from Japan focus on the colour and the size of the fruit in the box, which must be even and regular. Indeed, one grapefruit farmer with a 'pack' characterised by uneven colour and size was encouraged by a Japanese buyer to purchase an optic sizer, a piece of equipment that automatically sorts fruit into colours and sizes. Fruit size is also important for the Japanese market, and consumers are demanding larger and larger sized fruit. Internal quality is also important for Japanese fruit buyers and they usually insist on a brix (sugar acid ratio) of 10; anything higher than 10 is rewarded with much better returns for growers. According to one grapefruit farmer,

> The Japanese come here through the season. They cut the fruit, they taste it, and they use their hygrometers (to measure the sugar acid ratio). They want to know what the fruit is like before it arrives in Japan.

Many grapefruit producers in the Nkwaleni region have responded positively to these demands, although it has required significant new investments in trees and rootstocks. During the single channel era grapefruit farmers were encouraged to produce higher volumes – but not necessarily better quality – fruit. In response farmers used rough lemon rootstocks for the grapefruit trees which delivered higher yields per tree. Grapefruit producers were also instructed to plant orchards in high densities to encourage higher yields per hectare. Although these methods improved the yields of individual farmers they tended to result in smaller fruit with poor internal quality. When South African citrus producers gained access to the Japanese market in the 1990s, they found it necessary to shift away from volume and towards larger and better quality fruit. After the liberalisation of exports, the pressure to improve quality has increased and Japanese buyers, as noted earlier, play an active role in encouraging citrus farmers to produce better quality grapefruit. Farmers that have responded to these quality demands have had to make radical changes to production methods including moving away from high density planting and by replanting grapefruit on better quality rootstocks.

Shifting to production methods to meet the demands of Japanese consumers has demanded huge investments and this, together with the greater uncertainty, has led to numerous bankruptcies. In Nkwaleni valley almost half of the grapefruit farmers have been forced out of business. Farmers that continue to supply the difficult Japanese market face very difficult economic conditions even though this is a very lucrative market. Supplying larger fruit means that the overall export per centage on any farm is much lower.

Meeting the demands for internal quality and size also has implications for the frequency of re-investment in trees and infrastructure. Younger trees tend to producer larger and better quality fruit; as trees get older the size and quality of the fruit tends to decline:

> In the old days you could leave trees for 20 to 25 years before you needed to pull them out. Now because of the size demands for grapefruit the lifespan is down to 15 years.

A more positive development is that it has led to more careful farming practices, especially those associated with water use:

> We used to stop irrigating when it rained. Then we'd scratch the soil to see if it was dry. If it was, we'd start irrigating again. We don't operate like that any more. Now we have a neutron probe and if the tree needs water we irrigate. It is now possible for us to get a brix of 11 or 12 for grapefruit, and that is something that was never heard of before.

The Western Cape province and the Nkwaleni region of KwaZulu-Natal highlight a very important post-liberalisation development in the South African citrus commodity chain. Both regions are now tied very closely to specific international markets and farming practices are now shaped by the quality and phytosanitary demands of these markets. In both regions, but perhaps more noticeably in Nkwaleni, farming practices have changed dramatically in line with their specific market. As one Nkwaleni farmer said, 'Every decision I make on this farm is for Japan'. While these regions appear to have been successfully 'incorporated' into global chains, there has been differentiation both within and between regions as a result of this process. In both Citrusdal and Nkwaleni there are a significant number of farmers that have failed to meet the new 'market realities' of a deregulated industry and have been forced out of business. Many other farmers are not prepared to risk investing additional resources to produce fruit that meets the size and internal quality demands of Japanese or US citrus consumers. The phytosanitary regime associated with citrus black spot is also leading to the marginalisation of regions especially vulnerable to the disease. Although the European Union's CBS directive was issued in 1993, the pressure on citrus producers has been more intense in the last 4 years. There is now 'zero tolerance' for fruit found to be infected with CBS and all South African citrus producers outside of the Western Cape face the prospect of being excluded from the European Union, a region that accounts for between 60 per cent and 70 per cent of fruit exports. Not surprisingly, the South African Department of Agriculture and representatives of the citrus industry have reacted quickly to prevent infected fruit from reaching European markets. Farmers that consistently fail to control CBS are now placed on a 'black list'; fruit originating from black listed farms is diverted to markets which are not sensitive to CBS, including the Middle East, Indian Ocean islands and the rest of the African continent. Unfortunately for these producers, returns in these markets are far lower.

Citrus producers in regions more vulnerable to black spot can control the disease but it is extremely costly, both in the long and short term. In the short term, there are stronger chemicals, which promise more effective control of CBS. However, these are considerably more expensive than chemicals previously used to control the disease. Beyond chemical treatments, black spot can be managed by improved monitoring in the pack-house, although since the disease can manifest itself some time after harvesting this is not a fool-proof method of controlling CBS. A complicating factor is that it is very difficult to identify black spot in its early phases and fruit that has minor black blemishes, but is otherwise disease free, may be rejected for fear of problems at later stages in the chain. The result is that export percentages decline along with the revenues of citrus producers. Finally, since older trees are more vulnerable to CBS, farmers can plant more frequently, although this also affects the economic sustainability of citrus farming. This section of the chapter has suggested that the regional geography of the South African citrus industry is being reshaped in the post-deregulation era. Regions are now being tied to specific global chains citrus, which is in turn reshaping the structure of production in these regions as producers meet, or fail to meet, the high quality demands of international buyers. One of the drivers of this process is Citrus Black Spot, a phytosanitary disease that affects fruit produced outside of the Western Cape. As a black spot free region, Western Cape growers now have exclusive access to the United States market and production is being reshaped to meet the specific demands – both relating to phytosanitary and quality issues – required for this market. Regions outside of the Western Cape are finding it extremely difficult to participate in global chains as restrictions on the disease become tighter.

Conclusion

This chapter has explored the impact of liberalisation on South Africa's citrus export chain. The first section of the chapter examined the extent to which the chain could be described as 'producer driven' in the period when all exports were controlled by Citrus Exchange. I argued that the single desk agent was able to drive the chain in important respects, but that its power to govern the chain was always compromised by quality problems, competition from other citrus producing countries and, by the late 1980s, by the power of multiple retailers. The second sections of the paper explored the impact of 'buyer-driveness' in the citrus industry. The term 'fragmentation' was proposed a suitable word to describe the impact of liberalisation. There has been significant differentiation amongst citrus growers, between those who are willing to meet the new requirements of global markets and those who cannot, or who will not. Amongst cooperatives there has been a similar process of differentiation with older and larger cooperatives facing enormous challenges in meeting the quality and codes of conduct required for export markets. The last section of this chapter suggested that an important result of liberalisation has been the spatial reorganisation of the industry, with regions now identifying themselves closely with specific citrus chains. The phytosanitary regime associated with citrus production is playing an important, although not exclusive role in this process of spatial reorganisation. The consequence of this spatial reorganisation is uneven development in that some citrus producing regions are being actively incorporated into global chains, while others are being marginalised from export markets.

References

Barrett, H.R, Ilbery, B.W, Browne, A.W and Binns, T, (1999) Globalisation and the changing networks of food supply, *Transactions: Institute of British Geographers*, 24, pp. 159-174.

Cartwright, A.P. (1977) *Outspan Golden Harvest: A History of the South African Citrus Industry*, Cape Town: Purnell.

Citrus Board (various years) Annual Report, Citrus Board of South Africa, Pretoria.

Citrus Southern Africa, (2002) Newsflash Week 23, 13 June 2002, Citrus Southern Africa, Cape Town.

Coe, N., Hess, M., Yeung, H.W-C., Dicken, P., and Henderson, J., (2003) 'Globalizing regional development': a global production networks perspective, Paper presented at the 99th annual meeting of the Association of American Geographers, New Orleans, USA.

Dolan, C. and Humphrey, J., (2000) Governance and trade in fresh vegetables: the impact of UK supermarkets on the African horticulture industry, *Journal of Development Studies*, 37(2), pp. 147-176.

Friedland, W.H. (1993) The new globalization: the case of fresh produce, in A. Bonanno *From Columbus to Conagra: the Globalisation of Agriculture and Food*, University of Kansas, Kansas.

Friedman, H. (1993) The political economy of food, *New Left Review*, 197, pp. 29-57.

Gereffi, G. (1994) The organisation of buyer-driven global commodity chains: how US retailers shape overseas production networks in Gereffi, G. and Korzeniewicz, M. (eds) *Commodity Chains and Global Capitalism*, Praeger, Westport.

Gibbon, P. (2003) Value chain governance, public regulation and entry barriers in the global fresh fruit and vegetable chain into the EU, unpublished mimeo.

Greenberg, S. (2002) *Citrus production in the Eastern Cape's former homelands*, MA thesis, University of the Witwatersrand, Johannesburg.

Goodman, D. and Watts, M., (1997) *Globalising Food: Agrarian Questions and Global Restructuring*, Routledge, London.

Henderson, J., Dicken, P., Hess, M., Coe, N., and Yeung, H.W-C, (2001) *Global production networks and the analysis of economic development*, unpublished paper.

Kassier, W.E., (1992) *South Africa: committee of inquiry into the Marketing Act*, Department of Agriculture, Pretoria.

Le Heron, R. and Roche, M. (1995) A 'fresh' place in food's space, *Area*, 27(1), pp. 23-33.

Lesley, D. and Reimer, S. (1999) Spatializing commodity chains, *Progress in Human Geography*, 23(3), pp. 401-420.

Mann, R. (2001) Homeland agriculture in the era of liberalisation: high value fruit exports from parastatal projects, *South African Geographical Journal*, 83(3), pp. 214-220.

Marsden, T., Flynn, A., and Harrison, M., (1999): *Consuming Interests: The Social Provision of Foods*, London, UCL.

Mather, C. and Greenberg, S. (2003) Market liberalisation in post apartheid South Africa: the restructuring of citrus exports after deregulation, *Journal of Southern African Studies*, 29(2), pp. 391-410.

Mather, C and Rowcroft, P. (2004) (Re)presenting South African citrus, *Geographies of Commodities*, S.Reimer and A. Hughes (eds), Pearson, London.

McKenna, M.K. and Murray, W.E., (2002) Jungle law in the orchard: comparing globalization in the New Zealand and Chilean apple industries, *Economic Geography*, 78(4), pp. 495-514.

McMichael, P. (1996) Globalization: myths and realities, *Rural Sociology*, 61(1), pp. 25-55.
Outspan International, (1995) Outspan briefing document, unpublished report on the citrus industry, Outspan, Pretoria.
Rabobank, (1997) *The World of Fresh Fruit Trade*, Rabobank International, Utrecht.
Rabobank, (2001) Strategies implemented by the top four listed fruit traders, Industry Note, Food and Agribusiness Research, Rabobank, Utrecht.
Raikes, P. and Gibbon, P., (2000) 'Globalisation' and African export crop agriculture, *Journal of Peasant Studies*, 27(2), pp. 50-93.
Raynolds, L. (1997) Restructuring national agriculture, agro-food trade, and agrarian livelihoods in the Caribbean, in D. Goodman and M. Watts, *Globalising Food: Agrarian Questions and Global Restructuring*, Routledge, London.
Shepherd, A. and Farolfi, S., (1999) *Export crop liberalization in Africa: A review*, FAO Agricultural Services Bulletin, 135.
Stanbury, J. (1996) Unity is strength, letter to *Financial Mail*, 5 January 1996.
Van der Laan, H.L., Dijkstra, T., and van Tilburg, A. (eds), (1999): *Agricultural Marketing in Tropical Africa, African Studies Centre*, Leiden, The Netherlands.
Watts, M. (1996) Development III: the global agrofood system and late twentieth-century development (or Kautsky *redux*), *Progress in Human Geography* 20, pp. 230-245.

PART II
REGIONAL SYSTEMS AND GLOBAL NETWORKS: THE CASE OF INNOVATION

Chapter 3

Global Networks as Open Gates for Regional Innovation Systems

Ayda Eraydin

Introduction

Currently, there is widespread interest in regional innovation systems and regional innovation strategies as new tools for regional development. This new interest goes back to the 1970s, when traditional policies were replaced by a strong appeal for endogenous development and local and regional initiatives for sustained economic growth. The emphasis on the local introduced a new understanding and several competing and co-operating models of territorial development-evolution, which were primarily based upon proximity dynamics (Eraydın, 2002a). The new focus was on different types of newly emerging clusters referred to as innovative milieux (Camagni, 1991), industrial districts (Belussi, 1999), new industrial spaces (Scott and Storper, 1987) or learning regions (Florida, 1995; Morgan, 1997). However, analytical studies of these grassroots developments were not sufficient to provide clear policy recommendations. That is why in the 1990s regional innovation strategies became the core of an agenda that aimed to create territorial innovation systems with the ability to learn, diffuse and exploit knowledge (BITNET, 2001). The term 'regional innovation system' not only includes the collaboration of innovative activities between enterprises, but also captures the trend to build regional organisations and networks in order to strengthen the innovation capability of enterprises using the support mechanisms provided by regional and national governments (Cooke, 1998a).

Both the early debates on industrial clusters and the concept of regional innovation system have a certain bias, namely they analyse individual regions, which overemphasises local agents and local relations. In fact, it is possible to trace this tendency via the debates on local embeddedness, tacit knowledge, institutional thickness, social capital, local production culture and many others. In contemporary economic relations, however, it is obvious that none of the regions can achieve sustained growth dependent solely on endogenous development processes. They require networks with non-regional actors, which link regional economic systems to the global economy. There is essentially a consensus on the positive role of external networks and the need for external knowledge transfer. They are believed to be important for preventing a lock-in situation in regional clusters, which is usually faced after the initial phase of growth, following the exploitation of already available capacities. There are increasing numbers of studies that provide evidence to support this hypothesis.

This chapter is on global networks that act as open gates for regional innovation systems. It aims to discuss the increasing role of global networks for the sustained growth of regional clusters and to define how global networks and regional innovation systems are connected to each other. After a brief description of a regional innovation system in the first section, the second section of this chapter will attempt to re-identify this complex economic space with reference to networks. This manner of conceptualisation defines global networks as the integrated parts of innovation activities. In section three the role of global networks in the successful development of a regional innovation system will be discussed with reference to three key issues, (a) the characteristics of clusters and global networking, (b) the changing network structures and (c) the importance of network organisers in the arrangement of new networks. The concluding remarks basically refer to the findings of empirical studies on networks and emphasise the importance of global ties for the long-term benefits of regional clusters.

From Clusters to Regional Innovation Systems

The clustering of firms has been in the agenda since the 1950s, which marked the beginning of regional planning and regional policies in many countries of the world. In order to initiate agglomeration economies, the makers of regional policies identified core industries and encouraged the clustering of other industries around them. These clusters based on traditional vertically integrated production units, however, have proved too inflexible to adapt themselves to new circumstances, whereas clusters of small firms have demonstrated the capacity to adapt to changing conditions (Arzeni, 1996). The experience of successful clusters shows that they usually emerge and begin to grow because of local dynamics, and it is after this stage that governments reinforce and support this growth trend through various institutions and measures (Cooke, 2002). It is interesting to note that the success factors attributed to clusters form a long list, which increased considerably during the evolution of the discourse on new spaces of growth. A review of the key features of successful clusters shows that they tend to display certain common characteristics, which include mutual support, entrepreneurship, learning, creativeness, innovation, network development, new market relations, conflict solving, lower transaction costs, collaborative action, etc. However, it is possible to regroup them under three main headings: the power of networking, the socio-cultural and institutional context in which firms operate and learning opportunities. These characteristics are defined as essential for local clusters to sustain their competitiveness by learning, adaptation and innovation.

Recently, however, it became more obvious that not all clusters can sustain their earlier growth performance (Amin, 1999; Malmberg and Maskell, 1997;

Lyons, 2000; Staber, 1996). There may be different reasons for the downturn, ranging from local dynamics to external conditions (Kautonen, 1996; Amin and Thrift, 1995). If a cluster is not able to initiate substantial adjustment of its fundamental features, it is not possible to guarantee long-term sustained growth. Obviously, adjustment to changing conditions is not easy for any cluster. There are several factors that generate difficulties in transformation and degrade competitiveness.

Firstly, a cluster based on crafts and local small entrepreneurship defines a self-limiting model. In this model there are different things that obstruct the flow of local knowledge, such as problems arising at the boundary between one speciality and the next and the tendency to safeguard a stock of proprietary knowledge, in other words craft secrets (Sabel, 1996). Secondly, learning by learning by collaborative relations is also problematic. Mutual reliance has its price and it is difficult to draw a line between the zones of complementary responsibilities. Thirdly, the successful industrial districts of the 1970s faced a new set of pressures, such as more global patterns of subcontracting, problems in keeping abreast of new technologies in internationalised markets, as well as protecting existing inter-firm relations within the existing districts. These conditions have a negative effect on the internal coherence of districts and their model of learning based on interactive and collaborative networking. In fact, as the experience of several clusters indicates, agglomeration may be an asset for enterprises, but it can also lead to regional lock-in as a development path, which can turn out to be harmful in the long run (Glasmeir, 1991 and 1994). The main solutions to lock-in problems are related to innovation and innovative networking. As Kautonen (1996) suggests, to avoid lock-in, network relations that will lead to innovation are needed both within local milieux and in the global environment. Co-operation among local producers can create a division of labour among 'network innovators', who are able not only to diffuse but also to stimulate innovations (Bianchi and Bellini, 1991: p.488). However, there is also a need for other agents and measures in order to upgrade regional innovation potential.

What determines regional innovation potential? Spatial cumulativeness, which shows the degree of persistence of regional technological capabilities, is a fundamental explanatory variable of the higher innovative performance of certain regions (Breschi, 2000). The factors that contribute to regional potential can be defined at two levels, the regional infrastructural level and the superstructural level (BITNET, 2001). At the infrastructural level autonomous tax spending, available regional private finance, well-equipped infrastructure services and a regional industry-university strategy and many others provide a higher potential. At the superstructural level, the basic determinants of a region's innovation potential are defined as firstly, the institutional characteristics such as co-operative culture, interactive learning and associative consensus, secondly, the organisation of firms with harmonious worker relations and interactive learning, and thirdly, policies based on the principles of consultation, inclusion, networking and monitoring.

Obviously, not all regions have these characteristics. What can be done to promote growth and innovativeness in less developed regions without the above characteristics? The recent answer by regional scientists is to design '*a regional innovation strategy*' under the guidance of public institutions for creating '*a regional innovation system*'. In defining these new strategies and public policies the key issue is how to design an enterprise support system which resonates with the firms' immediate needs on the one hand and yet stimulates their long-term learning capacity on the other (Morgan, 1996).

Networks as the Basis of Regional Innovation Systems

In the 1990s, after a long period in which the characteristics of clusters and the factors of their success were described, the need to move from analysis to policy became more evident. Regional innovation strategies were formulated as a new policy framework in order to support the needs of less favoured areas. These strategies are based to a great extent on the theoretical debates on regional innovation systems. That is why in the literature 'a regional innovation system' concept is used in both theoretical discussions (Cooke, 1998a; Cooke et al., 1998; de la Mothe and Paquet, 1998) and empirical studies (Braczyk et al., 1998; de la Mothe and Paquet, 1998a; Asheim, 1997; Asheim and Isaksen, 2002) and also as the basis of several project reports (Leigh, 1995; Mc Kibbin, 2000). In these studies a regional innovation system is defined as 'collaborative relations between the actors that translate knowledge into regional income growth' (Landabaso, Oughton and Morgan, 1999). If the actors of the system are specified the definition becomes:

> a system in which firms and other organisations [such as research institutes, universities, innovation support agencies, chambers of commerce, banks, government departments] are systematically engaged in interactive learning through an institutional milieu characterised by embeddedness (Cooke et al., 1998: p.1581).

The aim of a regional innovation system is to combine traditional, context-linked, regional knowledge with codified knowledge available worldwide in order to stimulate the regional endogenous potential. In this process the *innovating firm* is defined as the focus of the innovation process, which is surrounded by various actors who are linked to each other within innovation activities, including their support industries (Asheim and Isaksen, 2002). An enterprise's innovative capacity is highly dependent upon the ability of the enterprise to come into contact with and co-operate with other actors, such as customers, suppliers and R&D organisations (Gregersen and Johnson, 1997). Therefore, a regional innovation system refers to how firms, institutions and government – jointly and individually – contribute to promote the innovation process within a regional context. There are contradictory views on the sectoral composition of innovative firms. According to Audrecht and Feldman (1996), the diversity in metropolitan areas generates a good medium for innovation. They support their argument by showing that 96 per cent of all registered innovations were generated from metropolitan areas. Similarly, in addition to the vertical co-operation partners, such as buyers and suppliers, producer services and other supporting firms have been found to be very important (Revilla-Diez,

2001, Eraydın, 2002b). On the other hand, there is also extensive literature indicating the benefits of specialisation and the availability of complementary firms in certain fields of specialisation.

In regional innovation systems *institutions* are very important, since they determine the rate and the direction of innovative activities (Lundvall, 1998). In the literature the term 'institutions' may refer to different things; e.g., institutions can be considered 'formal structures with an explicit purpose' or 'things that pattern behaviour' such as norms, rules and laws (Hassink, 2001; Edquist, 1997). Although considerable reference is made to the latter type in many theoretical debates, in the analytical studies formal structures are taken into account for practical purposes. Cooke (2002: p.137) groups all these components into twosub-systems, the knowledge application and exploitation sub-system and the knowledge generation and diffusion sub-system. Firms as customers, collaborators, contractors and competitors form the first group, whereas business consultants, technology centres, R&D centres, university departments, laboratories, technology transfer and R&D centres, development agencies, etc. constitute the supporting units for specialised knowledge creation and diffusion (Hassink, 2001). In innovation systems regional governments have several roles, such as catalyst, facilitator or broker in the articulation of innovative activities. Articulation means linking regional actors and matching these actors according to their innovation needs, initiating collaborative activities among different actors and designing policies to integrate all actors in a regional innovation system (Landabaso, Oughton and Morgan, 1999).

Lastly, *networks* constitute the major component of an innovation system, since the main character of this type of territorial systems is their efficiency at creating, diffusing and exploiting knowledge. Networks enable the transfer of knowledge among the different parts of the system. There are flows of intellectual resources between institutions, which are linked with formal and informal networks, and firms are the important parts of these networks whose activities initiate, support and diffuse new technologies and innovations (de la Mothe and Paquet, 1998). Such regional networks include firms, research and technology organisations, innovation support agencies, venture capitalists and local/central government bodies. Linkages between these various bodies are essential, since they turn knowledge into competitive economic advantage (Northern Ireland Economic Council, 2001). In regional innovation systems networks that bind the agents of a system also define the performance of these territorial systems, since the existence of well-defined institutions may make a limited contribution if they underutilise possible networks.

Redefining Regional Innovation Systems on Networks

Several studies have indicated the importance of networks and network partnerships in innovative activities[1] (McKibbin,[2] 2000; Hassink, 2001). Formal and explicit co-operation between firms in the innovation process (Asheim and Isaksen, 2002), links to new knowledge bases, access to venture capital (Cooke at al., 1998) and university-industry-government relations (Etzkowitz, 1994) are defined as the prerequisites for a regional innovation system. Having looked at a number of best practice systems, Cooke, Roper and Wylie (2001) suggest that, although the nature of regional innovation systems can vary, successful systems tend to display a number of common characteristics. These include intensive co-operation and knowledge spill-over via networks, which are very important in originating network externalities. There is no doubt that local/regional networks are fundamental to regional economic growth, especially for innovation-oriented regions (Granovetter, 1985; Grabher, 1993; Cooke and Morgan, 1993). In fact, in many studies a regional innovative system is characterised by a number of key local/regional networks. Figure 3.1 shows the possible local networks between the actors of a regional innovation system. A systematic definition includes the networks between firms (other firms, clients/customers, competitors and consultants), between firms and R&D infrastructure (government research institutes, private research institutions and universities) and between R&D infrastructure units (government-private R&D, university-private R&D, university-government R&D). It is true that many networks have a strong local dimension despite the emergence of global networks (van der Berg et al., 2001), since co-operation with a partner located in the same region is obviously the first step in building a network.

The existing models of RIS are more concerned with internal relations among firms and institutions and they treat external relations as residuals (Cooke, 2002; Landabaso, Oughton and Morgan, 1999). It is also interesting to see that in many empirical studies on RIS little reference is made to external networking. However, for certain activities other long distance relations may be even more valuable than local networking. For example, a study on the innovation systems of several metropolitan areas showed that spatial proximity is more important in horizontal linkages between manufacturing firms, research institutes or competitors, whereas vertical cooperation can be realised over longer distances (Revilla-Diez, 2001). The intensity of internal networking in many regional innovation systems does not mean they do not need knowledge, information and other opportunities provided by global networks. In fact, an increasing number of studies, such as Echeverri-Caroll and Brennan (1999), indicate the vital importance of external

---

1 The West Midlands survey results indicate that firms that engaged in joint activities were up to 6 times more likely to innovate than firms in isolation (Landabaso, Oughton and Morgan, 1999).

2 The paper evaluates the experience of 20 regions that participated in the RIS and RITTS programs of the EU.

networking for sustained development. The three issues described below provide a summary of discussions that emphasise the need to take global networks not as residuals of an innovation system, but as an integral part of it.

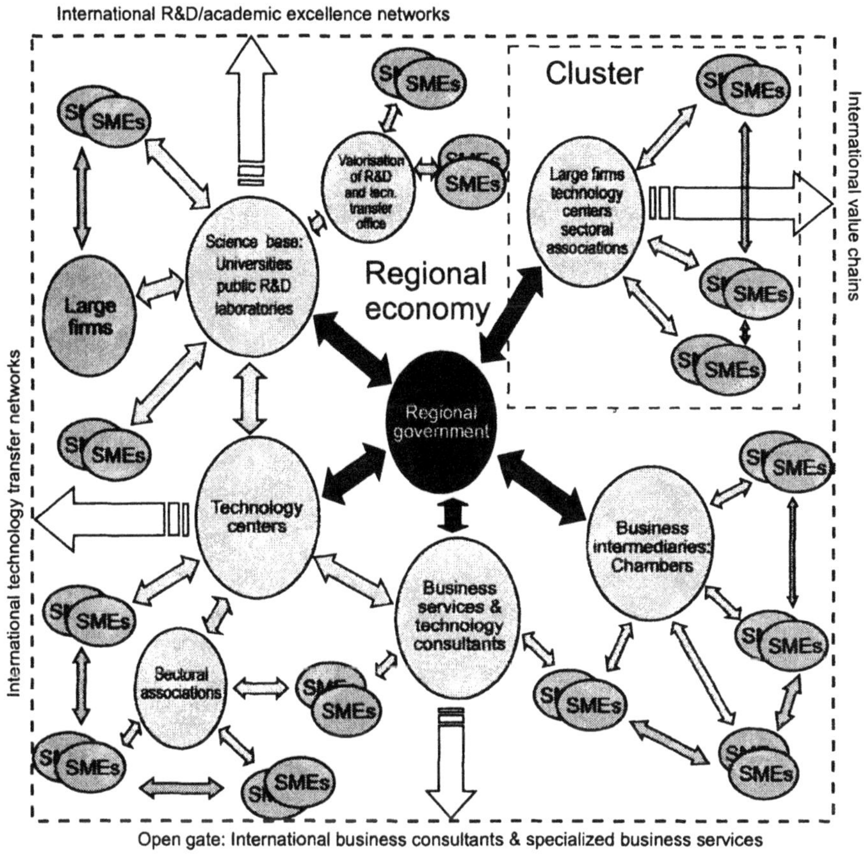

Figure 3.1 Local networks in a regional innovation system

Source: Landabaso, M. Oughton, C. and Morgan, K. (1999), Learning regions in Europe: Theory, policy and practice through the RIS experience, paper presented at the 3rd International Conference on Technology and Innovation Policy: Assessment commercialisation and application of science and technology and the management of knowledge, Austin, TX, USA, August 30-September 2, p. 7.

- Networks in different spatial categories ('within the region', 'rest of the country', 'abroad'), do not substitute for each other (Fritch, 2001).
- Global networks, which are not formed in a bounded regional space, enable firms and other actors to obtain up-to-date information and use a variety and mixture of competencies available in different places (Amin and Cohendet, 1999).
- In the contemporary economic conditions, where it is essential to gain dynamic competitiveness, there is a need to know what is happening in other regions and firms. The only way to obtain knowledge is through global networking.

The above points support the need to conceptualise a regional innovation system by taking both local/regional and global/national networks into the focus of analysis, instead of overemphasising relations between local agents. Accordingly, a regional innovation system can be defined as *an area of interaction (a medium of interplay) of networks where local innovative networks enable the exploitation of local capabilities and capacities and external networks function as open gates in order to recapture and adapt the different types of knowledge.*

Landabaso, Oughton and Morgan (1999: p. 7) define four types of global networks that perform as open gates for an efficient regional innovation system. These open gates are *international value chains, international R&D/academic excellence networks, international technology transfer networks and international business and consultancy services*. If these four types of global networks are combined with local networks, the number of possible networks to be used for knowledge transfer, dissemination and sharing of innovative activities and the complexity of the system increase enormously. Figure 3.2 shows that when global networks are fully utilised a regional innovation system not only becomes more open to external contributions, but can also become a part of the global economy. Moreover, they can have a greater chance to capture knowledge circulating in the world, even that embedded in other networks and regions. Under these conditions a regional system moves from being a bounded space towards an unbounded space, still with locally embedded knowledge accumulation.

Innovative Networking among Local Actors

In idealised models it is argued that regional innovativeness is rooted in networking that can exploit local capabilities and effectively combine them for innovative actions. In this system there are several possible networks, some of which are facilitated by interpersonal networks. It is believed that they enable their members to obtain 'sticky' knowledge embedded in social interactions (Asheim and Isaksen, 2002). These models need to be verified by empirical studies with the help of two main questions. Firstly, to what extent is the possible networking defined in idealised models actively utilised in the existing innovative clusters? Secondly, to what extent are networks between firms and institutions and among institutions themselves important in generating an innovative action? These questions encourage a search for a deeper understanding of networks between local actors.

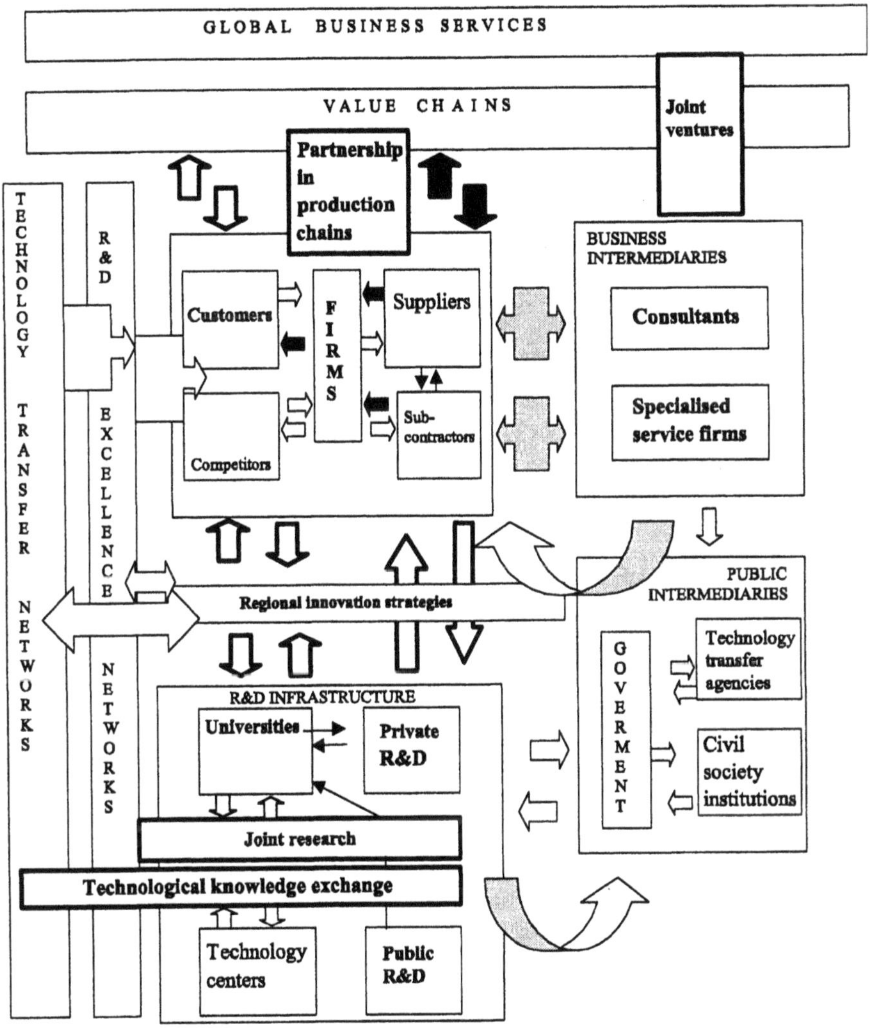

Figure 3.2 The interplay between local actors in RIS and global networks

Networking between Firms

Relationships between firms as customers and suppliers are a very important factor in their access to innovative resources. Many studies indicate the crucial importance of relations with customer firms in innovative activities (Eraydın, 2002c; Boekholt and van der Weele, 1998; Tödling 1999). Cooke (1998b), Boekholt and van der Weele (1998: p. 58), Cihnioglu, Oğuzsoy and Şen (2002) and several other authors[3] agree that a very large share of innovative resources utilised by innovative firms comes from interactive learning and exchange relationships between supplier and customer firms.[4] Similar evidence is provided by the REGIS study (Tödling and Kaufmann, 1999: p. 708). According to the findings of this study, the most important partners in the innovation process are firstly customers and secondly suppliers, which are not confined only to their region. The findings indicate that national and European partners are important. Similarly, the evidence from a study on transport equipment firms in North Carolina indicates that customer and supplier networks are equally important (Bergman and Feser, 2001). In general, it is possible to say that customers are always important in innovative process, but suppliers can also trigger innovation through better performing components and new raw materials.[5]

It is also interesting to note that in order to fully utilise the benefits of supplier and customer networks some changes in the organisation of networking are needed. Ottai (1998) reports that in Tuscan districts (Italy) 'the stage' firms were inclined to form 'enterprise groups'.[6] The relations between parent and supplier firms changed from purchasing to subcontracting in the South Brabant region (Netherlands) (Boekholt and van der Weele, 1998). Another form of changing

---

3 Dankbaar, B. and Hassink, R. 1995, Technology networking in border regions: Case study of the Euregion Maas-Rhine, *European Planning Studies*, Vol.3, Nos 163-84 and STEP Group (Nås et al) 1994.

4 Cooke (1998) reports that more than 60 per cent of innovative resources gained by innovative firms comes from interactive learning and exchange relationships between supplier and customer firms. As he also defines, these relations may be with firms located outside or inside the region. After providing supporting evidence on the importance of suppliers and clients in innovation, Boekholt and van der Weele (1998:58) provide the weights of local and non-local supplier networks. According to them, in Eindhoven districts 20 per cent of innovation generated by supplier relations are local, whereas the share of local clients is 37 per cent of the innovations generated by client networks. The same ratios for the Southern Netherlands are 40 and 51 per cent, respectively. Similarly in a study on Ankara covering defence, electronics, machinery and software, 36 per cent of the innovations were generated by relations with customers (Cihnioglu, Oğuzsoy and Şen, 2002).

5 Maskell (1998) provides evidence on the dissemination of information by IKEA, a Swedish global firm operating in the furniture industry, to its small supplier firms. Still, for SMEs the services provided by public institutions may be important. The evidence, however, reveals that SMEs have difficulties building up networks with public R&D infrastructure units and utilising the services of these institutions (Tödling and Kaufmann, 1999).

6 Ottai (1998:39) says that the firms in these groups need not be organised under the same ownership; they are at the same time legally independent.

relationship is defined as 'joint venture activities' within the region, as observed among Californian multimedia firms (Scott, 1998).

Obviously, innovative relations among firms in various clusters differ according to the size composition of the firms. In SME dominated areas the bulk of information on new products and processes as well as new ideas is circulated via informal and personal links among producers. The findings on the several regions of Europe covered by the REGIS study indicate that small firms have fewer relations to customers at larger spatial scales (national, European and global), but more with local/regional customers and suppliers (Tödling and Kaufmann, 1999). Sectorally focused networks (called local initiatives) promoted by a state government, however, may be more important as a vehicle for helping small and medium-sized firms to learn how to master new lean production techniques, as seen in the Baden-Württemberg case (Morgan, 1996).

Relations between Production and Service Firms

In the economic success of some regions, such as Cambridge, local linkages with business consultancy services are essential. Findings show that local/regional consultancy firms are more important in innovative processes than the services provided at national and global levels (Keeble and Lawson, 1999; Tödling and Kaufmann, 1999). The findings of Koschatzky (1999) are to some extent similar. In his study covering several regions,[7] he found that for advisory services spatial proximity is quite important. He also presented findings that indicate that there are regional differences in terms of service linkages, e.g., firms in central regions can use more external service firms, whereas the service networks of firms in rural areas are too highly focused on regions.

Evidently not all networks can play a positive role in innovative activities. In many British regions there has been a proliferation of top-down and short-term networks which have had little or no positive impact on long-term regional industrial development (Leigh, 1995). Similarly, although vertical co-operation facilitates the dissemination of information, Baden-Württemberg metal industry firms were not successful enough to increase the innovativeness of existing mature industries (Heidenreich and Krauss, 1998).

Relations between Firms and R&D Organisations

There is essentially a consensus in the literature on the importance of firm-university and firm-research institute relations. A lesser degree of importance is attributed to linkages with government research laboratories and public research institutes in innovative activities (Cooke, 1998a; Roper, 2000). The evidence on firms/industry-university relations, the existence and usefulness of these networks,

7 The study covered the following regions; Hannover-Nord, Hannover-Central, Hannover-South, Braunschweig, Goettingen, Leipzig, Oberlausitz, Dresden, Chemnitz, Zwickau-Plauen, Karlsruhe, Freiburg, Schwarzwald-Baar-Heuberg.

however, reveals a great difference between different clusters. At the one extreme it is possible to find an absence of solid relations between industry and university, as in the Catalan case (Bacaria and Alomar, 1998) and the Lyons health cluster (van der Berg, 2001), and at the other extreme very dense networks, as defined in the Cambridge case (Keeble and Lawson, 1999). There are other examples between these two extremes, such as in the Baden-Württemberg case study, in which 32 per cent of the firms declared that universities are a very important source of knowledge and 15 per cent of them stressed the important contribution of technology transfer centres for sustaining competitiveness (Heinemann et al., 1995 cited in Heidenreich and Krauss, 1998). Similarly, in the New York State medical devices production sector 61 per cent of all firms and 65 per cent of innovative firms have linkages with universities (McPherson, 2002). Variations in the use of R&D support institutions are also observed in the REGIS study regions, such as a relatively lesser role of supporting institutions in innovation in Baden-Württemberg, compared to inter-firm linkages, but more intensive use of these institutions by innovative firms in Styria, Tampere, the Basque Country and Wales (Tödling and Kaufmann, 1999). This study reports very low use of institutions in the Wallonia and Brabant regions.

Why do these differences appear? There are several reasons, including the characteristics of the firms, the composition and size of firms, etc. However, the difference in linkage patterns may partly have to do with the traditions, norms and co-operative culture of different regions (Cooke and Morgan, 1998). Another widely held belief is the crucial importance of linkages with research institutes for small enterprises. Nevertheless, several empirical studies do not support this belief. They provide evidence of limited co-operation between small firms and fewer links with several institutions in regional innovation systems, such as universities, research organisations, technology transfer and training, whereas they claim that larger firms interact more with support institutions (Tödling and Kaufmann, 1999). Few SMEs actually utilise *government-sponsored enterprise support services*. According to Morgan (1996) and Leigh (1995), in successful regions formal strategising by regional agencies and support of local research institutes can give an impetus in the initial phases of development. Later innovative-research activities are likely to be sustained mainly by voluntary and informal activities in various networks. The existing evidence also suggests that the quality of these linkages is the most important determinant of an SME's innovative performance. Being small is not necessarily a major barrier if SMEs are able to match their limited internal resources with complementary resources from other firms, business associations, technical institutes, or whatever (Morgan, 1996: p. 62).

One of the services/institutions, which are not used effectively, is 'technology transfer agencies'. While the ones that are available are usually not well known by the firms, the lack of appropriate *technology transfer institutions* to link industry and the science base may be a problem, as indicated by Landabaso, Oughton and Morgan (1999) for the West Midlands.

Networks among Institutions

Although many studies perceptively define the existing institutions of regional innovation systems, they are less informative on the relations between these institutions (Boekholt and van der Weele, 1998). There are some studies indicating interregional differences in co-operative behaviour among regions, such as the study that uses the data on 10 regions[8] included in the ERIS Project (European Regional Innovation Survey) (Fritch, 2001). According to this study enterprises located in certain regions, such as Baden, are characterised by a relatively co-operative attitude, while in certain regions the co-operation between firms and institutions and between the institutions themselves is rather weak.

Table 3.1 shows the generalised findings of various empirical studies on the character and density of local networks. Obviously not all of these networks are realised in a given regional innovation system. Some of them may be obsolete or, as the table indicates, there is very limited information related to some networks. The existing literature mainly concentrates on the bilateral relations between firms and other actors in RIS. Among them the relations between government departments and public technology transfer agencies are unilateral and basically depend on the provision of facilities to firms. Except for the relations between universities and universities and other technology centres, it is quite difficult to find dense connections between the local actors in RIS. In fact, depending on the networks observed in empirical studies it is possible to say that the innovation systems defined in theory are still at the infant level.

8 The regions included in this study were; Barcelona, Rotterdam, Stockholm, Vienna, Alsace, Gironde, Hanover, Saxony, Slovenia and South Wales.

Table 3.1 Possible versus actual types of local/regional networks defined in the literature

Actors in RIS		Firms customers, suppliers and competitors	Business inter-mediaries	Public-private R&D	Universities (U)	Technology centres	Public inter-mediaries
Production firms		Concentration of local networks			Still differentiated, in some clusters no relations	Highly differentiated by regions. Not effectively used	Usually defined as weak relation except dirigiste systems
Business intermediaries * Consultants * Specialised service firms							
R&D units	Research institutions government R&D-private R&D				Very different		
	Universities	In general a positive relation is expected			There is also high variation among university-university relations		
Public inter-mediaries	Technology centres	Highly differentiated by regions					
	Chambers, etc. government technology transfer agencies	Input flow from government and technology transfer institutions to firms					

- [vertical hatching] Significant variation between different examples
- [shading] Density of interaction (not applicable to all cases)
- [blank] Limited information, difficult to generalise

Global Networks as Open Gates for RIS

Open gates are crucial for an effective RIS since innovative firms, which are the main agents of this system, have to be linked with national and global knowledge sources. Via these networks it is possible to generate new ideas by using not only local knowledge and expertise, but also external expertise, and to transform them into marketable products (Revilla-Diez, 2001). As Camagni (1991) emphasises, in a technologically dynamic and highly uncertain world, the contribution of local milieux has limits and they must be supported by wider inter-firm networks as a way of access to information on rapidly changing technologies and market opportunities.

There are four main reasons why the involvement of local firms in wider national and global networks is absolutely essential for long-term regional growth. The first is to open up the milieu to external energy in order to avoid entropic death and lock-in in its own innovative capacity, which is particularly vital in those areas of production characterised by fast innovation and technological change (Camagni, 1991; Sternberg, 2000). Secondly, in regional clusters export oriented attempts are not enough to deal with problems generated by declining domestic demand, until they are supported by local networking and technology transfer via external networks (Kautonen, 1996). Thirdly, non-regional activities can increase the contribution of regionally oriented measures and therefore provide stronger support for innovation management and the competitiveness of regional firms. It is not possible for firms, especially SMEs, to rely upon localised learning and tacit knowledge and they must be supplemented by formal R&D competence and access to universal knowledge via different types of non-regional and global networks (Asheim and Isaksen, 2002). Fourthly, an actor in an RIS can have the chance to obtain both tacit and codified knowledge and can achieve benefits from possible synergies when it joins global networks.

Individual studies also support the importance of external networking in fostering innovative activities (Keeble and Lawson, 1999; Hassink, 2001), as do various comparative studies done for European regions (Braczyk, Cooke and Heindenreich, 1998) and projects such as REGIS (Tödling and Kaufmann, 1999) and the European Regional Innovation Survey[9] (ERIS) (Sternberg, 2000), besides many others. These studies provide an important base of knowledge on the existence and contribution of global networks in regional innovation systems. At this point, however, we need to mention a contradictory view raised by Tödling (1999: p. 695), who claimed 'only a small share of firms has truly global links in the innovation process, despite much emphasis on this phenomenon in the literature'. However, it should be noted that both views have a limited empirical basis, since network structures are usually defined by a simple count of the linkages of firms with other actors at different layers (Staber, 2001). Also, existing methods measure dyadic relations, but do not provide enough information on how all these linkages are integrated with each other. Obviously, it is very difficult to learn how the second-order relations are formed among the agents of a regional innovation system. That is why it is only possible to discuss some general trends. Despite these difficulties the clear point is the difference in the importance of global networks, firstly according to the characteristics of RIS, secondly with respect to mutual benefits that networks offer for the actors of a regional innovation system.

In this section of the chapter, the existing literature on four different types of global network is presented in order to discuss what kind of participation is

9 The regions studied were Alsace, Baden, Barcelona, Gironde, Hannover-Brunswick, Goettingen, Saxony, Slovenia, South Holland, South Wales and Stockholm.

dominant in the different types of RIS and what the expected benefits from these network activities are.

Networks Formed with Global Suppliers and Customers: Value Chains

In many regions business communities do not limit their operations to a certain region. As reported by Leigh (1995), during the regional innovation strategy workshops organised in Yorkshire and Humberside the business community stated that national and international linkages are very important in developing successful supply chains and markets, as well as research work. The members of the business community were looking beyond their regional boundaries in order to find suitable partners or purchasers and also sources of information.

Networks among global suppliers are the most important means to reach universal knowledge. Keeble and Lawson (1999) declare that it is important to emphasise the role of both customers and suppliers as part of wider innovation networks, rather than focusing on simply evolving arms-length market relations. Other studies indicate the importance of global value chains in innovative activities (Tödling and Kaufmann, 1999), which have different roles in the local-global interplay. The first role can be defined as *supplementary role.* They may complement a local/regional network and enable firms to reach external markets, the sources of knowledge and information on new global trends. However, the importance of global suppliers may differ according to the different types of products. Global suppliers of standardised materials or components as well as competitors in the same business have been found to be more important than local and national ones in innovative activities, whereas suppliers of customised materials or components, clients and customers at the national level contribute more than regional and global ones (Keeble and Lawson, 1999). The importance of supplier networks also differs according to the type of firms. There is evidence of a much higher level of collaborative or partnership arrangements with customers for both high technology manufacturing and service firms than for non-high technology SMEs (Hughes, 1998).

To actualise supplementary relations between local and global value chains is not easy. The difficulty is to make these two types of networks co-operate for their mutual benefit. There are examples that contradict the general belief on the complementary and supplementary nature of global-local networks. On the contrary, some networks that are thought to supplement local capabilities can weaken the potential of local clusters (Hassink and Wood, 1998). In the Jena region (Germany), which is well equipped with a broad expertise in the field of opto-electronics, research institutes, universities and high-tech niche-market companies want to have contacts with international commercial networks, but at the same time they need to be embedded in the intra-regional knowledge network. Therefore firms and organisations in opto-electronics in Jena collaborated beyond national boundaries, although national networks of research expertise[10] already existed in the region.

10 The Max Planck and Fraunhofer Institutes, universities, etc.

The second role of global networks is to assist in the *transformation* of regional structures. In the course of evolution, clusters need to restore their existing structure and relations. At this moment global value chains are important since they may force an urgent change on firms located within clusters. That is why mature high-tech areas look for global networking in order to overcome an unforeseen lock-in situation. In Baden-Württemberg, which is one of the best known example of a RIS model, the main firms in the region, such as Mercedes, formed new networks with East Asian suppliers and help them via the Centre of Industry and Commerce in Singapore. Additionally, firms in the textile industry initiated joint research projects and training centres to provide assistance to related sectors in East Asia (Heidenreich and Krauss, 1998). These attempts are all designed to increase the level of competitiveness of the region, since the in-region activities are not enough to sustain the competitiveness of the main production sectors. In this respect Baden-Württemberg is a very good example indicating that local learning and innovation capabilities may be inadequate to deal with newly emerging conditions and the necessity of various types of cooperative action.

Thirdly, global value chains are important since they help the commercialisation of existing and new products as well as innovative activities. Their role here can be defined as *to increase the value of innovative resources*. Many regions have to depend upon not only global linkages but also global expertise in order to increase the value of their existing assets. In biotechnology clusters Cooke (2002) found customers and venture capitalists are frequently global, although there are also some regional and national ones during the commercialisation of innovations.

Besides the firms in value chains some firms have to take on roles in the global/local interface, which is the nexus between the external conditions needed for successful competition and local conditions, regulations and regimes (Simmie and Kirby, 1998). In that respect export firms are the key transfer points between global and local for firms with innovative products that are looking for export markets (Wiig and Wood, 1995). After the initial contact is achieved with the help of these networks the agents of a regional innovation system can force regional actors to adapt themselves to the needs of the global system and work on new products and processes.

International R&D/Academic Excellence Networks and Collaborative Research Activity

Clear evidence exists of the increasing importance of national and global collaborative research and the new opportunities for collaborative research[11] and the increasing number of institutions that support international research activities, although much of the collaborative research still stays at the national level (Keeble and Lawson, 1999). The main reasons for this situation are the long tradition and dominance of the national institutions that support research activities and, to a certain extent, existing cultural barriers.

Global partnerships are more important for research institutions. The research and development institutions within a regional innovation system may prefer to form more intensive external linkages than the local networks. Based on the experience of Silicon Glen, Charles and Benneworth (2001) showed that Scottish universities pursue the commercialisation of their knowledge base with partnerships on a global scale, limiting their local involvement to technology transfer and technical assistance type of activities. Similar findings were reported by Revilla-Diez (2001), who emphasised that some research institutes are mostly engaged in international co-operation projects and connected to international science networks. They do not want to stay local. In fact, it is very difficult to generalise this situation, since some of the R&D centres have limited interest in and opportunity to form global links.

Some conditions accelerate the search for collaborative research actions; e.g. transformation to high-tech production techniques necessitates global knowledge and global R&D. Asheim and Isaksen (2002) showed that the firms in three industrial clusters in Norway (Sunnmore, Jaeren and Horten) not only exploit place-specific local resources, but also external, world-class knowledge in order to strengthen their competitiveness and achieve the transformation of their industrial basis. According to their study on Jaeren, the initial development was based upon technological co-operation between medium-sized, export-oriented firms mainly producing farm machinery with the help of an organisation called TESA. When the district became the centre of industrial robot technology, local and national R&D institutes proved inefficient to meet the needs of firms. The firms in this area followed two strategies to find R&D based competence. The first strategy was to use foreign R&D institutes and universities in Sweden and Germany and the second to utilise R&D departments within the co-operation (if the firms are part of a large

11 A study on global, national and regional research collaboration networks in Cambridge (Keeble and Lawson, 1999) shows that the distribution of collaborative research activity with other firms is as follows: with firms in the Cambridge region 14.4 per cent, the rest of East Anglia and South East England 12.4 per cent, the rest of the UK 35.6 per cent, the rest of Europe 19.4 per cent and the rest of the world 17.8 per cent. For high technology firms in the same region the situation is almost the same. For those that engage in such collaboration, no less than 48 per cent of collaborative activity, on average, is with other UK firms, and 37 per cent with foreign firms, compared with only 14 per cent locally.

international corporation) or to initiate research in co-operation with foreign strategic partners.

There are also other ways to transfer new technology and knowledge besides collaborative research, such as the recruitment of high quality labour. In many hi-tech industrial districts, notably Silicon Valley and Cambridge, significant flows of expertise are achieved via the high mobility of technical staff between firms. However, it is also well known that important amounts of recruitment of research and management staff take place nationally and even globally.[12] Wider global and national innovation and recruitment networks thus play a very important role in the functioning and growth of high technology SMEs and complement regional collective learning processes, especially in the initial steps of development, as exemplified in the Sophia-Antipolis region of France (Longhi, 1999).

The other way to obtain up-to-date knowledge is to become a partner in certain international projects. Participation in large international projects can give a region a high profile. The experience of working together in international projects initiates trust between partners and teaches them how to integrate work processes. It will also strengthen the position of a region among the others that are specialised in the same activities. Two very different examples given below show the possible benefits of participation in large international projects. The example provided by Grotte et al. (2002) is of the County of Sogn og Fjordane, which has benefited from specific regional activities as a working group on ICT operations and large international projects like ENCATA, TITAN and ODIN, in which the partners worked together and produced common applications and services. These international projects have created a positive feedback loop for the partners and have strengthened the role of the organisational structures in the region. Another interesting, but quite different, international network project shows the immediate benefits of taking part in international projects. The project aimed to form networks between peripheral book towns with the help of ITC technology. BookTown.net is one integrated internet based application, which linked local villages, the book towns, and operational enterprises, the book dealers, across five European regions. The project is quite informative as to how to face a set of challenges due to both cultural and language differences between the book towns and to potentially competing internet services. Table 3.2 gives a brief summary of the discussions on participants in global networks and the type of joint activities that these actors are usually involved in.

12 In the Cambridge region Keeble et al. (1998) found that 62 per cent of the research staff and 67 per cent of the management staff appointed by local technology based firms were recruited from outside the local milieu.

Table 3.2 The possible and actual global networks that are important for RIS

GLOBAL NETWORKS

THE LOCAL ACTORS IN RIS	Value chains	R&D excellence networks	Global business services	Technology transfer networks
Customers				
Suppliers		Collaborative research		Access to network embedded knowledge
Competitors			Commercial-isation of innovation	
Consultants + service firms				Mediation
Research institutions + technology transfer agencies				
Universities		Joint projects/research		Research projects + applied research
REGIONAL NETWORKS				
Between firms	Foreign trade companies			Participating international projects
Between firms and R&D infrastructure		Participating international projects		
Between R&D infrastructure units				

Areas of collaboration

The density of interaction (not applicable for all cases)

There is limited information, difficult to generalise

The Importance of Regional Characteristics, Agents and Motivations in Global Networking

The discussions above emphasise that in the contemporary economic conditions none of the regions can depend only on local resources and dynamics. There is a clear need to build different types of linkages with the external world. There are, however, still many issues that are not clear enough to define the role of interactive relations between networks. These issues and the points of emphasis given below will be the subject of the discussion in the remaining sections of the chapter.

- The importance of global networks varies highly by region. That is why there is a need to discuss regional innovation systems with different characteristics and the way they use these global networks.
- Yeung (2000: p.312) defines networks as 'both a governance structure and a process of socialisation through which actors and organisation are connected in a coherent manner for mutual benefits and synergies'. Obviously, during the course of the development process, the partners, structure and relational patterns of networks continuously change in order to maximise mutual benefits. That is why there is a need to define how networks evolve and to discuss the causes and consequences of changing network relations on regional performance.
- Theoretical debates point out that networking is not always an individual choice, but is authorised by certain institutions and organisations. Power relations are especially important in order to define how external networks are connected to regional innovation systems and also the actors of these systems (Eraydın, 2002d). The network organisers deserve further attention since they define the rules of the relations in global networks.

In this section of the chapter the existing literature will be reviewed on three issues related to global networks, namely *the importance of global networking in different types of RIS, the changing roles of global networks during the development process and the main actors that connect regions to global networks.*

The Characteristics of Clusters Matter in Global Networking

It is obvious that the importance of external networks is not the same for all clusters. In the formation and effectiveness of external networking, according to the existing literature, four characteristics of clusters strongly matter, the type of regional innovation system, the sectoral composition of industries, firm size and external ownership.

The Type of Regional Innovation System Defines the Role of External Networking

RIS are categorised according to the dominance of different networks. It is obvious that regional innovative systems have different types of critical networks between the actors that comprise the system. Moreover, the mechanisms of governance of these networks are quite different. It is possible to classify regional innovation systems according to the character of the public governance system. Cooke (1998a) and Asheim and Isaksen (2002) present three different types of RIS, namely grass roots systems, integrated systems and dirigiste systems.

The first type of RIS is designated as territorially embedded regional innovation network or *grassroots RIS*. The performance of this type of regional innovation system depends mainly on the localised learning process of firms without many interactions with knowledge organisations. The externalities of geographical, social and cultural proximity enable collaborative action and knowledge dissemination. In this type of regional innovation system, networks external to these regions are usually not very important. However, it is still possible to observe different versions of grassroots systems, such as in California, where clustered supply chains, which meet the needs of global corporations, dominate the interaction pattern of the system.

The second type of RIS is *regional networked innovation systems*, in which firms and organisations are still embedded in a specific region, but the system is supported by regional and institutional infrastructure, such as R&D institutions, technology transfer agencies, various producer services, training institutions and other local organisations. The firms in this type of cluster have external connections but they also utilise local institutions and networks between innovative actors. According to Cooke (1998a: p. 20), in this type of region access to information and technology transfer through networking is multi-level, varying from local to global levels.

The third type of RIS is defined as *regionalised national innovation system*, where exogenous forces and networking are more important. In these regions there may be firms and institutional infrastructures that are functionally integrated into national and international innovation systems. Asheim and Isaksen (2002) state that innovation activity takes place mainly in co-operation with actors outside the region in this type of RIS.

Cooke, Roper and Wylie (2001) introduced another dimension to this categorisation, namely a business innovation system.[13] The three varieties of business innovation system are localist-associative, globalised and non-associative. The idealised model of RIS falls into the category defined by an interactive business innovation system and public governance based on networked structure.[14] The examples in each matching category are given in Cooke (1998a) and the Report prepared by the Northern Economic Council (1991: p. 13). This categorisation is important since it enables us to show that the importance of local and global linkages varies in different governance and business innovation systems.

Sectoral Specialisation Makes a Difference in the Importance of External Knowledge Dependence and Networking The sectoral composition of clusters defines their main characteristics and their dependence on external knowledge transfer. Theoretically, in high technology activities and new sectors external links are more

---

13 Business Innovation System
 (i) Local and associative
 (ii) Globalised and non-associative
 (iii) Fully interactive and associative.
Public Governance of Innovation Systems
 (i) Grassroots and bottom-up
 (ii) Centralist and top-down
 (iii) Fully networked and balanced.

14 Baden-Württemberg (Germany), Bavaria (Germany), Cambridge (UK), Massachusetts (USA), Thames Valley (UK) are examples of idealised models of regional innovation systems.

important. The findings of several studies support this hypothesis. Using evidence from three important metropolitan areas of Europe, namely Stockholm, Vienna and Barcelona, Revilla-Diez (2001) found that innovative SMEs in biotechnology and communication have to compete internationally. The complexity of innovation process in these two branches forces firms to have immediate access to leading firms and research institutes, whereas firms in traditional sectors are less keen on external networking. Similarly, the Ankara RIS study (Eraydın, 2002b) also showed that the importance of external networks differs highly among different production sectors. Almost all firms in the defence industry have non-regional networks. 89 per cent of firms in this production sector have also formed global networks for product development and marketing; while in other manufacturing sectors, firms connected to global networks make up 56 per cent in machinery, 66 per cent in software and only 40 per cent in electronic products.

The Size of Firms Matters in Global Networking It is usually claimed that small business gains the most from integration into regional and successively into global networks (Tödling, 1994), but many empirical studies have found smaller firms to be less integrated into global networks due to certain deficiencies and obstacles. The lower rate of engagement of SMEs in innovation networks is well documented in the SMEPOL[15] (SME Policy and the Regional Dimension of Innovation) Research Project (Tödling and Kaufmann, 2001). According to the findings of this project, SMEs less often engage in research than large firms do, since they have weak boundary-spanning functions and a low labour capacity to establish and maintain interfaces to innovation networks. The REGIS project, which covered 11 regions in Europe,[16] also showed that for larger firms (200+) global co-operation is more important than for small and medium firms in order to sustain their competitive advantage (Tödling and Kaufmann, 1999). With respect to innovative actions, similar findings came from Northern Ireland, where large, companies with global reach have high innovation rates compared to smaller ones (Northern Ireland Economic Council, 2001).

External Ownership Capital transfer from the outside world is among the most critical global linkages of a regional innovation system. Capital flows and direct foreign investment are important in the emergence of effective value chains and technology knowledge transfer, especially in the initial years of development (Amin and Cohendet, 1999). The recent discussions and empirical studies on

15 The regions studied were Northern Norway, South-Eastern Norway, Upper Austria, Triangle region in Denmark, Lombardy, Apulia (Italy), Limburg (Netherlands), Wallonia (Belgium), Valencia (Spain), London-Lee Valley, adjacent outer metropolitan area (Hertfordshire, Essex) in UK.

16 Baden-Württemberg (Germany), Brabant (Netherlands), Styria (Austria), Tampere (Finland),Wales (UK), the Basque Country (Spain), Wallonia (Belgium), Aveiro and Fruili (Portugal), Fejer (Hungary), Lower Silesia (Poland).

regional development, especially the Irish examples, show the importance of foreign capital for motivating the growth process.

In the later stages of development one might expect another wave of capital flow via international firms to innovative territories in order to take over some successful local firms and utilise their increasing innovation capacities, but a study on Tuscany (Ottai, 1998) showed that a completely different strategy may work. What happened was the takeover of raw material supplier firms located abroad by local firms. A similar trend was observed in Baden-Württemberg (Heidenreich and Krauss, 1998), namely accelerated foreign direct investment from this region to abroad during the transformation of the regional structure from production to service economy. With respect to the issue of capital transfer, it is interesting to note that in many recently prepared regional innovation strategies, such as the Regional Innovation Strategies for Strathclyde, West Midlands and Yorkshire and Humberside, foreign direct investment did not play a role in encouraging exogenous innovation (Thomas, 2000).

Changing Actors and Patterns in Global Networking

Participation in a network may be based on different needs and aims. According to Capello (2000: p. 1931), participation should depend upon at least an efficiency goal, which should be followed by a synergy goal and even by a competence goal in the later stages of participation. This means that what you want to get and what you are eager to give may be quite different and, during the course of interaction, participation is expected to move from efficiency to a competence sharing type of relation. In order to increase the benefits of networking and strategic network advantages there is a need not only for serious participation in a network, but also flexibility with respect to organisational changes and an open attitude to network behaviour.

The above premises on networks help us to understand some of the observed patterns of evolution in networks. Firstly, many agents of various regional innovation systems are still reluctant to be a part of global networks, since when they participate in a global network they need to offer or share information related to their existing advantages, which they do not want to do. After participating in a network there is need for flexibility with respect to organisational changes. In local networking this may be less crucial, but in global networks, because of the differences in cultures, norms and practices, flexibility is more important for internal coherence. This situation may be hard for firms, since they usually consider conditions that enforce changes in their autonomous decision making as interference in their freedom of choice. Similarly an open attitude to network behaviour is difficult to develop due to problems of trust, etc.

Secondly, a review of the literature shows that among the four global networks that are important for RIS, namely international R&D/academic excellence networks, international technology transfer networks, international business services and international value chains, the last mentioned is easiest to establish. The main reason is that most value chains increase efficiency, but do not require sharing important amounts of knowledge and exchange of information. In R&D/academic excellence networks synergy may be the first step, but after building trust and confidence, these networks can be formed much more easily among specialised agents since they are expected to have a more open attitude towards networks. On the other hand, the other two types of global networks are less widely used. Usually business consultancy services are defined as a one-way exchange with more intimate relations. That is why the consultancy services are still relatively local, although there is a growing expertise in several fields.

Thirdly, the studies show that some unexpected conditions, such as an economic crisis, may enforce changes in the attitude of firms to networking. One of the strategies followed is restructuring of the system and improving external links. Restructuring necessitates participation in external networks in order to make sure that a 'milieu' or district does not stew in its own juice (Johannessen, Dova and Olsen 1997). Increasingly, these necessary ties are seen to be global (Amin, 1999) and co-operating and merging with a globally competitive firm are defined as possible strategies. As discussed above, due to the sceptical approach of many local firms, these strategies stimulate limited interest in business environments (Cooke, 1996).

Schmitz (2000) provides some evidence on changing network patterns in four regions that faced economic crises during the 1990s. The evidence from Pakistani, Mexican, Indian and Brazilian clusters shows that, as a reaction to the crisis conditions, the local horizontal co-operation between local actors increased little, whereas initiatives and joint actions aiming to carry firms into global value chains gained an important momentum of growth. According to Schmitz (2000), more than half of the surveyed enterprises reported increased co-operation with both global suppliers of inputs and subcontractors.

Fourthly, the role global networks play may be quite distinct in exogenous and endogenous models of development. Contrary to the endogenous growth models, in some regional innovation systems, such as Sophia-Antipolis, the development of a region begins with the help of firms connected to global chains and their existing networks. It is after the cluster evolves with the contribution of small and medium-sized enterprises that the globalised firms also induce local linkages between the actors of the system (Longhi, 1999).

Network Organisers

The Role of Multinational Corporations

In the local-global interplay multinational companies (MNCs) have specific roles. They are important as open gates since they frequently exchange information with their headquarters and other research and production sites worldwide (Revilla-Diez, 2001). As Malmberg et al. (1996) stressed, they can also act as the disseminators of new knowledge and information from abroad to a region, if they are rooted and integrated in the region and engaged in local networks. They can even initiate a successful regional innovation system (Longhi, 1999). Their contribution, however, may be considerably less if they are inward looking and less collaborative with local actors. The positive contribution of multinationals in the local-global interplay are defined in the literature by their roles as 'cluster engines' (van den Berg, 2001), 'network organisers' (Radosevic, 1999) or 'potential breeding grounds'.

Multinationals and large companies with a dominant position in a cluster can act as a leading engine of growth in regional clusters. Their main contribution to the growth process can be the product of combining two types of relations; firstly their linkages with universities and research institutions, which enable them to generate and transfer knowledge, and secondly their connections with the spin-off firms that have grown under their umbrella. In the literature there are several examples indicating the role of multinationals as the engine of an innovation system, such as Nokia (Helsinki, telecom), Novartis-Boehringer Ingelheim (Vienna, health) and Marieux (Lyons, health), which made important contributions to the performance of the clusters they are located in (van den Berg, 2001). Cooke (2002) also defined the vital role of internationally known biotechnology firms for innovative activities in both Cambridge, Massachusetts (USA) and Cambridge (UK). According to him, in both of these biotechnology clusters global linkages with other clusters and particularly 'big pharma' partners play important roles in external knowledge transfer.

Besides their flagship role, multinational companies can even play an additional role, defined by Radosevic (1999: p. 314) as *network organisers*. Their role as network organisers is especially important in countries that are currently in economic transition. The current modes and levels of integration of Eastern European countries into international production networks are mostly led by foreign enterprises (Radosevic, 1999: pp. 314-315). In transition economies the inherent advantages of multinationals in terms of access to finance, control over technology and organisational capabilities provide them first-mover advantages that domestically controlled firms are not able to match. That situation gives foreign enterprises the role as the main organisers of production networks.

The role of multinationals is also defined as important for regions/countries that want to transform their industrial basis. In Northern Ireland local innovation linkages have been developed by multinationals operating as they looked upstream to universities and start-up firms for fundamental R&D and innovation functions, and downstream to the supply chain for near-market applications (Roper, 2000). This situation means that they have more complex relations beyond supplier relations with domestic firms. It is known that via multinationals some clusters are involved in cross-border R&D projects, particularly in the development and implementation of the R&D outcomes (Charles and Benneworth, 2001). Moreover, multinationals are important open gates of the regional innovation systems, since they facilitate the links between several centres of excellence.

Obviously, not all multinationals have a leading role and very dense networking with regional actors. There are too many examples in the literature of foreign enterprises/multinationals with limited interest in forming local linkages and using local partners in their activities (Eraydın, 1993). This does not decrease their importance for regional innovation systems. They can still function as a source of knowledge via knowledge spill-over, help to improve the quality of labour and increase the potential of regions as a *breeding ground for spin-off firms*.

In order to evaluate relations between multinational and local firms and the contribution of multinationals, there is also a need to look how far the local system utilises knowledge and the opportunities it provides. Local capacities may not be adequate to utilise the opportunities provided by multinational enterprises. For example, recent research findings on knowledge transfers from multinational enterprises in Northern Ireland (Crone and Roper, 1999) point out that there is unexploited potential for local learning from these enterprises.

Leading Local Firms as Network Organisers

Although the literature focusses mainly on the relations between small and medium enterprises, the role of large enterprises in the organisation of both local and global networks is quite important (Eraydın, 1997 and 2002c). The Northern Ireland Economic Council (2001) identifies the importance of leading firms for organising innovative networks and indicates the potential of indigenous 'lighthouse' firms in Northern Ireland that display high levels of successful innovation. Learning from these firms will be the key building block in the design of a Northern Ireland RIS. The general expectation is that leading firms will help to generate new technologies to stimulate local regeneration by organising local networks and later global value chains. This situation is clearly observed in the development experience of the Denizli (Turkey) industrial cluster (Erendil, 1998). Some of these firms may even become multinationals (Harrison, 1994), as the Benetton case shows, which obviously can change their ties with the local production environment.

Special Organisations for Global Networking

The literature includes, besides multinationals and large local firms, different types of special organisations that are established especially for global networking. These examples can be classified as private firms jointly established by local enterprises, regional public institutions with responsibility for global networking and central state sponsored bodies and international organisations, such as the different sections of the EU, which support co-operative activities between regions.

Among them the classical examples are state-sponsored companies that are specifically designed for export promotion activities. Although they are not interested in initiating networking between local and non-local/global firms, they are useful in paving the way for further relations. In Turkey foreign trade companies (FTCs) played an important role at the initial stages of export-oriented production and showed local firms that it is possible to take part in global value chains (Eraydın, 1993). This initial step later fuelled regional joint efforts, which were not restricted to export activities. Special organisations were also formed by small enterprises in order to not only increase collaborative action among them but also to develop their external linkages. In 1993, a group of small enterprises (464 producers) located in the Aegean Region (Turkey) formed a co-operative institution (EGS), which was expected to provide support in the competition with the leading larger firms. While the member firms stayed autonomous, they first formed an export company and then several support companies. The services provided by the EGS Group were very influential in promoting exports, and, in 1996 and 1997, the EGS Foreign Trade Company became the leading export firm in Turkey. The recent economic crises, however, struck this organisation and the co-operative behaviour among its members. There are several examples that were formed in order to meet the need to strengthen global connections and knowledge transfer to regions. In Cambridge (UK) the biotechnology cluster Cambridge Network Ltd was set up in March 1998 to formalise linkages between the business and research community and to connect both local and global networks in a systematic manner (Cooke, 2002). The activities of this firm are mostly IT focussed due to the demand of biotechnology for IT equipment and opportunities for IT delivered patient and clinical services in the region. As discussed earlier, public institutions may also initiate strategies for the internationalisation of regional innovation systems. The Catalan innovation system has been supported by policies, new public institutes and also via several agreements with other regions and companies (Bacaria and Alomar, 1998). These types of institutions are especially important in dirigiste type of systems.

Concluding Remarks

The recent literature emphasises the impact of non-local relations and global networks on the performance of regional innovation systems. In the increasingly globalised world it is obvious that global networks are more important than in earlier decades and the opportunities to form networks with distant partners have expanded considerably. This is also true of regional networks, which became the basis of production and innovation systems. In fact the dominant narrative emphasises the importance of networking in the development and innovativeness of space.

The literature on regional innovation systems, following the same trend, tries to define agents and networks that comprise a regional system. The conceptualisation, however, has a major weakness, namely the strong focus of attention on internal relations between regional agents compared to the lesser degree of interest in the interplay between regional and non-regional/global networks. Although several models define the relation between regions with external world as 'open gates' (Landabaso, Oughton and Morgan, 1999) or under the heading of 'external influences' (Cooke, 2002), how global networks interact with local actors and networks needs to be depicted precisely. Obviously, this is a difficult task, but if we accept that regions are increasingly moving from being semi-closed systems to open systems, it is also indispensable.

This situation indicates the necessity of defining a regional innovation system by taking global networks into account, not as a residual but as an essential element of an innovation system. In this chapter based on several premises and theoretical concerns, it is argued that a regional innovation system should be defined not within the boundaries of a region, but as a hub of wider networks. It is obvious that proximity still matters in various kinds of relations and networking behaviour, but information on different opportunities supports increasing external networking, which is also encouraged by various institutions in different fields. To conceptualise regional growth and innovation as dependent on locally embedded relations may be true for certain clusters, but may be less explanatory for others. In fact, the evidence also shows that most semi-closed regions with limited external linkages are not innovative. Even to categorise them as a type of regional innovation system may be misleading (Cooke et al. 1998; Asheim and Isaksen, 2001; Cooke, 2002).

How far does the existing evidence justify a model, which tries to integrate global network relations into an innovation system? The review of the literature on regional innovation systems with respect to interactions between local-global actors and local-global networks indicates important deficiencies in the existing work. Firstly, most of the studies concentrate on the internal structure of regional systems and focus mainly on the features of different agents and institutions. The studies on networks are usually restricted to defining bilateral relations without giving adequate information on the qualitative nature of these relations. Since most them depend on interviews of one partner in a relationship, it is not possible to trace a whole network. Additionally, most of the studies are based upon information provided by firms and there are very limited numbers of studies that concentrate on the relations between different institutions and their partners. Secondly, some connections between regional actors and global networks are well documented, but there is no information on other possible networks and the relations between local and global networks. There may be different reasons for this outcome; either the studies are inadequate for or uninterested in analysing existing relations or there are no relations between different actors.

This evaluation needs further elaboration, since it is not fair to assume that every connection between the agents of a system defined in theoretical studies is possible in various regional structures. The work by Suarez-Villa (2000) on network segmentation and barriers to network structures, which he describes for Southern California, is quite useful for understanding the difficulty some actors have in taking part in global networks. He defines the isolated parts of the network's structure territorially or within the same activity and the differences in the chances of each actor to participate in or form a network. Barrier structures impede or resist flows towards certain parts of their networks and even close a part of the network to others. As this study shows, there is a need for a detailed analysis of the structures and properties of existing networks and of barriers to networking, instead of assuming equal chances of formation of different networks. Unfortunately, the material at hand does not provide any chance to discuss the network properties and factors that influence the formation of networks in different types of territorial units. The literature only enables us to trace the role of global networks in regional innovation systems, which can be summarised in four different points.

Firstly, the characteristics of clusters matter in establishing global networks. While the territorially embedded regional innovation systems have less connections with global networks, in regionally networked systems global connections act as open gates for external partnerships. It is also possible to say that innovation systems in which small firms are dominant are usually less enthusiastic about global networking, because of their lower capacities to benefit from these networks with their unequal power relations. There is some evidence, though it is not very trustworthy, to indicate that mature clusters are more interested in global networking, after the exploitation of internal sources and advantages.

Secondly, global networks differ in terms of their structure and weight. Networks formed with global suppliers and customers are the main gateways to a regional innovation system. These networks both play a supplementary role to local networks and act as the transformers of degrading economic structures by making other opportunities possible. The second type of global networking is R&D and excellence networks, which take place between local actors of RIS, global R&D units and international universities. They are crucial for obtaining new technologies and becoming familiar with new innovations. Two other types of networks need further attention, namely joint project networks and export linkages. These networks have a crucial role in information dissemination and technology transfer, although their motivations fall into more classical economy norms. On the other hand, it is interesting to see that technology transfer networks and some business services take place implicitly in value chains and are still less important in global economic relations.

Thirdly, there is limited information on the evolution of networks. Although changes in the patterns of networking are obvious, case studies do not provide sufficient information for a detailed analysis. One thing that can be generalised is that global connections increase when regions achieve high rates of growth and during and after periods of economic transformation and economic crisis.

Fourthly, the search for intermediaries and network organisers for establishing and sustaining relations between RIS actors and global networks indicates the important role of multinationals. The other network organisers, which include large regional firms, special organisations designed for increasing global networking and also several national and international institutions that support increasing interactions, are less influential.

The above points explain why any regional innovation strategy should take global networks into account as an integral part of the system. The evidence on the role and impact of global networks on regional strategies, however, is not very reliable and more evidence is needed for a lively theoretical discourse. Moreover, all of these findings show that any regional innovation strategy has to have at least one, or if possible more than one, open gate in order to grasp externalities in different territorial systems. Obviously in some regions local actors can manage to form non-local networks and get into contact with other global networks, but if they do not have this capability, various institutions can play an intermediary role. To define institutions and organisation that can devise the best strategy is the main duty of policy-makers.

References

Amin, A. (1999) 'The Emilian Model: Institutional Challenges', *European Planning Studies*, vol. 7, pp. 389-405.

Amin, A. and Cohendet, P., (1999) 'Learning and adaptation in decentralised business networks', *Environment and Planning : Society and Space*, pp. 87-104.

Amin A. and Thrift N. (1995), 'Globalisation, Institutional Thickness and the Local Economy', in P. Healey, S. Cameron, S. Davoudi, S. Graham and A. Madani-pour (eds), *Managing Cities: The New Urban Context*, John Wiley and Sons, London, pp. 91-108.

Arzeni, S. (1996) 'Introduction', *Networks of Enterprises and Local Development: Competing and Cooperating in Local Systems,* OECD, Paris, pp.13-18.

Asheim, B.T. (1997) *'Towards a learning based strategy for development: structural limits and new possibilities?'* Paper presented at the Regional Studies Association "Regional Frontiers" EURRN European Conference 20-23 September 1997 Frankfurt (Oder) Germany.

Asheim, B. T. and Isaksen, A. (2002) 'Regional innovation systems: The integration of local 'sticky' and global 'ubiquitous' knowledge', *Journal of Technology Transfer,* vol. 27, pp. 77-86.

Audrecht, D. and Feldman, M. (1996) 'Innovative clusters and the industry life cycle', *Review of Industrial Organisation*, vol. 11, pp. 253-273.

Bacaria, J. and Alomar, S. B. (1998) 'The Catalan innovation system: governing Rapid changes', Braczyk, H-J., Cooke, P. and Heidenreich, M. (eds) *Regional Innovation Systems: The Role of Governances in a Globalized World*. London, UCL Press, pp. 72-98.

Belussi, F. (1999) 'Policies for the development of knowledge-intensive local Production systems', *Cambridge Journal of Economics*, vol. 23, pp. 729-747.

Bergman, E. and Feser, E. J. (2001) 'Innovation system effects on technological adoption in a regional value chain', *European Planning Studies*, vol. 9, no. 5, pp. 629-648.

BITNET Project (2001) *Analysis of the REGIS report and other EU research*, Report 7, 2001-10-04, Deliverable 4-Work package 2.

Bianchi, P. and Bellini, N. (1991) 'Public policies for local networks of innovators', *Research Policy*, vol. 20, pp. 487-497.

Boekholt, P. and van Weele, E., (1998) 'Southeast Brabant: a regional innovation system in transition', Braczyk H. J., Cooke, P. and Heidenreich, M. (eds) *Regional Innovation Systems: The Role of Governances in a Globalized World*, London: UCL Press.

Braczyk, H-J., Cooke, P. And Heidenreich, M. (eds) (1998) *Regional Innovation Systems: The Role of Governances in a Globalized World*. London, UCL Press.

Breschi, S. (2000) 'The geography of innovation: A cross –sector analysis', *Regional Studies*, vol. 34, no 3, pp. 213-229.

Camagni, R. (1991) 'Local Milieu, Uncertainty and Innovation Networks: Towards a New Dynamic Theory of Economic Space' in R. Camagni (ed.), *Innovation Networks*, Belhaven, London, pp. 121-144.

Capello, R. (2000) 'The city network paradigm: measuring urban network externalities', *Urban Studies*, vol. 11, pp. 1925-1945.

Charles, D. and Benneworth, P. (2001) 'Clustering and economic complexity: ICT Clusters in the United Kingdom', in E. Bergman, D. Charles and P. den Hertog (Eds) *Innovative Clusters: Drivers of National Innovation Systems* (OECD: Paris) pp. 327-346.

Cihnioglu, S., Oguzsoy, Şen, S. and Tırpanlı, O. (2002) *Ankara Bölgesel Yenilik Çalışması*, Middle East Technical University: Department of Urban and Regional Planning, Ankara (unpublished report).

Cooke P. (1996) 'Building A Twenty-First Century Regional Economy, In Emilia-Romagna', *European Planning Studies*, vol. 4, No.1, pp. 53-62.

Cooke, P. (1998a) 'Introduction', in H.-J. Braczyk, P. Cooke and M. Heidenreich (eds) *Regional Innovation Systems: The Role of Governances in a Globalized World*, London, UCL, pp. 2-27.

Cooke, P. (1998b) 'Global clustering and regional innovation: systemic integration in Wales, in H.-J. Braczyk, P. Cooke and M. Heidenreich (Eds) *Regional Innovation Systems': The Role of Governances in a Globalized World*, London, UCL, pp. 245-262.

Cooke, P. (2002) 'Regional innovation systems: General findings and some new Evidence from biotechnology clusters', *Journal of Technology Transfer*. Vol .27, pp. 133-145.

Cooke, P. and Morgan, K. (1993) 'The network paradigm: new departures in corporate and regional development', *Environment and Planning D: Society and Space*, vol.11, pp. 543-564.

Cooke, P. and Morgan, K. (1998) *The Associational Economy: Firms, Regions And Innovation*, Oxford, Oxford University Press.

Cooke, P., Roper, S., and Wylie, P. (2001) *Developing a Regional Innovation Strategy For Northern Ireland*. Northern Ireland Economic Council, Belfast.

Cooke, P., Uranga, M. G., Etxebarria, G. (1998) 'Regional systems of innovation: An evolutionary perspective', *Environment and Planning A*, Vol. 30, pp. 1563–1584.

Crone, M. and Roper, S. (1999) *Knowledge Transfers from Multinational Enterprise Plants in Northern Ireland*, Northern Ireland Economic Research Centre, Belfast.

de la Mothe,J. and Paquet, G. (1998) *Local and Regional System of Innovation*, Kluwer, Boston.

Echevelli-Caroll, E. and Brennan, W. (1999) 'Are innovation networks bounded By networks and proximity', in Fischer, M., Suarez-Villa, L. and Steiner, M. (eds) *Innovation, Networks and Localities*, Springler-Verlag, Berlin.

Edquist, C. (ed.), (1997) *Systems of Innovation: Technologies, Institutions and Organisations.* Pinter, London.

Eraydın, A. (1993) 'The new international relations, restructuring in the economy and the emerging changes in the business behavior', in C. Rogerson, E. Schamp and G.J.R. Linge, (eds), *Business Behavior, Markets, Finance and Industrial Change,* Gruyter, Berlin.

Eraydın, A. (1997) 'LDC Industrial Districts: the Challenge of the Periphery', in K. I Westeren (ed.) *Cross Border Cooperation and Strategies for Development In Peripheral Regions,* Nord-Trondelags, Forskning, Oslo, pp. 411-436.

Eraydın, A. (2002a) 'Building up competence, institutions and networks in order to catch up in the knowledge economy', in R. Hayter and R.LeHeron (eds), *Knowledge, Territory and Industrial Space,* Ashgate, Aldeshot, pp. 67-87.

Eraydın, A. (2002b) 'Network externalities: Preliminary evidence on Ankara Metropolitan innovation system', Paper prepared for EU-COST A17 Action Cambridge (unpublished).

Eraydın A. (2002c) 'The Roles of Central Government Policies and the New Forms of Local Governance in the Emergence of industrial districts', in M. Taylor and D. Felsenstein (eds), *Promoting Local Growth: Process, Practice and Policy*, Ashgate, Aldershot, pp. 81-104.

Eraydın, A. (2002d), 'The local embeddedness of firms in social networks in Turkish industrial districts: the changing role of networks in local development', in M. Taylor and S. Leonard (eds), *Social capital and the Embedded Enterprise: International Perspectives,* Ashgate, Aldershot, pp. 269-289.

Erendil, A., (1998) 'Using Critical realist Approach in Geographical research: An Attempt to Analyze the Transforming Nature of Production and Reproduction in Denizli', Unpublished Ph.D Thesis. Middle East Technical University, The Departmcnt of City and Regional Planning.

Etzkowitz, H (1994) *'The Triple Helix: A North American Innovation Environment'*, Paper prepared for the opening session of the NAFTA Institute on Innovation, August 14-21, 1994, British Columbia, Canada.

Florida, R. (1995) 'Towards the Learning Region', *Futures,* vol. 27, pp. 527-536.

Fritsch, M. (2001) *'How And Why Does Efficiency Of Regional Innovation Systems Differ?'*, Kiel Institute Of World Economics Institute Of Regional Research Organisation or Economic Co-Operation And Development, Local Economic And Employment Development Programme International Workshop Innovation Clusters And Interregional Competition, November 12-13, 2001.

Glasmeir, A. (1994) 'Flexible Districts, Flexible Regions? The Institutional And Cultural Limits To Districts In An Era Of Globalization And Technological Paradigm Shift', in A. Amin, and N. Thrift (eds.) *Globalisation, Institutions And Regional Development In Europe,* Oxford University Press, Oxford, pp. 118-146.

Glasmeir, A. (1991) 'Technological Discontinuities and Flexible Production Networks: The Case Of Switzerland and The World Watch Industry', *Research Policy*, vol. 20, pp. 469-485.

Grabher, G. (1993) 'Rediscovering the social in the economics of interfirm relations', in G. rabher (ed.), *The Embedded Firm. On Socioeconomics of Industrial Networks,* Routledge, London, pp. 1-33.

Granovetter, M. (1985) 'Economic action and social structure: the problem of embeddedness', *American Journal of Sociology,* vol. 19, pp. 418-510.
Gregersen, B. and Johnson, B. (1997) 'Learning Economies, Innovation Systems and European Integration', *Regional Studies*, vol. 31, pp. 479-490.
Grøtte, I.V., Skogseid I., Strand, G. and Jansen, A. (2002) *A Holistic Approach to Regional Infrastructure: Theoretical Perspectives and Examples*, http://www.vestforsk.no.
Harrison, B. (1994) 'The Italian Industrial Districts And The Crisis Of Cooperative Form: part II', *European Planning Studies*, vol. 2, no. 2, pp. 159-174.
Hassink, R. (2001) 'Towards Regionally Embedded Innovation Support Systems in South Korea? Case Studies from Kyongbuk– Taegu and Kyonggi', *Urban Studies,* vol. 38, no. 8, pp. 1373–1395.
Hassink, R. and Wood, M. (1998) 'Geographic 'clustering' in the German opto-Lectronics industry: its impact on R&D collaboration and innovation', *Entrepreneurship & Regional Development*, vol. 10, no. 4, pp. 277-297.
Heidenreich, M and Krauss, G. (1998) 'The Baden-Württemberg production and Innovation regime: past successes and new challenges', in Braczyk, H.J., Cooke, P. and Heidenreich, M. (eds) *Regional Innovation Systems: The Role of Governances in a Globalized World,* London, UCL Press, pp. 214-244.
Hughes, A. (1998) *'Hi-tech firms and hi-tech industries: Finance, innovation and human capital,* paper presented at "SMEs and Innovation Policy: Networks, Collaboration, and Institutional Design", ESRC Center for Business Research, University of Cambridge, Cambridge.
Johannessen, J.A., Dolva, J.O. and B. Olsen, (1997) 'Organising innovation: Integrating knowledge systems', *European Planning Studies*, vol. 15, no. 3, pp. 331-349.
Kautonen, M. (1996) 'Emerging innovative networks and milieux: the Case of furniture industry in the Lahti region of Finland', *European Planning Studies*, vol. 6, no. 4, pp. 439-456.
Keeble, D. and Lawson, C. (1999) 'Collective Learning Processes, Networking and 'Institutional Thickness' in the Cambridge Region', *Regional Studies,* vol. 33, no. 4, p. 319.
Keeble, D., Lawson, C., Smith, H.L., Moore, B., and Wilkinson, F., (1998) 'Internationalisation process, networking and local embeddedness in technology intensive small firms', *Small Business Economics*, vol. 11, pp. 327-342.
Koschatzky, K. (1999) 'Innovation networks of industry and business related services-relations between innovation intensity of firms and regional inter-firm cooperation', *European Planning Studies*, vol. 7, no. 6, pp. 737-757.
Landabaso, M. Oughton, C. and Morgan, K. (1999*) 'Learning regions in Europe: Theory, policy and practice through the RIS experience'* paper presented 3rd International Conference on Technology and Innovation Policy: Assessment commercialisation and application of science and technology and the management of knowledge Austin USA August 30-September 2.
Leigh, C.M. (ed.) (1995) *A Regional Innovation Strategy: towards a blueprint for Yorkshire and Humberside*, Yorkshire and Humberside Regional Research Observatory.
Longhi, C. (1999) 'Networks, collective learning and technology development in innovative high-tech regions: The case of Sophia-Antipolis', *Regional Studies*, vol. 33, no. 4, pp. 333-342.
Lundvall, B. A. (1998) 'Why study national innovation systems and national styles of innovation', *Technology Analysis and Strategic Management,* vol. 10, no. 4, pp. 407-421.

Lyons, D. (2000) 'Embeddedness, Milieu, and Innovation Among High Technology Firms: A Richardson, Texas Case Study', *Environment and Planning A*, vol. 32, pp. 891-908.

Malmberg, A. and Maskell, P. (1997) 'Towards an Explanation of Regional Specialisation and Industry Agglomeration', *European Planning Studies*, vol. 5, pp. 1997.

Malmberg, A., Sölvell,O. and Zander,I. (1996) 'Spatial clustering, local accumulation of knowledge and firm competitiveness', *Geografiska Annaler B*, vol. 78, no. 2, pp. 85-97.

Maskell, P. (1998) 'Learning in the village economy of Denmark: the role of institutions and policy in sustaing competitiveness', Braczyk, H-J., Cooke, P. and Heidenreich, M. (eds) *Regional Innovation Systems: The Role of Governances in a Globalized World*, London, UCL Press, pp. 190-213.

Mc Kibbin, S. (2000) *Network building and institutional environment: A case study of innovation strategies of the European Union*, Institute of Social Studies, Working Paper 314.

Mc Pherson, A. (2002) 'The contribution of academic-industry interaction to Product innovation: The case of New York State's medical devices sector', *Papers of Regional Science*, vol. 81, pp.121–129.

Morgan, K. (1997) 'The Learning Region: Institutions, Innovation and Regional Renewal', *Regional Studies*, vol. 31, pp. 491-503.

Morgan, K. (1996) 'Learning-by interacting: Inter-firm networks and enterprises support', *Networks Of Enterprises And Local Development: Competing And Cooperating In Local Systems*, OECD:Paris, pp. 53-66.

Northern Ireland Economic Council (2001) *Developing a Innovation Strategy for Northern Ireland: A Statement by the Economic Council on research conducted by Philip Cooke, Stephen R Roper and Peter Wylie*, Occasional Paper 14: September 2001.

Ottai, G. (1998) 'The remarkable resilience of the industrial districts in Tuscany', in Braczyk, H. J., Cooke, P. and Heidenreich, M. (eds) *Regional innovation systems: the role of governances in a globalized world*, London, UCL Press, pp. 28-47

Radosevic, S. (1999) 'Transformation of science and technology systems into systems of innovation in central and eastern Europe: the emerging patterns and determinants', *Structural Change and Economic Dynamics*, vol. 10, pp. 277–320.

Revilla-Diez, J. (2001) 'Metropolitan innovation systems- a comparison between Barcelona, Stockholm and Vienna', *International Regional Science Review, vol. 21, no. 1, pp. 63-86.*

Roper, S. (2000) *Benchmarking Regional Innovation: A Comparison Of Baden-Württemberg, Bavaria, Northern Ireland And The Republic Of Ireland*, Northern Ireland Economic Research Centre, Working Paper Series No. 56.

Sabel, C. (1996) 'Learning-by-monitoring: The dilemmas of regional economic policy in Europe', *Networks Of Enterprises And Local Development: Competing And Cooperating In Local Systems*, OECD, Paris, pp. 23-49.

Schmitz, H. (2000) 'Does Local Co-operation Matter? Evidence from Industrial Clusters in South Asia and Latin America', *Oxford Development Studies*, vol. 28, no. 3, pp. 323-336.

Scott, A. (1998) 'From Silicon Valley to Hollywood: growth and development of The multimedia industry in California', in H.-J. Braczyk, P. Cooke and M. Heidenreich (eds) *Regional Innovation Systems: The Role of Governances in a Globalized World*, London, UCL, pp. 136-163.

Scott, A.J. and Storper, M. (1987) 'High Technology Industry and Regional Development: a Theoretical Critique and Reconstruction', *International Social Science Journal*, vol. 112, pp. 215-232.

Simmie, J. and Kirby, M. (1998) 'Innovation and the theoretical base of Technopole planning', *Progress in Planning*, vol. 49, pp. 159-198.

Staber, U. (1996) 'Accounting for Variations in the Performance of Industrial Districts: The Case of Baden-Württemberg', *International Journal of Urban and Regional Research*, vol. 20, pp. 299-316.

Staber, U. (2001) 'The structure of networks in industrial districts', *International Journal of Urban and Regional Research*, vol. 25, no. 3, pp. 537-552.

Sternberg, R. (2000) 'Innovation networks and regional development evidence from the European Regional Innovation Survey (ERIS): Theoretical concepts, methodological approach, empirical basis and introduction to the theme issue', *European Planning Studies*, vol. 8, no. 4, pp. 390-407.

Suarez-Villa, L. (2000) 'Southern California as a gateway region: policentricity and Network segmentation as competitive advantages', in A.E. Andersson, and D. Andersson (eds) *Gateways to the Global Economy*, Edward Elgar, Cheltenham.

Thomas, K. (2000) 'Creating regional cultures of innovation? The regional Innovation strategies in England and Scotland', *Regional Studies*, vol. 34, pp. 190-198.

Tödling, F. (1994) 'The Uneven Landscape Of Innovation Poles: Local Embeddedness And Global Networks', in Amin, A. and Thrift, N. (eds) *Globalisation, Institutions And Regional Development In Europe*, Oxford University Press, Oxford, pp. 68-90.

Tödling, F. (1999) 'Innovation networks, collective learning and industrial policy In Europe' Guest Editorial, *European Planning Studies*, vol. 7, no. 6, pp. 693-697.

Tödling, F. and Kaufmann, A. (2001) 'The role of the region for innovation activities in SMEs', *European Urban and Regional Studies*, vol. 8, no. 3, pp. 203-215.

Tödling, F. and Kaufmann, A. (1999) 'Innovation systems in Europe-A Comparative perspective', *European Planning Studies*, vol. 7, no. 6, pp. 699-717.

van den Berg, L., Braun, E and van Winden, W. (2001) 'Growth clusters in European Cities: An integral approach', *Urban Studies*, vol. 38, no. 1, pp. 185-205.

Wiig, H. and Wood, M. (1995) 'What Comprises a Regional Innovation System? An Empirical Study', Paper prepared for *Regional Association Conference. 'Regional Futures: Past and Present, East and West'*, Gothenburg, Sweden, 6-9th May 1995, STEP Group (Nås et al,), Studies in technology, innovation and economic development.

Yeung, H.W.C, (2000) 'Organising 'the firm' in industrial geography I: networks, institutions and regional development', *Progress in Human Geography*, vol. 24, no. 2, pp. 301-315.

Chapter 4

Local and Global Networks of Innovation

Sam Ock Park

Introduction

In this era of knowledge-based economy and globalisation, industrial networks are important not only to save transaction costs but also for the promotion of innovation through the process of knowledge creation and transfer. During the last decade geographers have been witnessing globalisation as a crucial driving force in the dynamics of economic spaces on the one hand, whereas on the other hand localisation has concurrently been progressing in production and innovation activities (Park, 2003).

Innovation is critical for gaining and maintaining the competitiveness of firms and for promoting economic development in a knowledge-based economy (Felsenstein and Taylor, 2001). Innovation can be defined as the transformation of knowledge into novel wealth creating products, services, and technologies (Cooke et al, 2003). It is not only a technological process but also a social process within economic spaces, where networks are important. In a knowledge-based economy, networks go beyond inter-firm relations and encompass inter-personal relations, supplier-customer relations for transactions of materials and services, relations with diverse organisations such as producer service firms, universities, and public institutions. Accordingly, networks are regarded as relational processes among diverse agents of economic activities over space (Dicken et al, 2001). These networks contribute significantly to learning processes, promote knowledge spill-over, and provide a context for knowledge creation and innovation (Bunnel and Coe, 2001; Malecki, 1997; Yeung, 2000).

The process of developing innovation networks reveals interactions among various economic actors and individuals in a regional context, reflecting the fact that spatial proximity is important. There is no disagreement that the local dimension is important for social processes in innovation networks. Innovation networks are, however, not limited to local networks. Innovation networks and innovation systems can also be developed from non-local and global networks of innovation, suggesting the development of spatial systems of innovation that integrate local, national and global dimensions of networks (Oinas and Malecki, 1999). In the spatial systems of innovation, networks among the clusters of innovation, networks of knowledge intensive manpower, and interactions among different cultures have become increasingly important. Dominant innovation networks in a global context are stimulated by global commodity chains, global financing, and ethnic networks (Dicken et al, 2001; Gereffi and Koreniezwicz, 1994; Saxenian, 2000). Global commodity chains and the ethnic networks in global society facilitate the flow of knowledge, information, capital, products, services, etc., which are required for innovation.

The purpose of this study is to examine types and dynamics of innovation networks in a globalised world with regard to regional innovation systems. Theoretical notions of the development of regional innovation systems and regional advantages of industrial clusters are analysed with relation to the local and global networks. Basic supporting sectors required for the promotion of innovation clusters are examined. Types and dynamics of innovation clusters and their global networks are discussed. The final section concludes with some policy implications in a networked world.

The Spatial Dimension in the Networks of Innovation

Development of a National Innovation Systems Approach

Economic growth and development differ from one nation state to another and one could ask why these differences between nation states exist. There are three major approaches for explaining the national differences in growth and development, namely: approaches dealing with national production systems, business systems, and innovation systems (Park, 2001; Lundvall and Maskell, 2000).

Traditionally, the production structure or national systems of production were regarded as a major factor for explaining the differences in growth and development between countries. In the structural approach of the 1960s and 1970s it was assumed that different sectors affected growth differently. Later, in the 1970s, more dynamic approaches to vertical linkages in the production system were introduced as an important element in the development processes. For example, Anderson (1992) introduced more dynamic elements in the production systems, such as the importance of backward linkages in the form of flows of information, learning by doing, and life-cycle perspectives in the development of industrial complexes. In this approach, in addition to structural aspects, the system of production was considered. That is, Anderson regarded interactions between user sectors and producer sectors in the production systems and the quality of demand to be important elements for the development of new technology.

However, the differences in production systems can only explain a part of the national differences in economic growth and development. Historically, there have been considerable differences between countries in the ways in which they have organised and sustained the development, introduction, improvement, and diffusion of new products and processes within their national economies (Freeman, 1995). That is, the national education system, industrial relations, technical and scientific institutions, government policies, cultural traditions, and other national institutions have been regarded as important for explaining the differences in economic development between countries. In the national production system approach, the role of institutions in economic growth was not dealt with, because it focuses on the organisation of production (Lundvall and Maskell, 2000).

Accordingly, beyond the differences in the national systems of production, national differences in the organisation of firms and markets are regarded as important elements of national differences in growth and development. This aspect was clarified in more detail by Whitley (1996) with the introduction of national business systems. The national business systems approach is interdisciplinary, combining elements of economic reasoning with sociological perspectives, because differences in culture and formal institutions are regarded as significant factors for explaining how firms and markets are organised in different countries. The fundamental reasoning of the national business systems approach is that 'in nations where the institutional set-up supports the establishment of trust between no-kin agents, and especially where this is combined with credit-based financial system, firms will tend to be involved in cooperative networks' (Lundvall and Maskell, 2000). In such a nation, firms can evolve a high degree of specialisation and division of labour because they can share their risks with collaborative partners, and their expansion will be based on incremental growth in terms of capabilities and assets.

The national business systems approach contributed to our understanding of the differences in growth and development of countries by spelling out the institutional dimension that was neglected in the national production systems approach. However, the national business systems approach has its limitations, in that it neglects the production structure of nations. The national business systems approach disregards the fact that the organisation of firms and markets also differs according to production sectors. Accordingly, it is necessary to integrate the national production systems and business systems to understand the differences in growth and development of nations. The integration of the two approaches represents the integration of structural and institutional aspects. Innovation is regarded as an important factor for national growth and development in a knowledge-based economy.

The national innovation systems approach may represent 'a kind of synthesis of national production systems and national business systems in the sense that it tries to capture the co-evolution of structural and institutional characteristics in a systemic perspective' (Lundvall and Maskell, 2000: p. 359). A national innovation system can also be regarded as 'a system of interconnected institutions to create, store, and transfer the knowledge, skills, and artefacts that define new technologies' (Hotz-Hart, 2000: p. 437). Favourable development of national innovation systems can provide firms with valuable capabilities and favourable conditions for innovation not available to competitors located in other countries, even under the open market conditions of the globalisation era. Even in this era of globalisation, while global networks are certainly growing in importance, national systems of innovation are regarded as fundamental for understanding changes in economic spaces.

Traditionally, innovation was regarded as a linear process that can be conducted separately in R&D labs. In the national innovation systems approach, however, innovation is regarded as an interactive learning process that requires exchange of knowledge, interaction, and cooperation among various actors in a production network. Various actors of interaction and networking include other firms, contractors and subcontractors, equipment and component suppliers, customers or users, competitors, private and public research labs, universities and other higher education institutions, providers of consultancy and technical services, state authorities and regulatory bodies (Hotz-Hart, 2000).

Because of these interactive processes of innovation in our knowledge-based economy, the production of goods and services is becoming more knowledge intensive but not necessarily more R&D intensive. For example, rapidly growing new advanced service activities such as software and venture capital funds employ highly qualified manpower and are highly intensive in immaterial investment but not in formal R&D (OECD, 1999). Innovation requires more than R&D and a too narrow focus on R&D overlooks the importance of other types of innovative efforts such as design or market analysis. The breakdown of innovation expenditures of the OECD countries reveals that R&D has an average share of 33.5 per cent; product design, 24 per cent; market analysis, 6.6 per cent; patents and licenses, 4.6 per cent; and external spending, 22.4 per cent (OECD, 1999: p. 18). There are, of course, variations in the R&D content of innovation and innovative performance between various countries and types of industries.

Regional Innovation Clusters (RICs)

It is true that the differences in the national innovation systems are an important factor in national growth and development even in the era of globalization. It is also true that the impact of national boundaries on the flow of capital, knowledge, products, services, etc., has been weakened by the progress of globalisation and the development of information and communication technology (ICT). Regions have become a more important identity for shaping the economic spaces and economic growth and development of a nation in the era of globalisation and a knowledge-based economy.

Because of the importance of the role of 'regions' in the era of globalisation and a knowledge-based economy, the concepts from the national innovation systems approach can be applied to regions, thus creating a regional innovation systems approach. Saxenian's (1994) identification of the major source of regional advantages of Silicon Valley compared to Route 128 in the United States contributed significantly to making regional innovation systems an important policy issue in regional development in many advanced countries. Saxenian (1996: p. 2) described Silicon Valley as a 'regional network-based industrial system that promotes collective learning and flexible adjustment among specialist producers of a complex of related technologies'. It is well accepted that 'knowledge spill-overs are geographically bounded within a limited space over which interaction and communication is facilitated, search intensity is increased, and task coordination is enhanced' (Feldman, 2000: p. 389).

There are several reasons why regions are important for innovations in firms (Park, 2001). Firstly, collaborations and inter-firm networks play an important role in innovation within firms. Especially for SMEs, collaboration with other firms and organisations makes it possible to gain access to a diverse specialist expertise to meet customer needs and to share risk and costs (Keeble, 1997). Because of the significant impacts of networks, the design of technology networks has been regarded as an important policy issue at a local and national level in some countries like Denmark. The main strength of networks of SMEs is derived from the ability that small firms have shown to learn from each other and from an intermediary system which has acted as a successful catalyst for cooperation (Huggins, 1996). In the inter-firm networks, relationships between large and small firms as well as those between small and small firms are also important for innovation (Park, 1996; Young, Francis, and Young, 1994).

Secondly, innovation is an interactive process that is shaped by a variety of institutional routines and social conventions. Innovation is an interactive process between different functions within firms, between suppliers and customers at the inter-firm level, and between firms and the wider institutional milieu (Morgan, 1997). Routines and conventions may help to regulate economic life by reducing uncertainty and facilitating coordination and cooperation for mutual benefit. The interactive process can be regarded as a process of interactive learning in which institutional routines and conventions can play a role (OECD, 1992).

Thirdly, regional clustering and specialisation are important for innovation. In recent years, regions have been rediscovered as an important source of competitive advantages and organisation of global space economy (Scott, 1995). Clustering and specialisation in a region are regarded as some of the key elements of growth and competitiveness, due to the reduction of transaction costs, agglomeration economies, and technological and skill advantages (Krugman, 1995, Porter, 1994). Beyond this regional competitiveness based on the economies of clustering, geographical studies of recent years suggest the importance of locally embedded social, cultural and institutional arrangements as a source of knowledge creation and learning (Amin and Thrift, 1995; Storper, 1996). Informally constituted knowledge and information environments, derived from local tacit knowledge and face-to-face exchange, the quality of local institutions, long-standing social habits and local conventions of communication and interaction contribute to a firm's learning-based competitiveness. In this institutional perspective, Silicon Valley and Italian industrial districts are regarded as 'learning regions', which display intensive inter-firm interaction, shared know-how, spill-over expertise, and strong firm support systems (Amin, 1998). In RICs, accordingly, organisation in terms of inter-firm networks and collaboration, learning dynamics with interactive processes, and the territorial dimension with regional clustering are regarded as three important dimensions.

Global Networks of Innovation

In this era of globalisation, both global and regional networks are important for innovation. The process of creating skills and important influences on the rate of the innovation process are intensively local (Hotz-Hart, 2000). Local networks and inter-firm interactions in local areas are important for the exchange of tacit knowledge and the generation of new knowledge that can serve as a basis for an innovation. Paradoxically, more open global competition makes the home base more important and the multinational firms' objectives are to be global and local simultaneously (Porter, 1990). Regionally embedded agglomeration economies based on intensive local competition and cooperation networks can promote incremental innovation. However, if a situation of lock-in occurs, because the local networks are too strong and closed and small and medium sized firms are accordingly unable to change their technological trajectory, the region may lose its competitive advantages and become incapable of technological development. Especially when the existence of a strong industrial atmosphere in a lock-in situation is used to squeeze wages, in order to remain competitive, this would not result in continuous innovations (Glasmeier, 1994). Because of this lock-in effect, the non-local networks in the global economic space are also important for new technological development. Extending the competition and cooperative networks from local to global suggests the progress of spatial innovation systems from regional, to national, and to global.

That is, RICs should have innovation networks with other regions in a nation on the one hand, and they should have links with global innovation centres on the other hand. Especially networks with the original innovation centres in global economic space are critical for sharing information and transferring knowledge and technology. The regional innovation clusters have networks with other innovation clusters through flows of information, knowledge, technology, capital, products, and services in all spatial dimensions, regional, national, and global. Networks with global innovation centres or original innovation centres are sometimes regarded as more important than networks with other national innovation centres. Because of the open characteristics of the networks of innovation, Oinas and Malecki (2002) suggest a major role for spatial innovation systems in the evolution of technologies over time and space. However, it should be considered that the spatial dimension of innovation networks varies according to the type of innovation cluster.

Innovation Clusters: Basic Supporting Sectors, Types and Global Networks

Basic Supporting Sectors for Regional Innovation Clusters

As discussed in the previous section, in RICs, inter-firm networks and collaboration, collective learning with interactive processes, and the territorial dimension with regional clustering are regarded as three important dimensions. In order to evolve or build such regional clusters of innovation, supporting environments and processes are required, because 'a regional innovation system consists of firms that innovate and a variety of support organizations such as banks, investors, lawyers, consultants, grant providers, technology agencies, incubators, science parks, research bodies and training bodies' (Cooke et al, 2003: p. 373). In this chapter dealing with the variety of support organisations, six basic supporting sectors are considered important for the evolution of RICs. First , agglomeration of production activities and inter-firm networks of production are necessary for the development of RICs. Innovation of new products or new processes is related to the production activities. Accordingly, production networks can be regarded as a basic sector for innovation networks. Traditionally, the inter-firm networks of production are regarded as an important process in the clustering and development of production systems. Production is a primary function in regional industrial systems (Hamilton and Linge, 1983).

Second, services for supporting production and business activities are critical for the development of RICs. Financing, accounting, design, advertising, consulting, and legal services are all knowledge-intensive services and are required for the formation of RICs. Especially a supply of venture capital is a critical factor for successful innovation clusters. Most of the industrial parks in developing countries are merely agglomerations of production activities and are deficient in producer services. The provision of producer services and advanced producer services in Silicon Valley is regarded as one of the important factors in the development of a venture habitat and innovation environment (Lee et al, 2000).

Third, innovation clusters should supply new innovative firms in order to sustain regional competitive advantages. New venture firms and high tech start-ups and spin-offs are important sources of dynamic evolution for the innovation clusters. For the generation of new high tech start-ups, abundant high quality manpower with entrepreneurship is necessary. Entrepreneurs and engineers are important manpower for the generation of new start-ups and spin-offs. Continuous training and retraining of manpower is also important. Thus, the supply of high quality manpower is another basic supporting sector for innovation clusters.

Fourth, successful innovation clusters should continuously conduct R&D activities for new product innovations. Continuous development of new products is a source of the evolution of innovation clusters. Accordingly, university and R&D institutions can be regarded as a basic supporting sector for the innovation cluster. Creation and spill-over of knowledge from the universities and R&D institutions are important sources for improving the regional innovation potential. Most of the innovation centres in the world are closely linked with universities and R&D institutions.

Fifth, in addition to the formal R&D activities, a collective learning process is also important for creation and transfer of knowledge in the RICs, as examined in the previous section. Social relations with formal and informal meetings are significant mechanisms for transferring tacit knowledge and they contribute to the new generation of tacit and codified knowledge. Inter-firm relations, business associations, small groups for common interests, ethnic groups, etc. are all significant contributors to the creation and transfer of knowledge and then to innovation.

Lastly, provision of supporting infrastructure is also important for the development of innovation clusters, especially in newly industrialised countries. Provision of high quality information and communication infrastructure is a primary condition for the successful development of innovation clusters in an era of globalisation. A supply of high quality office buildings and living environments is also important, especially in non-Western countries, to attract innovative firms and foreign direct investments in knowledge-intensive sectors.

In addition to the six basic supporting sectors, there are many other aspects in the supporting sectors of innovation clusters. Accordingly, the six basic supporting sectors can be regarded as core sectors for the successful development of innovation clusters. It should be noted that the six sectors are not independent of each other, but interrelated with each other. For example, on the one hand, the university and R&D sector can contribute to the supply of high quality manpower and training and on the other hand, they can be sources and participants of active collective learning and advanced services such as consulting activities.

Types and Dynamics of Regional Innovation Clusters

Cooke (1998) considered two key dimensions for the typology of regional innovation systems – the governance infrastructure and the business superstructure. Based on the public governance systems, three types of regional innovation systems were proposed, namely grassroots, networked, and centralist, while based on the business innovation system, localist, interactive, and globalist regional innovation systems were suggested (Cooke et al, 2003). Cooke's classification of regional innovation systems is meaningful because it is derived from theoretical and empirical scrutiny. However, since this study focus on the networks, the types examined here are based on the degree of networks and innovation. That is, innovation clusters can be classified by both the level of innovation and the degree of local networks and embeddedness in a region. Even though this classification is very simple, it helps to understand the dynamic evolution and networks of innovation. As seen in Figure 4.1, four types of clusters can be identified.

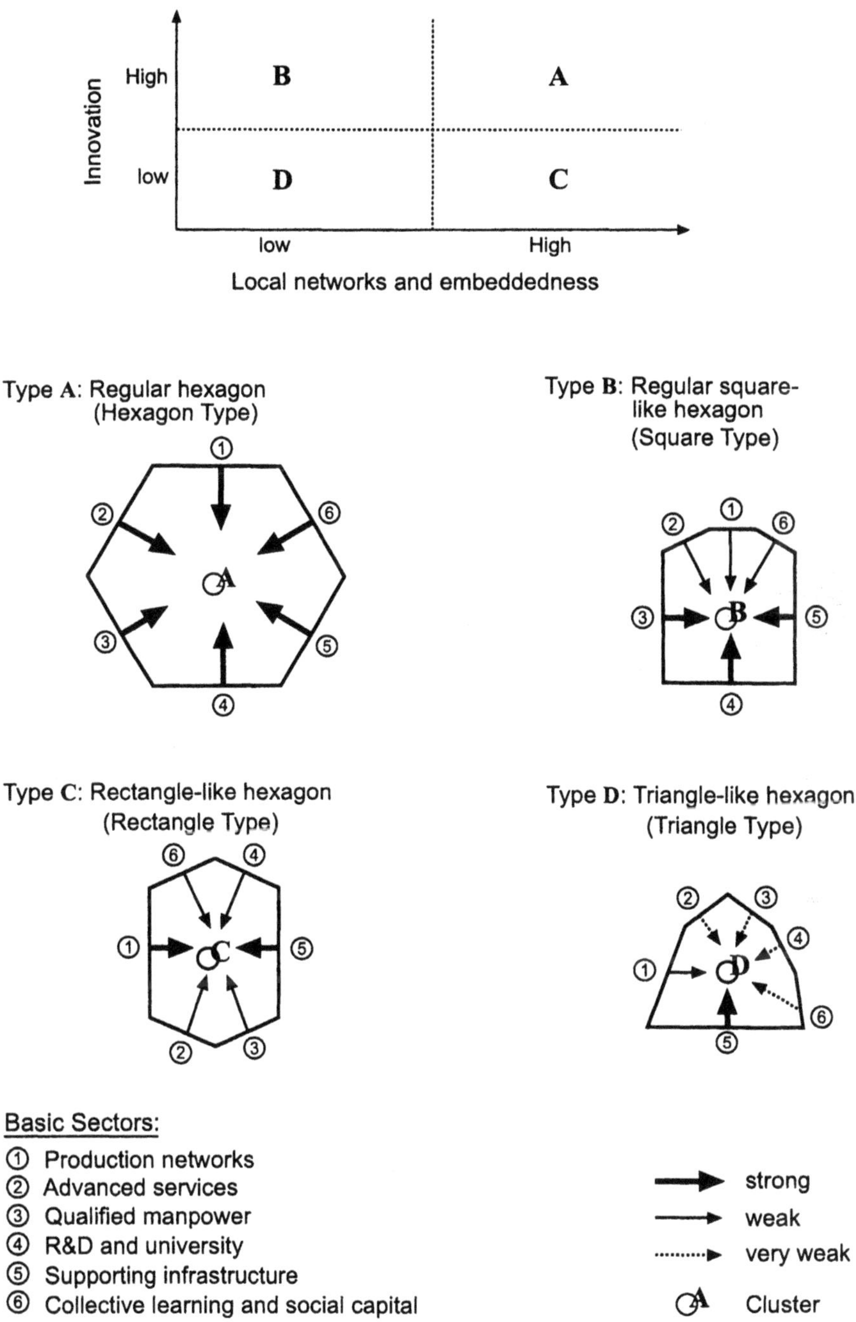

Figure 4.1 Types of innovation clusters

Type A represents a high level of innovation and a high degree of local networks and embeddedness. It is a successful innovation cluster. The six basic supporting sectors are all sufficiently provided: strong inter-firm production networks; a supply of diversified advanced services; a supply of mobile and highly qualified manpower; strong formal R&D activities in universities, firms, and public sectors; qualified supporting infrastructure; and strong collective learning processes and social capital. These six sectors are all balanced with each other in terms of importance, creating a regular hexagon type of innovation cluster (Hexagon type). There exist strong processes of collective learning, creation and transfer of knowledge, and development of new products and technology through strong local networks among diverse economic actors. Silicon Valley is a typical example of this model (Lee et al, 2000; Gray et al, 1998).

Type B represents a high degree of innovation but a low degree of local networks and embeddedness. It is a relatively research oriented innovation cluster. It has a sufficient supply of qualified manpower, strong R&D activities and research universities, and sufficient supporting hard infrastructure. However, production networks, advanced services, and collective learning processes are not strongly embedded locally. Because it has only three strong basic sectors, the shape of the hexagon looks like a regular square (Square type). There are strong R&D activities, but most of the R&D institutions are branch type research centres. Research Triangle Park in USA is a typical example of this type (Luger and Goldstein, 1990).

Type C shows relatively strong local networks and embeddedness, but a relatively low level of innovation. It is a production oriented innovation centre. It has relatively strong local production networks and good supporting infrastructure, but insufficient processes of collective learning in the local area. In this type, sectors in advanced services, qualified manpower, and R&D and university are not strong. Since two sectors are strong and the other four sectors are relatively weak, the hexagon model looks like a rectangle (Rectangle type). Hsinchu, Taiwan, and Kumi, Korea, are typical examples of the Rectangle type (Saxenian and Hsu, 2001; Park and Markusen, 1995).

Type D represents a low level of innovation and a low degree of local networks and embeddedness. The supporting infrastructure is relatively well developed, but other basic sectors are rather weak. The shape of the hexagon model now looks like a triangle (Triangle type). There is an agglomeration of production activities, but networks and innovation potential are not yet sufficiently developed. This type cannot be regarded as an actual innovation cluster. Most of the industrial parks in developing areas can be regarded as the Triangle type.

It should be noted that each of the types of innovation cluster is not constant but dynamic over time. The Hexagon type (Type A) can be further developed by intensifying innovation potential and local networks. The Square type (Type B) can be developed into a Hexagon type by increasing local networks and embeddedness. Local production networks and local entrepreneurship seem to be the critical factor for the evolution to the Hexagon type. The Rectangle type (Type C) can also be developed into a Hexagon type by supporting innovation networks, R&D and collective learning processes. Type D can possibly be developed into a

Rectangle type by increasing local networks. The four types can also be dynamically developed by increasing the degree of global networks in terms of flow of capital, manpower, knowledge, advanced services, etc. Due to the dynamic aspects of the types of innovation clusters, however, there are diverse transitional or hybrid types of innovation clusters in the real world, which are not clearly covered by the four types.

Global Networks of Innovation Clusters

1) International Flows of Network Factors

On the one hand, successful innovation clusters have strong local networks and embeddedness integrating the six basic supporting sectors. On the other hand, they also have strong global networks in this era of globalisation and an information society. The global networks of innovation clusters result from various flows in the global space economy, such as knowledge and technology, manpower, capital, products, advanced services, and strategy and power. These flows are related to the diverse logic of global networks, which is reviewed briefly in this section.

First, the Internet and training systems encourage a global inflow of information, knowledge and technology. Information about new products, markets, and technology can be transferred easily through the Internet in recent years. For example, large Korean firms have R&D centres in Silicon Valley and conduct R&D activities employing local engineers and scientists. Local tacit knowledge can be converted into codified knowledge through the R&D activities and transferred to headquarters in Seoul (Park, 1996). Localised codified knowledge can also be combined into systemic knowledge and transferred to parent firms. In addition, Korean firms send their employees to Silicon Valley for training and their tacit knowledge acquired in Silicon Valley can be naturally transferred when they come back from their training. Accordingly, direct investments in global innovation clusters and training systems in the innovation clusters can be sources of flows of knowledge and technology in global networks.

Second, the international flow of qualified manpower can be promoted through ethnic networks. Advanced countries have offered opportunities of higher education for students from developing countries. Some of them went back to their home country after their advanced education, but the rest of them got a job in the advanced countries. However, later when their home country offers better opportunity to work, they come back. In addition, second generation immigrants in the advanced countries can be given the opportunity to work in their parents' country. They have close networks of technology and knowledge within the advanced country where they were trained. Many engineering and high tech societies of ethnic groups in the global innovation centres can be the source of networks. Accordingly, at first, there is a flow of manpower from developing countries to advanced countries for higher education and jobs, but later when the home country becomes industrialised there is a good possibility of return-migration of engineers and entrepreneurs. More than 15,000 engineers and entrepreneurs returned to Hsinchu from foreign countries during the period of 1983-1998 (Hong, et al, 2000). Chinese immigrant networks within Silicon Valley and their connection with Hsinchu clearly demonstrate global manpower networks based on ethnic networks (Saxenian, 2000; Saxenian and Hsu, 2001). Indian ethnic networks are also an important factor in software networks between Bangalore and Silicon Valley (Saxenian, 2000). This flow of manpower is a very important source for the transfer of tacit knowledge and technology.

Third, international flows of capital have been greatly increased by foreign direct investments in this era of globalisation and open-door policy. Foreign direct investments from advanced countries to developing countries have increased since the 1980s. Direct investments to a developing country can be a source for transferring technology and knowledge. In recent years, there have been direct investments from newly industrialised countries to the global innovation centres in advanced countries, which can be an important source of technology and knowledge transfer from the Hexagon type of innovation cluster to other types of innovation clusters. Ethnic networks can also facilitate international flows of capital, as seen in Chinese overseas networks of capital investments (Saxenian, 2000). Venture capital has concentrated in the global innovation clusters, which has activated more innovation in the Hexagon type innovation clusters. Overall, the international flow of capital, especially from foreign direct investment, has contributed significantly to the international transfer of knowledge converting locally embedded tacit knowledge to codified knowledge.

Fourth, the international flow of products has also contributed to the formation of global innovation networks. Global commodity chains (GCC) have been a mechanism for promoting the flow of products through international trade and inter-firm networks. According to Gereffi and Koreniezwicz (1994), there are two types of global GCC: producer-driven GCC and buyer-driven GCC. In the producer-driven GCC, the producer of final products is the key actor for organising production networks in the global economic space, while, in the buyer-driven GCC, the buyer who sells the final products to consumers is the key actor for organising production networks. Automobile and semiconductor producers control design, R&D, marketing, subcontracting, etc., and can be regarded as a producer-driven type. In the shoes and apparel industries, the buyer has the power to organise and control such activities as design, R&D, marketing and contracting products. Along the chains of production networks, knowledge of marketing, product innovation, and process technology can be transferred. Accordingly, GCC provide not only networks of production and trade, but also networks of innovation.

Fifth, international flows of advanced services are another important source of global innovation networks. In recent years, producer service firms have been more active in expanding their global market. Multinational firms dealing in legal, design, consulting, advertising, and technical services are now global service networks with opening service markets under the WTO system (Daniels, 1998). Most of the innovation clusters in newly industrialising countries have difficulties in product and process innovations due to the lack of advanced producer services. With the opening of the service sectors in the newly industrialising countries, advanced services from the global innovation clusters help to improve the regional innovation potential. In addition, direct investments in the advanced service sector from the global innovation clusters usually concentrate at the national innovation centre and contribute to the knowledge spill-over effect with inter-firm interaction and various informal meetings.

Sixth, and lastly, strategy and power flow along the global networks of innovation. On the one hand, firms in the global innovation clusters have the power to organise their global innovation networks along their chains of investments throughout the world. On the other hand, firms in the Square type or Rectangle type of innovation cluster in the newly industrialising countries establish strategies for marketing and innovation through direct investments in the global innovation clusters as a base for their global networks. In general, however, power and strategy flow along the hierarchy of GCC and along the international networks from Hexagon type to other types of innovation clusters.

2) Global Networks According to Types of Innovation Clusters

Global networks between types of innovation clusters through international flows of knowledge and technology, manpower, capital, products, advanced services, and strategy and power can be depicted as in Figure 4.2. The pattern of flows in the global space economy differs according to the type of innovation cluster.

Global innovation centres (Hexagon Type) have a close network with Rectangle Type innovation clusters. International innovation networks between Silicon Valley and Hsinchu, Taiwan, are a good example of Hexagon-Rectangle type networks. Especially ethnic Chinese networks in the flow of capital and manpower are a critical factor for the strong global networks of innovation between the two regions (Saxenian, 2000). There was a 'brain drain' from Taiwan in the 1960s and 1970s because Taiwan's best and brightest engineering students went to the USA for graduate studies and chose to stay to pursue professional careers in the USA, especially in Silicon Valley. However, during the last decade, engineers began returning to Taiwan as a result of an active government recruitment policy and opportunities for good jobs created by rapid economic development (Saxenian, 2000).

Hexagon type innovation centres make up global networks with Square type and Triangle type ones, but their intensity of interaction is lower than that with the Rectangle type. Rectangle type innovation clusters also have relatively strong networks with Triangle type ones. Relatively strong networks between innovation clusters in the newly industrialised countries and industrial clusters in China and south-eastern Asia are mostly regarded as Rectangle-Triangle type global networks. There are strong networks along the hierarchy of Hexagon – Rectangle – Triangle type clusters due to the complementarities along the hierarchy. GCC and ethnic networks facilitate the hierarchical global networks of innovation. In general, products flow from the lower order of the hierarchy to the higher order, but knowledge and technology, manpower, capital, advanced services, and power flow from the higher order to the lower order of the hierarchy. However, there are also flows of manpower and capital from lower level to higher level innovation clusters.

The reverse flows are possible because of education and training in the advanced innovation clusters and investment strategies to acquire local embedded knowledge and information from the advanced innovation clusters.

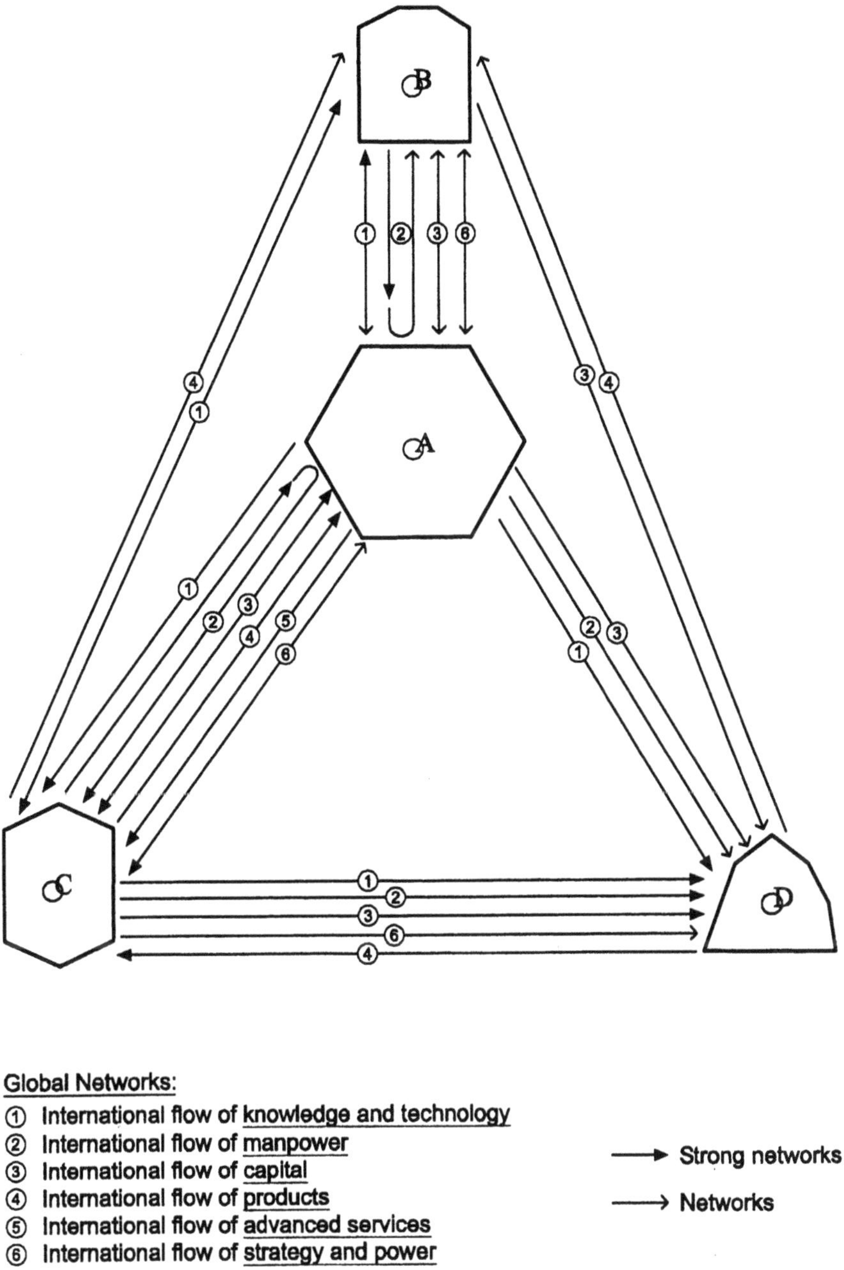

Figure 4.2 Global networks between types of innovation clusters

Conclusion

The purpose of this study is to analyse networks of innovation and to understand their spatial dimensions in a networked world. On the one hand, even in an era of globalisation and the Internet, regions are critical to the innovation process, because organisation in terms of inter-firm networks and collaboration, learning dynamics with interactive processes, and the territorial dimension with regional clustering are important. On the other hand, globalisation, development of ICT, and opening markets under the WTO system promote the international flow of products, knowledge and technology, qualified manpower, advanced services, capital, and strategy and power, creating global networks of innovation.

Collective learning processes through formal and informal meetings, inter-firm cooperation and competition, intra-firm networks, and inter-organisational collaborations are important for local networks of innovation. Global commodity chains and ethnic networks in global society have significant impacts on the formation of innovation clusters and networks of innovation at an international scale. The pattern of spatial innovation networks differs, however, according to the type of innovation clusters, which is classified on the basis of the level of innovation and the degree of local networks and embeddedness. Different types of regional innovation clusters have different basic supporting sectors and different international flows of network factors. However, it should be noted that each of the types of innovation clusters is not constant but dynamic over time. Due to the dynamics, there are diverse transitional or hybrid types of innovation clusters in the real world.

The formation of local networks of innovation and collective learning processes are critical for the evolution of regional innovation clusters. However, the development of regional innovation clusters or systems is not only related to the regional dimension and the closed regional system. The local, regional, national, and global scales are all interlinked in networks of innovation. Innovation clusters are linked to each other beyond regional or national boundaries, suggesting that spatial innovation systems beyond the national or regional innovation systems are evolving in the global space economy. Even global innovation systems are evolving in this era of the Internet and a knowledge-based economy. The dynamic evolution of spatial innovation systems suggests that diverse spatial scales should be considered for regional development and regional competitive advantages in a networked world. Policy makers should consider the regional characteristics in terms of the degree of networks and embeddedness before implementing regional innovation strategies.

References

Amin, A. (1998) An *institutionalist perspective on regional economic development.* Paper presented at the Economic Geography Research Group Seminar, 'Institutions and Governance'. July 3, 1998, Department of Geography, UCL, London.

Amin, A. and Thrift, N., (1995) Institutional issues for the European regions: from market plans to socioeconomics and powers of association. Economy and Society 24 (1): pp. 41-66.

Anderson, E. S. (1992) Approaching national innovation systems, In Lundvall, B. A. ed., *National Innovation Systems: Towards a Theory of Innovation and Interactive Learning*, London: Pinter Publishers.

Bunnell, T. G and Coe, N. M., (2001) Spaces and scales of innovation. *Progress in Human Geography* 25: pp. 569-589.

Cooke, P. (1998) Introduction: origin of the concept, In Braczyk, H., Cooke, P., and Heidenreich, M., eds., *Regional Innovation Systems*, Reading, UK: Arroehead books Limited: pp. 2-27.

Cooke, P., Roper, S., Ramos, R., and Royuela, V., (2003) 'The golden thread of innovation and Northern Ireland's evolving regional innovation system', *Regional Studies* 37 (4): pp. 365-380.

Daniels, P. (1998) Economic development and producer service growth: the APEC experience, *Asia Pacific Viewpoint* 39 (2): pp. 145-160.

Dicken, P. Kelly P., Olds K., Yeung H. W. C, (2001) Chains and networks, territories and scales: towards a relational framework for analyzing the global economy. *Global Networks* 1: pp. 89-112.

Feldman, M. P. (2000) Location and innovation: the new economic geography of innovation, spillovers, and agglomeration. In Clark, G. L., Feldman, M. P., and Gertler, M. S., Oxford, pp. 373-394.

Felsenstein, D. and Taylor, M., (eds.) (2001) *Promoting Local Growth. Process, practice and policy,* Aldershot, Ashgate.

Freeman, C. (1995) The 'national systems of innovation' in historical perspective. *Cambridge Journal of Economics* 19, pp. 5-24.

Gereffi, G and Koreniezwicz, M. eds. (1994) *Commodity Chain and Global Capitalism,*Westport, Conn: Praeger.

Glasmeier, A. (1994) Flexible districts, flexible regions? The institutional and cultural limits to districts in an era of globalization and technological paradigm shifts. In Amin, A. and Thrift, N., eds., *Globalization, Institutions, and Regional Development in Europe.* Oxford University Press, Oxford, pp. 118-146.

Gray, M., Golob, E., Markusen, A., and Park, S. O., (1998) New industrial cities? The four faces of Silicon Valley, Review of Radical Political Economics 30. (4): pp. 1-28.

Hamilton, F. E. I. and Linge, G J. R., (eds.) (1983) Spatial analysis, Industry and the Industrial Environment. Volume 3- Regional Economies and Industrial Systems, Chichester and New York, Wiley.

Hong, S. B., Lim, D. S., Lee, M. J., Lee, J. H., and Chang, S. K., (2000) Policy Research on Overseas emerging innovation clusters and Global Open Strategy, Science & Technology Policy Institute, Seoul.

Hotz-Hart, B. (2000) Innovation networks, regions, and globalization. In Clark, G, L., Feldman, M. P., and Gertler, M. S., eds., *The Oxford Handbook of Economic Geography.* Oxford University Press, Oxford, pp. 432-450.

Huggins, Robert, (1996) Technology Policy, Networks and Small Firms in Denmark. *Re*

gional Studies 30 (5): pp. 523-526.
Keeble, David. (1997) Small Firms, Innovation and Regional Development in Britain in the 1990s, *Regional Studies* 31 (3): pp. 281-294.
Krugman, P. (1995) *Development, Geography and Economic Theory*, London, MIT Press.
Lee, C.M., Miller, W., Hancock, M., and Rowen, H. Eds. (2000), *The Silicon Valley Edge. A Habitat for Innovation and Entrepreneurship*, Stanford, Stanford University Press.
Luger, M. and Goldstein, H., (1990) *Technology in the Garden: Research Parks and Regional Economic Development,* Chapel Hill, NC: University of North Carolina at chapel Hill.
Lundvall, B. and Maskell, P., (2000) Nation states and economic development: from national systems of production to national systems of knowledge creation and learning. In Clark, G. L., Feldman, M. P., and Gertler, M. S., eds., *The Oxford Handbook of Economic Geography,* Oxford University Press, Oxford, pp. 353-372.
Malecki, E. J. (1997) Entrepreneurs, networks, and economic development: a review of recent research, *Advances in Entrepreneurship, Firm Emergence, and Growth* 3: pp. 57-118.
Morgan, K. (1997) The learning region: institutions, innovation and regional renewal, *Regional Studies* 31 (5): pp. 491-504.
OECD (1992) *Technology and Economy - the Key Relations*. OECD, Paris.
OECD (1999) *Managing National Innovation Systems,* OECD, Paris.
Oinas, P. and Malecki, E. J., (1999) Spatial innovation systems, in Malecki, E.J. and Oinas, P., eds., *Making Connections. Technological learning and regional economic change,* Aldershot: Ashgate: pp. 7-34.
Oinas, P. and Malecki, E. J., (2002) The evolution of technologies in time and space: from national and regional to spatial innovation systems, *International Regional Science Review* 25, (1): pp. 102-131.
Park, Sam Ock, (1996) Network and embeddedness in the dynamic types of new industrial Districts, *Progress in Human Geography* 20 (4): pp. 476-493.
Park, Sam Ock, (2001) Regional innovation strategies for regional development in the Knowledge-based economy, *Geojournal* 53, pp. 29-38.
Park, Sam Ock, (2003) Economic spaces in the Pacific Rim: a paradigm shift and new dynamics, *Papers in Regional Science* 82 (2): pp. 223-247.
Park, S. O. and Markusen, A., (1995) Generalizing new industrial districts: A theoretical agenda and an application from a non-Western economy, *Environment and Plan ning A* 27: pp. 81-104.
Porter, M. E. (1990) *The Competitive Advantage of Nations,* Free Press, New York.
Porter, M. E. (1994) The role of location in competition, *Journal of Economics of Business* 1 (1), pp. 35-39.
Saxenian, A. (1994) *Regional Advantage: Culture and Competition in Silicon Valley and Route 128.* Harvard University Press, Cambridge, Mass.
Saxenian, A. (1996) *Regional Advantage: Culture and Competition in Silicon Valley and Route 128,* 2nd edition, Harvard University Press, Cambridge, Mass.
Saxenian, A. (2000) Networks of immigrant entrepreneurs, In Lee et al eds., *The Silicon Valley Edge. A Habitat for Innovation and Entrepreneurship,* Stanford, Stanford University Press: pp. 248-268.

Saxenian, A. and Hsu, J. Y., (2001) The Silicon Valley-Hsinchu connection: technical communities and industrial upgrading, *Industrial and Corporate Change* 10, (4): pp. 893-920.

Scott, A. J. (1995) The geographic foundations of industrial performance, *Competition and Change* 1 (1): pp. 51-66.

Storper, M. (1996) Institution of the knowledge-based economy, In OECD, 1996: *Employment and Growth in the Knowledge-based Economy,* Paris, OECD, pp. 255- 283.

Whitley, R. (1996) The social construction of economic actors: institutions and types of firm in Europe and other market economies, In Whitley, R. ed., *The Changing European Firm,* Routledge, London.

Yeung, H. W. C. (2000) Organizing "the firm" in industrial geography. Part 1: Networks, Institutions and regional development, *Progress in Human Geography* 24: pp. 301-315.

Young, Ruth C., Joe D. Francis and Christopher H. Young, (1994): Flexibility in Small Manufacturing Firms and Regional Industrial Formations. *Regional Studies* 28 (1): pp. 27-38.

Chapter 5

From Technology Transfer to Knowledge Transfer: An Institutional Transition

Norbert Buzás

Introduction

Knowledge has always been a crucial resource in economy, but the traditional theories have not recognised it as a factor of production. Formerly knowledge was only considered as a source of increasing returns, because once created, it can be reused without cost. During industrialisation, learning became the ubiquitous process, and due to emerging information and communication technologies (ICT), in the post-Fordist era knowledge became essential for the economy. ICT has dramatically reduced the cost of gathering and spreading information, and the primary problem is no longer the lack of relevant data but competence in mining the information flow. In this new era called 'the knowledge-based economy', information and knowledge became the key factors of production.

Prior to the knowledge-focused era, it was primarily the tangible assets that were to be managed. The strategy was to operate equipment and production lines so as to provide maximum value to the owners. In a knowledge-driven economy, however, knowledge is the asset to be managed. 'Knowledge production' requires special knowledge management tools, such as monitoring the knowledge flow in the same way as the flow of capital and raw material is managed in a company. Customer relationship management is also transformed into customer knowledge management (Gibbert et al. 2002), in which the new focus is on customer success rather than customer satisfaction.

The accelerated development of technology and the shorter life-cycle of products make the learning ability of companies essential for creating and keeping competitiveness. It becomes necessary to alter the organisational structure in order to increase the learning capacity and open up horizontal communications at all levels of firms (Lundvall/Johnson 1994). On the other hand, the diversity of knowledge sources makes it indispensable for firms to enter into co-operations in order to gain access to a more diversified knowledge base.

The above-described changes also influenced the processes of technology diffusion. The traditional model of technology transfer, in which technology moves from a well-characterised economic unit to another well-defined one, has changed decidedly (Amesse/Cohendet 2001). In a knowledge-based economy, technology transfer must be considered as a complex knowledge-transfer/knowledge-sharing process emphasising the importance of intellectual assets and research and development management.

According to the traditional approach towards technology transfer each activity of the company is maintained within the firm or externalised according to transaction-cost criteria. In a knowledge-based economy, however, it is no longer possible to control each part of the value chain on the same level; hence companies have to focus on some special expertise, called 'core competences'. Such favoured competences became limiting factors, and knowledge-based firms rank their activities according to the distance from core activities determining the relationship (acquisitions, networks, alliances, etc...) between the firm and the external organisations.

The altered ways of transferring technology result in a serious institutional transition. The role of the traditional institutions has transformed into new organisations, which can be characterised as knowledge transfer/knowledge-sharing platforms. This kind of organisation can overcome various fundamental obstacles to the evolution of an effective environment for a knowledge-based economy, such as the shortage of private money in the R&D process, entrepreneurial deficits in the knowledge-driven sectors or weak university-industry relations.

First, we will summarise the traditional context of technology transfer by featuring the internal and external processes, respectively. Then we will propose a new knowledge matrix underlining the importance of complementary knowledge in networking. To illustrate the structure and function of knowledge-based networks, finally, two different examples will be described.

The Traditional Technology Transfer Context

In the traditional technology transfer contexts, problems related to the process of transfer used to be solved with different contractual or organisational arrangements. In order to summarise the traditional technology transfer context, we use a matrix (Amesse/Cohendet 2001) based on

- the channel of transfer as internal (within organisations) or external (between organisations); and
- the life cycle of technology as pertaining to the creation or the diffusion process (Table 5.1).

In the cell labelled 'managing innovation', the main problem of technology transfer has traditionally been how to manage technology efficiently during its way from research phase to development phase or prototype turnout resulting in a market-oriented product (Souder 1987). The lack of integration of customers' value into the process line has an unfavourable influence on the success of the product/technology. As a rule, the customer-focused approach towards R&D solved this problem.

In the situation 'transferring to divisions/subsidiaries', the main issues in technology transfer are related to the emitting and receiving capacity of the source and target, respectively. Successful diffusion, on the other hand, requires of the firms appropriate organisational arrangements depending strongly on equipment, documentation, skills and management routine (Behrman 1976).

Table 5.1 The technology transfer activity matrix

	Within organisations	**Between organisations**
Creating technology	Managing innovation	Contracting out R&D; outsourcing
Diffusing technology	Transferring to divisions/subsidiaries	Buying/selling embodied technologies

Source: Based on Amesse/Cohendet (2001).

In the case of 'contracting out R&D, outsourcing', technology transfer problems are related to the well-known 'make-or-buy' decision, which depends on the alternative cost. Firms traditionally kept innovation projects in house as long as they could, in order to maintain full control over them. Beforehand only an extensive co-operation with universities slightly modified this situation (Bozemann 2000). Nowadays, however, there is a shift from contractual to relational arrangements, such as alliances.

Last, but not least, in the cell labelled 'buying/selling embodied technologies', technology transfer can mainly be characterised as licensing processes in developed countries and technology buying/selling between firms in developed and developing countries. Relating to the latter phenomenon there were several problems with exploitation. Firms from developed countries applied technologies patented by firms located in 'poor' countries and when these technologies proved successful, they offered much less compensation as a 'license fee' than the real market value of the technology.

Towards a New Technology Transfer Concept in a Knowledge-Driven Economy

Over the past 10 years development time has been reduced extremely and globalisation has forced 'front-runners' to introduce new technologies on a global scale as fast as possible instead of the old sequential diffusion from the birthplace first to the domestic market and then to foreign markets. Time pressure forces the innovation process out of the boundaries of firms and terminates its strong localisation within R&D departments. The wide variety of modes of application (the number of 'codebooks', Cowan/David/Foray 2000) reshaped the technology transfer process. The number of available codebooks for a single firm is very limited now. Thus technology transfer has to be managed in a reduced space and new organisational arrangements are needed.

In a knowledge-driven economic context, the focus of attention is on the core knowledge and related competence, which determine the character of a firm's activities. If firms move away from their core business, their management becomes unable to efficiently monitor employees or evaluate their human capital (Langlois/Foss 1996). Because core competences are the principal sources of competitive advantages, core knowledge has governance priorities (Prahalad/Hamel 1990) and is no longer considered tradable on the market.

Based on the feature of knowledge, a matrix with three categories can be developed (Table 5.2), which determines the relationship between the firms involved in the knowledge transfer process.

The principal category is the core itself. Since it is the 'goose with the golden eggs' that is at stake in this class, knowledge is exchanged only with companies which produce strategic high-value components of the core business. Because they require very similar core activities, mergers and acquisitions are only typical actions between firms positioned in each other's principal category. The core also serves as a basis for spin-off companies, but in pursuance of spin-off formation the mother company may lose its core knowledge entirely or in part. It is important to note that the essential competences demanded by customers or required to defend the core must also be placed into this class.

Table 5.2 The knowledge matrix

Essentiality of knowledge	Knowledge strategy	Typical actions according to knowledge drift	
		in	out
Core	Safekeeping	Mergers, acquisitions	Subsidiaries, spin-offs
Complementary	Sharing, scattering	Networking (consortia, alliances, supplier-buyer interdependence etc.)	
Marketable	Trading	Buyer-seller interactions (in-licensing, buying know-how)	Seller-buyer interactions (out-licensing, selling know-how)

The next class can be characterised as 'complementary'. Firms here actually hold a significant amount of knowledge, but they need to access complementary forms of knowledge from other firms to ensure the most efficient exploitation of their internal knowledge (Arora/Gambardella 1990). This interpretation has its roots in the literature on strategic alliances, which recognises complementary knowledge as the driving force of international strategic alliance formation (Geringer 1988). Such networks 'can be seen as economic clubs acting to internalize the problem of effective knowledge transmission' (Metcalfe 1995) and can reduce the risk of overspecialisation.

Considering the degree of essentiality, the 'marketable' class represents the knowledge periphery, where the firm has no specific advantages in terms of knowledge. Here typical actions include seller-buyer relations. As for technology transfer, embodied knowledge (e.g. objects) is more likely to be exchanged than tacit knowledge. Activities related to the knowledge ranked here are very often outsourced in the long run (Quinn 2000).

Networks as Knowledge-Sharing Platforms

According to Eurostat's data (Eurostat 2002), in 2000 the share of R&D expenditure in GDP was 1.90 per cent in the EU-15 countries. This percentage is the same as it was in 1995, and it has actually decreased by 0.02 per cent since 1999. At the same time the gap has been growing; the USA and Japan spent 2.64 per cent and 3.04 per cent of their GDP, respectively, on R&D in 1999. Considering research at the national level, only two member states (Sweden, 3.80 per cent and Finland, 3.19 per cent) equalled and one (Germany, 2.44 per cent) approached the contribution made by the principal competitors of the EU.

In order to keep pace with the leaders in knowledge 'production', the EU defined a new policy approach, the European Research Area, to reinvigorate research in Europe. The project's aim is to strengthen the coherence of research activities and to help increase R&D funding in real GDP terms. Support to projects within the priority themes is provided through Networks of Excellence, Integrated Projects and joint participation in research programmes, i.e. Collective and Cooperative Research. All of these are joint programmes of activities leading to increased networking and creating R&D cohesion on a European scale. This initiative is not unprecedented, because since launching its first Framework Programme for research and technological development at the beginning of the 1980s, the EU has helped fund more than 80,000 collaborations by research bodies in member states (Gusmão 2001). These interventions also encouraged the formation of various internal networks that even altered national innovation systems.

Because of their different economical characteristics and variant social traditions, the different European countries have to face different challenges in building their knowledge-based networks to provide the appropriate environment for knowledge management. In those developed economies with satisfactory levels of R&D expenditures, the focus must be on the optimum distribution of financial appropriations, ensuring that smaller enterprises are sufficiently supported. In countries with lower expenditures, however, the principal aim is to encourage entrepreneurial activity in knowledge-based industry and to increase the R&D expenditures of private companies.

In order to demonstrate how various knowledge-sharing networks can overcome the different obstacles, we shall discuss two examples. First, Finland with its technology clinics programme, which supports enterprises by providing common technology services to avoid redundant R&D activities, represents one of the leading countries in European research. Hungarian Cooperation Research Centres (CRCs), in an EU-associated country, on the other hand, help to increase the rate of private money among R&D expenditures.

The common feature of these examples is that they raise awareness of external R&D sources. The Finnish technology clinics, in a developed economy with high GERD[1] and high ratio of private money in R&D expenditures, promote the adaptation of specified technologies for solving problems in small and medium-sized enterprises (SMEs). CRCs in Hungary, however, take only the first step in the technology-focused promotion of enterprises by boosting private-public partnerships in R&D.

Clinics without Medical Attendance: Sharing Technology Services

Because of the day-to-day pressures of running their business, only few SMEs can devote enough time and necessary resources to explore and develop technological issues. In addition, SMEs traditionally have approached technology development in an ad hoc way, accumulating learning-by-doing experiences rather than explicit activities. For these reasons, SMEs have little or no tradition of using external technological services. As a response to these difficulties, the concept of a Technology Clinic was developed (Osman 2001). The basic idea is to pre-select a technology issue which is strategically important for a group of firms from the same sector and to support these SMEs in understanding the issue and implementing responses within their firms. In this way it represents a sophisticated supply-side support measure, rather than a response to specific problems of individual SMEs.

The clinic initiative was started in 1992 by TEKES (National Technology Agency in Finland) in order to bring together SMEs, technology service providers and financiers of technological support. TEKES was founded primarily as a Technology Development Centre to assist Finland in the economic recession of the 1970s. Formed in 1983, TEKES was intended to work with the major areas of technology, because the Finnish government recognised that improved technology would play a key role in economic resurgence. Public technology funding and the

1 Gross Domestic Expenditures on R&D.

programme for additional financing in 1997-1999 created a significant competitive technological advantage for Finland.

Now TEKES is the main financing organisation for applied research in Finland and hence the principal body for implementing national technology policy. Its mission is to promote the competitiveness of Finnish industry and the service sector by technological means. In 2002, TEKES received about 28 per cent of government R&D appropriations. Public funding for an individual technology clinic is provided through TEKES. The financial support of the technology clinics can cover up to 50 per cent of occurring costs; the remaining 50 per cent must be covered by the customer SMEs. In addition, TEKES also pays a management fee to the clinic coordinator and allocates a budget for marketing activities.

During the first five years 15 clinics were built up and over 250 assignments carried out by SMEs (Autio 1998). Based on the number of customers, the most popular clinics were the Food Hygiene Clinic (62 customers) and the Surface Coating Clinic (50 customers). In each technology clinic four types of organisations are involved: customer SME(s), technology service provider(s), the clinic coordinator and TEKES. Sometimes their roles may overlap. In the above-mentioned Food Hygiene Clinic, for example, the co-ordination responsibility was assumed by the main service provider. The role and connections of involved organisations are illustrated in Figure 5.1. The role of technology clinics is to act as catalyst and also as agents of change, meaning that the topics of technology clinics should be new to the SME sector (at least in Finland). Topics for a technology clinic can be proposed by anyone. Most proposals came either from TEKES experts or from technological service providers.

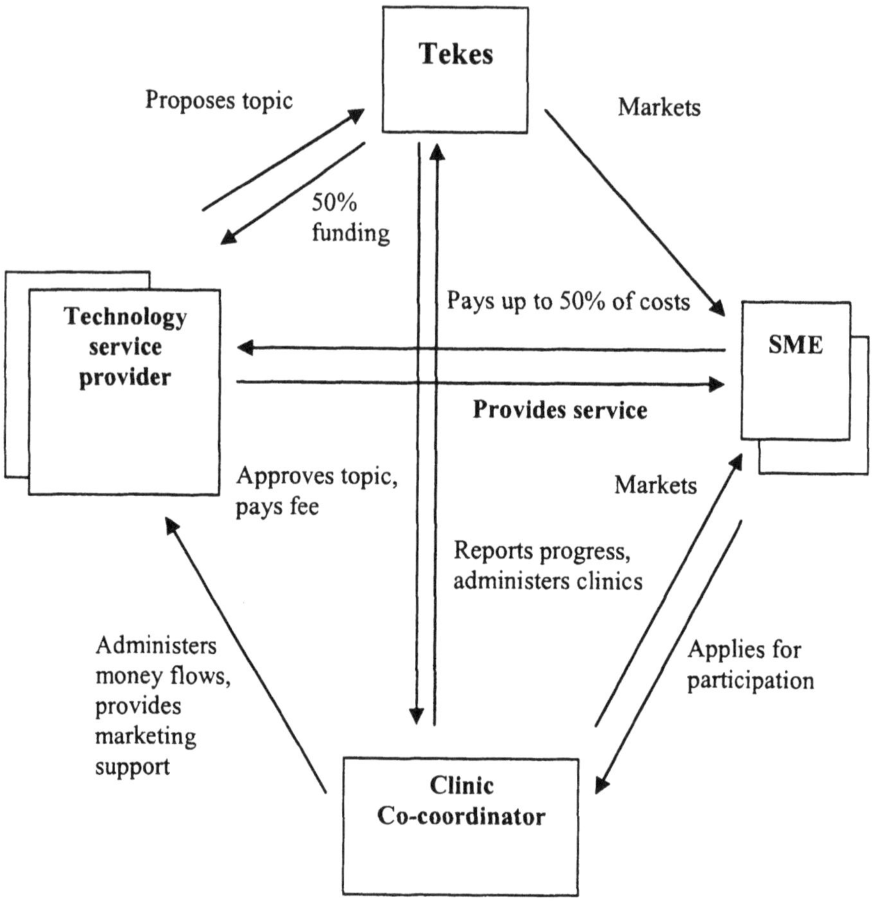

Figure 5.1 Schematic model and role of participants in technology clinics

Source: Based on Autio (1998).

Once a topic is selected for implementation, the clinic facilities are marketed to the targeted SMEs by the regional offices of TEKES, which serve as marketing interfaces. Based on the databases of potential customers built up by the clinic coordinator or service provider(s), participants are recruited not only by direct contacts, but also by mass mailing and marketing seminars. A technology clinic is usually open for up to three years, and clinic assignments may be proposed by SMEs anytime during the working period. The coordinator of the clinic is a specialised technology-driven company like Finntech Ltd., and does not participate in the provision of technological services.

The clients of technology clinics are supposed to be SMEs, according to the broad definition. This means that the definitive restrictions concerning the ownership of an SME are not usually applied in the clinics, but large companies are only exceptionally accepted as costumers. If they are, they receive a lower rate of subsidy.

In terms of objectives, six classes of technology clinics can be identified (Rhisiart/Roberts/Thomas 1999) (it is important to note that this is only an illustrative classification and these categories may overlap):

- *Technology-based clinics* focus on a specific technology with the objective of diffusing the specific technology into the SME sector. A typical technology-based clinic is the Surface Coating Clinic.
- *Theme-based clinics* promote awareness of and provide solutions to a specific theme or regulatory change. Such problems may be solved by applying several alternative technologies or methods. A typical example is the Noise and Vibration Clinic.
- *Cutting-edge clinics* aim at keeping SMEs at the forefront of technological development on an international scale, or at helping firms enforce get-ahead strategies and in this way improve their leading position in the international competition. An example of this is the Composite Material Clinic.
- *Catching-up clinics* are intended to assist 'latecomer' SMEs to reach the standards already achieved by the international competition by providing a chance to catch up in some areas of technology. A good example of such clinics is the Electro-magnetic Compatibility Clinic.
- *Methodology clinics* disseminate good management practices and methodologies into the SME sector. Among such clinics the Rapid Prototyping Clinic is a good example.
- *Demonstration clinics* offer demonstrations of new technologies and processes to a selected group of SMEs in a particular sector. One typical example of such clinics is the Recycled Materials in Road Construction Clinic.

The technology clinic initiative was evaluated in 1997 by experts. The diagnosis was completed based on in-depth interviews with experts and on a customer feedback survey employing fill-in questionnaires. The survey findings suggested that technology clinics had a surprisingly strong learning impact. The organisational and the technical implementation of these clinics also proved to be successful. According to the evaluation, service providers demonstrated a high-degree of professionalism and commitment. Only the marketing of the clinics, together with after-sales support, was given poorer ratings (Autio/Wicksteed 1998).

Cooperation Research Centres as Prep Schools for European R&D

In the 1970s and 1980s, in Hungary about 2 per cent of the GDP was spent for R&D, mainly by state-owned enterprises. There was regular contractual co-operation between public research institutions and firms, which more or less assured the transfer of the results from the research institutes to industry. After the transition, the ownership structure of business changed substantially, resulting in a dramatic decrease in R&D expenditures. In 2000 this value was still below 1 per cent. The number of researchers employed in the knowledge intensive sector is still low and the knowledge transfer between the academic and business spheres is incidental.

Based on its institutional development towards a market-driven economy, Hungary is characterised as one of the most progressive countries, with a share of the private sector over 80 per cent (Tihanyi/Roath 2002). Besides private ownership and commercial banking, limiting government intervention was a high priority in new emerging institutions.

Accordingly, in Hungary the major industrial investors (the big market players) were to finance high-risk scientific research projects on the basis of their long-term interest in:

- results coming from the research that would become the financer's property,
- a persistent supply of human resources by using the research project as 'training site',
- cost-effective research facilities because of the shared use of devices.

The integration of private firms into the joint research projects is also essential because the rate of private money among R&D expenditures is much lower than in the EU. While in Western Europe the average rate of private money among R&D expenditures is about 65 per cent, and in Sweden approaches 75 per cent, to date in Hungary the contribution of private companies to research costs is only 40 per cent (Eurostat 2002).

In spite of preliminary expectations, traditional bigger firms had limited interest in research networks and showed little willingness to build up any R&D infrastructure outside their 'fences' for fear of losing control over their investments. For these reasons, the public sector still has to play a particularly important role in creating a business environment which can serve as a model of the European practice of scientific research in order to enable SMEs to keep their competitiveness after integration. Such governmental interventions, on the other hand, result in an increased integration of the academic, public and private sectors.

In order to reinforce Hungarian scientific and technological excellence through the integration of R&D capacities and activities, in 2001 the Hungarian Ministry of Education (R&D Secretary) gave financial assistance through an application process to establish thematic Cooperation Research Centres (CRC). Host institutions had to be universities with a certified Ph.D. program, and the criteria for the participation of business partners also included involvement in the centre's activities. As 'seed capital', the government provided half of the budget for the first three years, in the range of USD 200,000 to 1 million. During this initial period centres have to develop a cooperative way of doing business in order to make returns which will cover at least the burn rate of CRC for the next 3-6 years.

As a result of the governmental intervention, the approach of private companies changed abruptly. On one hand, the framework guarantees through its organisation (votes in the CRC's board) that the project can be directly influenced by all partners. The private companies, on the other hand, were benefited by various cost-based advantages. They gained access to low-cost university facilities and budget-priced Ph.D. students as well-trained human resource. Occasionally, private companies can use the 'no-cost' side within universities or any other public institutions (e.g. hospitals). Last, but not least, special taxation benefits relating to the cooperative research would also be enjoyed. In the first round, five CRCs were established:

- The *Inter-University Centre for Telecommunications and Informatics* was established in Budapest by three departments of two universities along with Hungarian subsidiaries of some multinational informatics companies (Ericsson, Compaq and Sun) and the market leader, the Hungarian mobile communication company Westel, among others. The main purpose of this CRC is to develop Internet-based and mobile communication systems for the 21st century.
- The *CRC for Industrial and Medical Application of Lasers* was set up in southern Transdanubia by five partners from the academic sector in association with 13 private companies. The research centre, with headquarters in Pécs, covers not only local knowledge, because both academic and private partners include expertise groups in laser technology located in Budapest or Szeged.
- The University of Miskolc and 30 industrial partners gave life to the *CRC for Mechatronics and Material Science.* Of all the five CRCs, this centre has the most diverse geographical coverage including representatives of the steel industry located at Dunaújváros or the subsidiaries of the household devices giant Electrolux in Jászberény.
- Similar to the info-communication research centre above, the *CRC for Rational Drug Design* was also established in Budapest. The academic sector is prominently represented in this research centre, because two academic, one non-profit and four private pharmaceutical and biotech founder partners are complemented by 10 other academic departments as associated partners. The latter take part without financial involvement, but give freely of their knowledge.

- The *CRC for Chemical Industrial Development Based on Green Technologies for Sustainable Development*, was established in Veszprém by the Institute of Chemical Engineering of the University, the chief Hungarian oil company (Mol Rt.), and five other chemical and pharmaceutical firms. This research centre is based on the long-time tradition of chemical industry existing in the city.

Due to the multiplicative effects, the first group was followed by other CRC-like organisations (for example the Renewable Energy and Green Technologies R&D Programme Centre in Szeged) working under similar conditions to those of the supported CRCs.

All such organisations develop and use a joint research infrastructure and jointly manage any knowledge produced. Sharing of all rights related to products, technologies or intellectual property is established by contract prior to the first step in joint development. A CRC offers the possibility to develop an excellent way of cooperative, interdisciplinary research for the parties involved. Last, but not least, the formation of CRCs and similar organisations greatly increased the proportion of private capital in innovation processes.

Figure 5.2 shows the schematic structure of a cooperative project, namely biogas production and use in the framework of joint research activities. The purpose of this project is to produce biogas for various regional users. The academic sector is represented by Bay Zoltán Applied Biological Research Institute, Szeged, which takes part in the optimisation of fermentation technology. A chemical company (Medikémia, Szeged) plays a role in the scale enhancement of the fermentation process. As a by-product of pig farming, a big meat-packing company (PICK) provides the main portion of the stock for the bioprocess as liquid manure. Through the selective collection of public waste, a regional public waste management firm (MÉH) also contributes to the raw material.

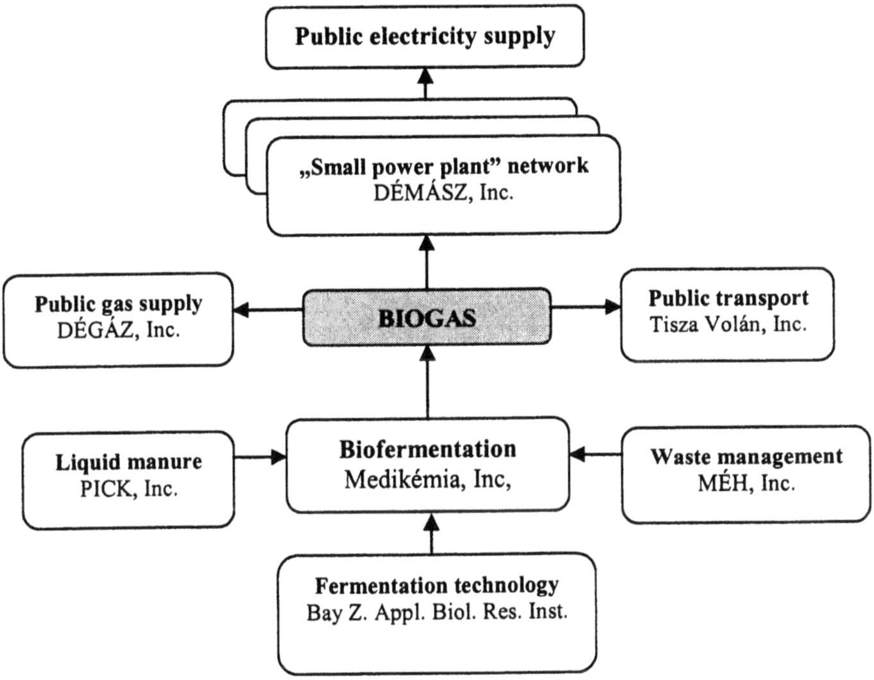

Figure 5.2 Biogas production by a joint research project in Szeged, Hungary

The users of biogas include the main public service provider companies (except for the drinking water supplier). The regional public transport company (Tisza Volán) has recently introduced buses with environmentally friendly gas engines. The public gas supplier (DÉGÁZ) can utilise biogas directly by introduction into the pipeline, while the electricity supplier (DÉMÁSZ) transforms the fermentation product into power with the help of its small power plant network. The latter devices supply the segregated farms and small settlements of the Great Hungarian Plain.

Concluding Remarks

In today's changing world knowledge-sharing communities have become the principal places for technology transfer. Such organisations help to overcome the various obstacles to effective technology and knowledge transfer and integrate handicapped SMEs into the innovation processes. TEKES and the Hungarian government have both developed a new approach to bringing private companies and the research community into an effective interaction. Through technology clinics new

SMEs have been reached as R&D customers, and technology suppliers have significantly increased their ability to work with SMEs. CRCs in Hungary also proved to be an effective way of technology co-operation providing a preparatory phase for acquiring European R&D practice. In both cases the use of governmental financial support and joint research organisations resulted in a huge increase in private capital in R&D expenditures (Hungary) or kept it on a higher level (Finland). On both sides there is clear evidence for maintaining the joint work on a full cost basis after the governmental support is terminated. Technology clinics can provide a knowledge-intensive environment specialised to fit a narrow field of technology. Hungarian CRCs can serve as a solid human resource and infrastructure basis for the companies and academic institutions involved.

The support of specialised services can serve as a model for further steps in governmental technology management in transition economies. If, by using CRCs, among other measures, the governments of the latter countries (such as Hungary) can enhance their GERD to 2-3 per cent and reverse the structure of expenditures towards private predominance, they can then extend their practice by more sophisticated elements, such as technology clinics.

Acknowledgement

Financial support from the Hungarian Research Fund (OTKA F043043) is gratefully acknowledged. A Travel Award of the IGU Commission on Dynamics of Economic Spaces for its Annual Residential Conference (Johannesburg, 2002) enabling me to present this paper is also warmly appreciated.

References

Amesse, F., Cohendet, P. (2001) Technology transfer revisited from the perspective of the knowledge-based economy. *Research Policy* 30, pp. 1459-1478.

Arora, A., Gambardella, A. (1990) Complementarity and external linkages: the strategies of the large firms in biotechnology. *The Journal of Industrial Economics* 38, pp. 361-379.

Autio, E., Wicksteed, B. (1998) *Technology Clinic Initiative.* TEKES, Helsinki.

Bozemann, B. (2000) Technology transfer and public policy: a review of research and theory. *Research Policy* 29, pp. 627-655.

Cowan, R., David, P., Foray, D. (2000) The explicit economics of knowledge codification and tacitness. *Industrial and Corporate Change* 9, pp. 212-253.

Eurostat (2002) *Eurostat Yearbook.* Office for Official Publications of the European Communities, Luxembourg.

Geringer, J. M. (1988) *Joint venture partner selection: Strategies for developing countries.* Quorum, New York.

Gibbert, M., Leibold, M., Probst, G. (2002) Five styles of customer knowledge management, and how smart companies use them to create value. *European Management Journal* 20, pp. 459-469.

Gusmão, R. (2001) Research networks as a means of European integration. *Technology in Society* 23, pp. 383-393.

Langlois, R., Foss, N. (1996) *Capabilities and Governance: the Rebirth of Production in the Theory of Economic Organization*, Working Paper, University of Connecticut/Copenhagen Business School.

Lundvall, B.-Å, Johnson, B. (1994) The Learning Economy. *Journal of Industry Studies* 1, pp. 23-42.

Metcalfe, S. (1995) The Economic Foundation of Technology Policy: Equilibrium and Evolutionary Perspectives, in: P. Stonemann (ed.), *Handbook of the Economics of Innovation and Technological Change*, Blackwell, Oxford.

Osman, P. (2001): A közösségi és a magánszféra együttműködése a vállalkozások fejlődése és technológiai fejlesztés érdekében (Cooperation of the public and private sector for the enterprise- and technology development) *Marketing & Management* 2, pp. 41-47.

Prahalad, C. K., Hamel, G. (1990) The Core Competence of the Corporation. *Harvard Business Review* 68, May/June, pp. 79-91.

Quinn, J. B. (2000) Outsourcing Innovation: The New Engine of Growth. *MIT Sloan Management Review* 41, 4, pp. 13-28.

Rhisiart, M., Roberts, G., Thomas, M. (1999) *Technology Clinics*. Report produced for the INNOREGIO project, Cardiff Business School, Cardiff University.

Souder, W. E. (1987): *Technology Transfer: from the Lab to the Customer in managing New Product Innovations*. Lexington Books, Lanham, USA, pp. 218-238.

Tihanyi, L., Roath, A. S. (2002) Technology transfer and institutional development in Central and Eastern Europe. *Journal of World Business* 37, pp. 188-198.

PART III
THE FIRM IN AN INTERLINKED TERRITORIAL WORLD: THE WORLD OF PRODUCTION

Chapter 6

Internal Networking in the Globalising Firm: The Case of R&D Allocation in German Automobile Component Supply Companies

Martina Fuchs

Introduction

The internationalisation of manufacturing industries is often discussed with reference to labour-intensive production, especially in the context of foreign direct investments in economically peripheral regions. But some tendencies point to the growing internationalisation of Research and Development (R&D) as well. And if it is true that multinational companies' know-how is going global, we may have to face a new division of skills between 'cores' and '(semi-)peripheries', i.e. between the centres of manufacturing industries on the one hand, and Newly Industrialising Countries and regions in Central and Eastern Europe on the other. This chapter is not concerned with the internationalisation of R&D in general, as observed by economists since the 1970s (see Cantwell 1995), but examines the internationalisation of R&D to traditionally low-cost regions. The chapter will present an empirical example of R&D in German automobile components supply companies that perform R&D for electronic components and systems, and produce such parts and modules as well. This sector is becoming increasingly expertise-intensive as a result of structural changes in automobile value chains, which have led to growth in this sector of the value chain. Thus, the sector has good preconditions for the analysis of change with new challenges for management. This study concentrates on 'medium size' companies in the automobile components supply sector. The companies are medium-sized not in terms of the number of their employees, but with reference to the automobile supply sector: they are companies with only a few product lines, and they are not the big global players (such as Siemens VDO Automotive and Bosch, for example). Such companies follow less heterogeneous R&D strategies than the global players in automotive supplies where 'everything happens' (a manager of Siemens VDO Automotive about the organisation of R&D).

The objective of this chapter is to examine selected aspects of the internationalisation of R&D. The focus is on the internal networks of business organisations. In contrast to other network studies that discuss networks between firms, this chapter analyses the inner network of the organisation of a specific function (R&D) in international companies. Companies that have already internationalised their production often also have to address the issue of internationalising R&D as well. Some managers already have decided in favour of going global, while others have kept their R&D capacities in the core plants. There appears not to be unavoidable 'pressure' to internationalise; there are many 'productive models' that lead to success (see Boyer/Freyssenet 2002).

Managers' decisions depend on their common ideas and aims for the company, their view of the general ongoing process of globalisation, and of the problems and potential involved in the internationalisation of R&D. Such 'shared visions' in management neither depend directly on the requirements of the value chain nor can they be seen as a simple reflection of circumstances. 'Reality', that is the basis for decision-making with regard to the internationalisation of specific functions, is not identical with 'location factors' that lead directly to a clear strategy; this 'reality' is a mental construction, a set of imaginations and deliberations. The managers communicate their subjective impressions to create a shared vision for their common strategies and actions. The management's 'mental maps' of globalisation are important for a company's R&D internationalisation strategy. There is no unique rationality and no best way, quite different strategies may lead to efficient internal organisation of R&D in the company. Schoenberger (1999) refers to the influence of the management's mental maps and the power relations inside of the company as 'the region in the firm', and thus she emphasizes the contrast between such aspects of cultural approaches (in a broader sense) and perspectives which are interested in the 'firm in the region'.

Management decisions are not arbitrary but depend, for example, on the requirements of clients. Often such requirements are not clear, and can be contradictory: the clients want low costs and costs could (in principle) also be reduced by internationalisation and the use of cheap labour in R&D. Low-cost labour is a commonly acknowledged motive for the internationalisation of production plants; and there are also examples of the internationalisation of R&D for reasons of labour costs (see Fromhold-Eisebith 1999, Fuchs 2003). As the case studies will show, some German automobile component suppliers have already gone global with their production. This could lead to internationalisation of R&D too. But to maintain high standards, managers often tend to keep R&D in the core region where they know how to find qualified workers. The location of the client is not a simple determinant either. If the clients of the German automobile supply companies are foreign firms that are located in the USA, Japan or elsewhere, for example, the supply company sometimes locates its R&D close to these foreign OEMs. This less for the need to have access to foreign expertise, as general studies of the internationalisation of R&D have shown (see Serapio/Dalton 1993), but rather results from the need for close contacts with the client. If the client is an OEM in Germany, the tendencies are contradictory for the supply companies: The

German automobile industry still tends to keep R&D in the German core plants, but there are recent trends towards going global as well. The 'Big Three' in the German automobile industry (BMW, DaimlerChrysler and Volkswagen) concentrate their R&D in Germany. But there is also a new tendency to organise R&D on a 'de-centralised' or 'multi-centred' basis. A good example is the design of Volkswagen's New Beetle, where headquarters did not control the R&D process entirely but gave important tasks to the foreign plant in Mexico (Pries/Schweer 2004). But as far as general tendencies are concerned, up to now German OEMs have concentrated their R&D in Germany, and want close cooperation between their R&D and the suppliers' R&D in this core region. Thus, this article discusses management's interpretations of space and R&D location strategy with regard to factors originating with the client, such as high-quality production, labour costs, and proximity to the client's R&D departments.

In order to analyse expertise in R&D with regard to globalisation or centralisation in the core regions, it is important to distinguish between different types of R&D. Typically, R&D is divided up according to the product cycle, i.e. in the automobile industry in basic research, applied research, experimental design, and the design of applications and routines (see Jürgens 2000, p. 164, Malecki 1997, pp. 51-53). The R&D process is supported by technological and organisational tools, such as computer simulation and simultaneous engineering, which integrate activities which had been done separately and in sequence (Jürgens 2001, p. 61f). However, the processes are interlinked, forming an interactive system rather than a cycle, although separate tasks can still be identified. There is a general assumption related to this differentiation of R&D. The hypothesis is that basic research, applied research, and experimental design are located close to headquarters, while the design of applications and routines is carried out near the production plant (Jürgens 2000). This is a very rough model and does not systematically take the forces for and against internationalisation into account.

R&D engineers in automobile components supply companies must have specific skills in order to calculate and draft designs as well as be able to use CAD (Computer Aided Design) programmes; this means that the actors in a company know the rules that are important for R&D and are able to apply the rules within a specific context. Thus they use instrumental knowledge.

Technological knowledge is not the only important skill in this context; steering skills are also very important (see Sugiyama/Fujimoto 2000, p. 182); the process of R&D must be competently organised. R&D takes place within a social system; the actors of this system are usually the director of R&D and the engineers, sometimes also managers of other company branches, and technicians (Figure 6.1). As trade unions tend not to be strong in R&D departments, members of trade unions or works councils are not usually important actors for locational strategy in R&D. Important actors outside the components supply company are the actors in the client's R&D. In the medium-sized firms examined here, there was no significant contact with external R&D suppliers. Further actors of the macro-network, such as those in politics, could not be identified in this study, which concentrates on the micro level, although they might influence the innovative environment more indirectly and in the long term.

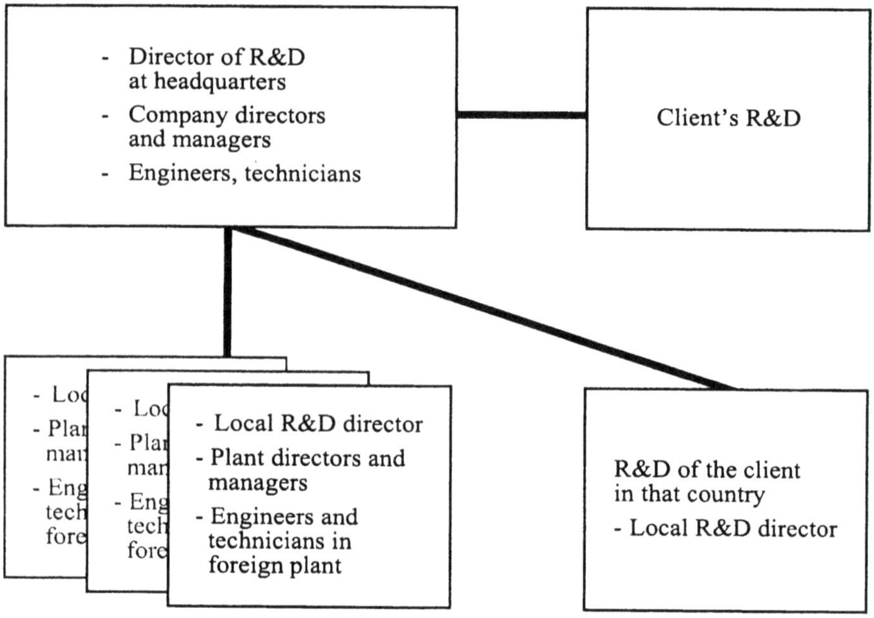

Figure 6.1 Relevant actors of R&D in the automobile components supply industry

As these actors do not act alone but as members of a social system, they need more than merely instrumental skills. They must act according to the rules of the social system, which in the main is the company. Furthermore, they must also communicate, develop and follow rules. Thus they try to create a shared vision of what is to be done in the future. Segrestin/Lefebvre/Weil (2002, p. 68) speak of R&D in the automobile industry as 'design, or the art of compromise'. According to Messner (2002), this skill can be described as the ability to direct the process of communication and to arrive at consensus within a social system. As in modern societies, there is no simple division between specific actors who steer and other actors who have to apply the rules in the globally organised R&D departments of manufacturing companies. On the other hand, there are interdependent relations between the plants and headquarters, centres and peripheries. Messner (1997) found a similar phenomenon in his analysis of the 'network society'. According to Messner (1997) in modern societies steering skills mean the ability to coordinate, and sometimes to correct or initiate, processes that are sensitive and difficult to guide. This is also true of R&D: managers usually cannot control the work of the engineers, which is largely based on tacit knowledge and organised in social networks between the engineers. The tasks of R&D call for cooperation between colleagues and the exchange of information. There is no formal hierarchy between R&D at headquarters and foreign R&D, unlike for example in the controlling of accounting, which is usually hierarchically organised in multinational enterprises. Thus, the central management has to convince the foreign partners. Top managers have to solve the problems if this is not possible, but that is the exception. Most processes consist of coordination in with the hierarchy in the background.

Thus R&D requires not only instrumental skills and the ability to direct the process of communication and to arrive at consensus in a social system, but also to transform that consensus into new institutions (as in codes of conduct) or into new organisations (such as new departments in the enterprise). This skill of transforming consensus into rules of conduct or new organisational structures can be interpreted as the ability to create structures as a framework for future processes.

This leads to the question of governance within the company. The discussion of governance has its roots in research by social scientists into globalisation on the macro and meso level, but is now spreading to research into multinational companies (see Lacertera 2001). As a company is integrated – and often globally organised – in value chains, these value chains influence the managers' shared visions of global chain governance (see Humphrey/Lecler/Salerno 2000, Humphrey/Schmitz 2000). In our case, as mentioned above, the forces of the value chain for internationalisation or centralisation in the core region that influence the shared visions are often contradictory forces such as labour costs, quality/standards, and the location of the client. If R&D is divided into 1) instrumental skills, and 2) steering skills, and again into 2a) ability to direct the process of communication and to arrive at consensus within a social system, and 2b) the ability to create structures (Figure 6.2), the hypothesis can be postulated that instrumental skills may internationalise earlier than the ability to direct the process of communication and create consensus within a social system, and the ability to create structures. The last two skills remain in the core companies and grow there. But such largely tacit forms of knowledge may grow later in the foreign plants, as tacit knowledge can become explicit and thus transferable (see Malmberg/Maskell 1999a, b), or may develop in the foreign plants as a necessary precondition for the efficient organisation of production. But before discussing the internationalisation of different R&D skills in depth, a better understanding of the sector is necessary.

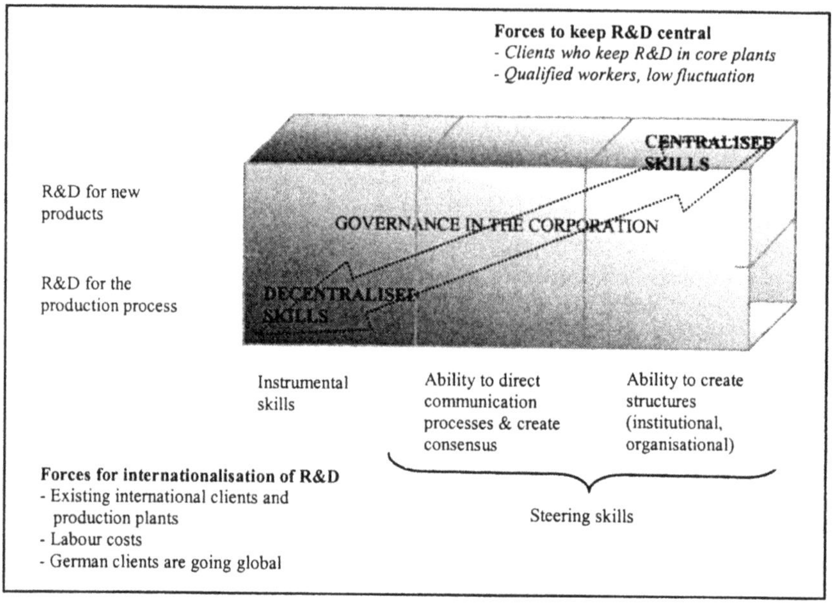

Figure 6.2 R&D skills in the automobile components supply industry

Specific Characteristics of the Sector: Important Trends in Automobile Components Supply Companies

The automobile industry is a very important sector of the German economy. About one-third of all R&D investments are made in this sector. About 770,000 people are employed in the German automobile industry, and about half that figure is in the components supply sector (VDA 2002). As the sector is broad, it is difficult to analyse the automobile sector as a whole. Automobile supplies include very heterogeneous sub-sectors, such as textiles for the seats, steel and plastics, electronic equipment, and so on. For a better understanding of the internationalisation of R&D, it is advisable to examine a dynamic sub-sector in detail. Rapid growth processes can be identified in the sub-sector of electronic components, parts and systems. By concentrating on one sub-sector, an important factor can be kept quite stable: the type of production process which strongly influences the need for the international division of labour and skills.

The expansion of the electronics sector was mainly initiated by the trend towards cars with many electronic components and systems, especially in Europe. Cars have become increasingly technology-intensive. Many parts which were formerly mechanical components became electronic or 'mechatronic'. Associated with this is an increasing complexity of components and sophistication of electronic systems, and this leads to the need for higher specifications. Instead of single electrical or electronic components, more complete systems are required. Also associated with this is the trend that first-tier suppliers in particular must offer more complete systems instead of single components. To do so they need more specialised knowledge.

This dynamic development is integrated in some structural changes in the automobile components supply sector. From the 1990s on, the traditional supplier relations of German car manufacturers, shaped by an industrial structure consisting of many medium-sized, family-owned companies, were fundamentally transformed. The OEMs introduced lean production strategies and restructured their process chains, trying to shorten the time-to-market span and bring new products on to the market faster. Most OEMs define themselves more as logistics organisers than as integrated factories. They have therefore decided to concentrate on the general design and systems integration of the product as their core tasks (Jürgens 2001, p. 61f, Kaplinski 2000, p. 27ff). Thus, a growing share of R&D in the automobile industry is not carried out by the OEM but by the components supplier. In many cases, the OEM and the components supplier work together closely, and the supplier is required to visit the R&D departments of the OEM for several face-to-face contacts.

Thus automobile component suppliers, especially in electronics, have to expand their R&D activities. In Germany, the suppliers to the automotive industry already carry one-quarter of the costs for R&D in the automobile supply chain (1999; VDA 2001, p. 57). According to Spohr (2002, p. 79), this increasing importance of R&D in the electronic components supply industry could change the relations in the value chain: the hierarchical structure from the second-tier of electronic semiconductors and software to the first-tier of system producers to the OEM could be replaced by a polycentric structure in which the OEM, the first and the second tiers become partners. But this is still a heuristic model for a better understanding of the processes of change, and not yet a reality because the OEMs are still maintaining their position as lead firms (Berger et al. 1999).

The transfer of tasks from the OEMs to the suppliers intensifies business rivalry. While in the early 1990s there were about 30,000 first-tier suppliers in Germany, there are now only about 8,000. The fierce competition not only leads to pressure on prices, but also to the need to produce very good quality. Quality standards have become a factor of increasing importance. In terms of safety, the 'quality' of electronic parts for automobiles not only means greater reliability compared to parts used in the computer industry and other consumer products. 'Quality' also means greater resistance to physical stress factors, such as temperature variations, vibrations, and electromagnetic interference (Spohr 2002, pp. 66-71). In the software industries, CMM levels (Capability Maturity Models) are an international quality standard. This emphasis on quality *may* lead to a strategy that avoids internationalisation (as case study I will show), because the managers orientate their strategy towards qualified personnel in the core plants and regions. This is a safety-orientated strategy: they are familiar with the engineers' general qualifications, and so they rely on these employees to maintain the high standards required by the OEMs. But it can also lead to a specific strategy of internationalisation, searching for specific regions that already have experienced engineers and technicians (as case study 2 will illustrate). These different possibilities for finding specific solutions make it necessary to take a closer look at the shared visions of the managers.

Methodology

Shared visions can only be identified by qualitative and explorative studies, and not by standardised methods. Most qualitative approaches regard communication between members of a specific social system as an essential element. 'Reality' is continuously produced and reproduced. But realities are not split up into different worlds, because communication leads to shared visions. The managers' perceptions and their decisions must be reconstructed, because they are not *ex ante* a rational choice. The shared visions can be reconstructed hermeneutically, not analysed using positivist methods. This study aspires to a deeper insight into the adaptation and alteration of reality by management. Managers' internationalisation strategies are not influenced directly by external place and space *per se*, but rather by their shared visions of place and space.[1] Pries/Schweer (2004) refer to automobile companies as being 'action units of the double transformation of reality': The companies not only transform raw materials and other inputs into goods and services; the reflexive transformation of reality refers to the recursive process of considering and deciding strategy.

It should be borne in mind that shared visions are not created and then exist for a long time, but that they are the outcome of discussions, and negotiations often continue. They are criticized, and the consensus must sometimes be questioned. Thus, a shared vision is not a fixed opinion, but is dynamic and open to change. Shared visions are a step in the learning process. With reference to Pries/Schweer (2004) it can be said that companies – as 'action units of the double transformation of reality' – realise innovations in the products and the process, but they also make innovations real in their own minds: they are learning.

This study of German suppliers of electronic components and systems for the automotive industry was carried out in 2002. The shared visions were analysed by in-depth interviews with the actors in the companies studied, especially the director of R&D and company managers. The companies were selected from the data base of the German Association of the Automotive Industry (VDA), taking into account the difficulty of separating electronic supply companies from other suppliers. Various multinational automobile supply companies were chosen for their different sizes and internationalisation strategies. Because of the explorative character of the study, six companies were chosen for in-depth interviews. In this chapter, three case studies are presented.

---

1 This approach could also be called 'de-constructivist', but as this term refers to a very specific and sophisticated research method in cultural geography, it should not be applied to the method used in this study.

Case Studies: Reasons for Internationalisation and Persistency

The following examples demonstrate the openness of the decision-making process and thus the importance of the shared visions of management in three companies who have chosen different ways of dealing with the constraints of *proximity to the R&D of the client, the need for experienced employees to guarantee high quality,* and *labour costs*. The first case study illustrates the locational stability of R&D for product and process design as a strategy to maintain the quality required by the clients. Here, the shared vision of the management is led by the idea that quality production by engineers and technicians in Germany is the key to success. It is worth noting that this strategy of locational stability in R&D is combined with a strategy of internationalisation for the production plants. The second example shows an internationalisation strategy for product design based on the search for 'competent regions'. This shared vision, to which 'competent regions' are central, leads to a cautious strategy to combine labour costs with the factor of experienced employees that the management sees as a guarantee for high quality. The third case study portrays a company where labour costs are the most important part of the shared vision of management. This company tries to bring process design to foreign plants, including instrumental skills and some steering skills which are necessary to organise the R&D process.

Case Study I: Locational Stability as a Strategy for Maintaining Quality

The company in this case study develops and produces locking systems for the international automobile industry. Huf Hülsbeck & Fürst was founded in Velbert (in the Ruhr, Germany) in 1908, a region with a tradition of locksmith skills. Parts for the automotive industry were first made in 1920, and today the most important customers are DaimlerChrysler, Volkswagen, BMW, GM/Opel/Vauxhall, and Ford. The company has about 5,000 employees worldwide, of which 1,750 are in Velbert. Another plant, which also includes R&D, is located near Munich, but the centre of R&D is still in Velbert.

The motive of proximity to the client leads to top management's shared vision that it is important to keep skills and responsibilities centralised. As mentioned earlier, the German OEMs still locate a great deal of their R&D in Germany, and they expect close cooperation between their R&D and the R&D of the supplier. Face-to-face contacts and personal meetings usually take place in the OEM's plant. Thus, the director of R&D and some engineers visit the R&D departments of the OEMs on a regular basis. While R&D is kept centralised, the international organisation of the OEM's *production* made it necessary for the company in this case study to internationalise its production. The company managers searched for highly skilled workers on a global level. Figure 6.3 shows the international production plants. The managers' shared vision is that qualified labour is important for production, as the company finds in the Velbert area with its long metalworking tradition.

Since the early 18th century, Velbert has been an important location for the production of keys and locks in a European context. This early production cluster created a well-qualified labour force, and in the shared vision of management this is elementary for generating high quality. While R&D was kept at the home base, internationalisation of production set in about 20 years ago. In 1983, Shiroki (Japan), a supplier to Toyota, received a licence from this company. Then, plants were built in South Korea, the United Kingdom (Willenhall and Tipton), and in the USA (Germantown, Detroit, Greenville). The Willenhall location was chosen because company management saw the location as a 'British Velbert' with regard to the metalworking tradition. The managers have the same opinion about the new locations in Saxony (Germany). Further plants (50 per cent ownership) are located in Sittard (Netherlands), in Katowice (Poland), in El Burgo de Osma (Spain), in Tondela (Portugal) and in Atibaia (Brazil). Furthermore, there are two plants in China (Yantei und Shanghai), in India (Delhi) and a new plant in South Korea (Chonan). Thus, the internationalisation of production placed much emphasis on the availability of qualified workers, and R&D was kept at the German headquarters.

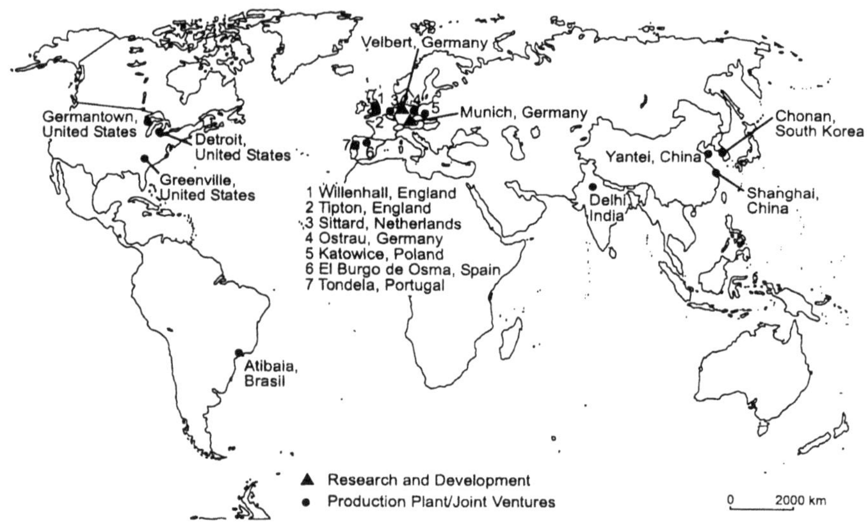

Figure 6.3 Locations of Huf Hülsbeck & Fürst, 2002

Data compiled by: M. Fuchs, *Cartography*: T. Eckhardt.

Huf Hülsbeck & Fürst generally keeps R&D for new products at the location of the parent company. But there are indications that some instrumental R&D skills necessary to improve the manufacturing process are beginning to internationalise, although this aspect was not originally included in the shared vision of the top management. The background to this is the fact that the company already has international production plants: although the parent company tends to keep R&D in the German plants, it was decided for reasons of efficiency to decentralise some maintenance and repair functions and tool design for the production processes to the foreign plants. The procedure in this company is as follows: the foreign plants buy the first machines and the know-how to operate the equipment from specific firms, and begin to assemble the parts. Already at that stage the factory needs decentralised skills for repair and maintenance of the equipment. Local expertise is gradually supplemented by skills in production planning and maintaining standards. Thus, on the whole, some upgrading processes can be identified in the foreign plants that were not planned by headquarters. Such unexpected upgrading processes cannot be interpreted as random phenomena, because they follow the internal logic of growth and of learning requirements in foreign direct investments (see de Meyer 1992). In contrast to such unplanned upgrading processes, the following two case studies show examples of planned internationalisation strategies: one which is a more cautious strategy and closely connected with the search for experts, the other a more offensive strategy. The last two case studies bear out the assumption that, in the case of internationalisation, instrumental skills internationalise earlier than the steering skills and responsibilities.

Case Study II: Cautious Internationalisation Linked to the Search for Experts

The second company studied, Cherry, was founded in 1953 in Illinois. Its first European company started up in Germany in 1963 (in Bavaria), and two years later the US parent company began developing automobile parts. Since 1979, the German headquarters and R&D have been located in Auerbach. Another plant has been operating in Bayreuth since 1969. The management of Cherry follows the shared vision that it is important to internationalise R&D very actively. In this case, there is a combination of decentralisation of skills and the search for cheaper labour. This is taking place in a company which also has its headquarters in a region with long years of experience in manufacturing and thus wants to retain this asset. The shared vision of management is that it is necessary to actively concentrate steering skills at headquarters in the central R&D department, and to decentralise instrumental know-how in the technological implementation of R&D. The company develops, produces and sells switches and steering systems for the automotive industry on a second-tier level (and, in a smaller sector, parts for personal computers). Because the US part of the company experienced economic difficulties, the German section acquired a strong position within the whole company. In 2002, there were 160 employees in engineering in Auerbach, and only about 60 in the USA. The corporation employs about 4,000 all told. Ever since the position of the German section gained in importance, the management in Auerbach has learned to acquire steering skills. At first, these were of a general nature and related not only to R&D, but later skills in guiding the process of finding consensus, and then creating new institutions based on this, helped to organise R&D on an international level, as will be shown later on.

The US parent company in Illinois (Figure 6.4) is linked to a labour-intensive plant in Ciudad Juárez, Northern Mexico and a factory in Klasterec Nad Ohri (Czech Republic) near the border with Germany (founded in 1993). The plants in Mexico and the Czech Republic do not have their own capacities for product development; another factory with a small R&D department is located in Harpenden (United Kingdom). Further plants are two joint ventures (Yokohama, Japan, and Vallaripatti Madurai, India).

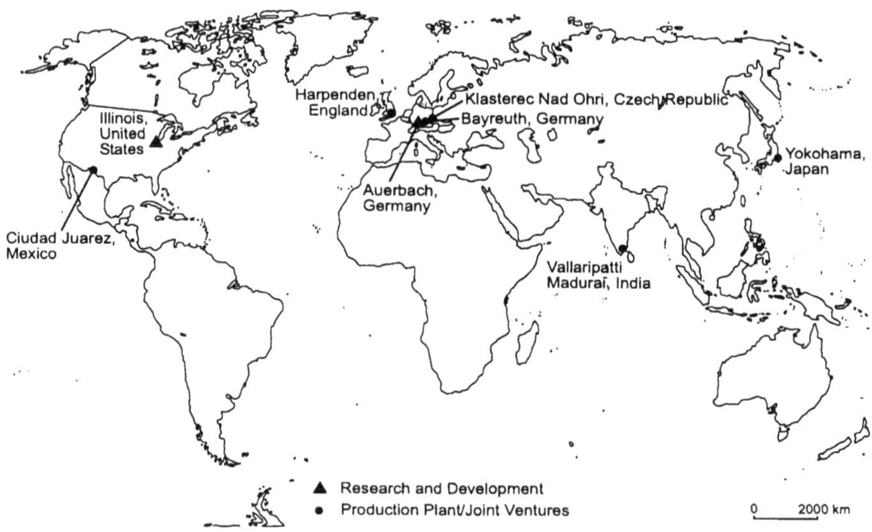

Figure 6.4 Locational structure of Cherry, 2002

Compiled by: M. Fuchs, *Cartography*: T. Eckhardt.

Although most of these plants do not have departments for product development, some elements of decentralised R&D are in evidence, especially with regard to the company plant and further R&D supplier firms in India. This is an effect of the deliberate concentration of steering skills in the central engineering department which made the decentralisation of instrumental skills possible, as the company had already regard to the US headquarters. In the German company, coordinating the R&D process requires precise planning and close cooperation with other plants. The specification and control of the steps is considered to be very important in the central engineering department. Before the start of a project, the engineers from the various countries meet face-to-face. They define the technological requirements and draw up the necessary documents. After that, each of the partners performs his tasks, and then they all meet again face-to-face. By stepping up such international co-ordination, the company in Auerbach is beginning to concentrate its steering skills. That is the basis for decentralising further standardised activities.

The management is, however, still seeking the best practice for the international division of R&D, and there is no clear and homogeneous shared vision. The advantages of internationalisation that the managers share in their common vision are the highly motivated engineers in India combined with cheaper labour and a flexible labour market. But, according to the managers, the other side of the flexibility coin is the high fluctuation of engineers in India. In contrast, the managers regard stable labour relations as a positive aspect of this system of regulation. Because in Germany employees usually stay for a long time – or their whole working life – with one company, it is possible to ask an engineer after a long time span how something had been organised or done in the past. Although the highly regulated German labour market with all its rigidities is much criticised, least in this respect the stability of labour relations is a locational advantage for R&D. This implicit knowledge is retained in the company. Only with regard to standardised tasks does the management prefer a quite unambiguous strategy of internationalisation, because the plant in India can meet the highest standards; it has attained the best CMM level. And this can be done with lower labour costs. As a vision for the future, the management suggests a division of tasks: basic product research and development should take place at the German parent company, while development and design, which is closely related to the production process, should be carried out in the USA, and other standardised functions are to be assigned to the plant in India. Production will take place in the plant which can do so most profitably. Thus, steering skills are growing at the headquarters in Auerbach; there a very deliberate management strategy exists to improve their ability to direct the processes of communication and consensus finding, as well as the ability to create institutional structures. The example shows that labour costs are an important factor in the shared vision of management, but not the only one. None of the companies analysed dares to ignore quality, but if a combination of low cost and quality is possible this would seem to be an attractive strategy to top management.

Case Study III: Internationalisation of R&D Competences as a Factor for the Reduction of Labour Costs

Compared to the other two companies interviewed, the shared vision of the following company attaches more importance to the labour cost factor. W.E.T. Automotive Systems AG (formerly 'Wärme und Elektrotechnik') was founded in 1967 and has been a supplier to the automotive industry since the early 1970s. The most important products are car seat heaters, but they also make cable harnesses and related products. Because the car manufacturers all have different seat designs there are a lot of heater types – about 1,000 in all. This leads to high production costs, and the competition is strong. Although the company is a world leader for car seat heaters, it has to struggle to maintain this position. One strategy for success in competition is innovation-based: W.E.T. invests an above-average amount for the automobile supply sector in R&D (9 per cent of turnover in 2002). The other strategy is to work to reduce labour costs, which are still high because production of the heaters is labour-intensive. Seat heaters are integrated in textiles, and textile production as a biaxial process sets limits to automation. Labour costs account for about 70 per cent of production costs. This has led to the internationalisation of production and also of engineering, as shown in the following.

The headquarters and centre of R&D are located near Munich. The company has about 300 to 350 employees at headquarters and about 2,600 in the whole multinational enterprise. The production operation, which was formerly located close to headquarters and employed about 100 to 150 workers, was shifted bit by bit, without redundancies, to Pilisszentivan in Hungary. There is a production plant in Valletta (Malta) as well, which developed as a labour-intensive plant. There is a similar structure in North America. The location in Windsor, Canada (50 employees), became a Customer Service Centre for North America, and production was shifted to Acuña, Mexico (Figure 6.5):

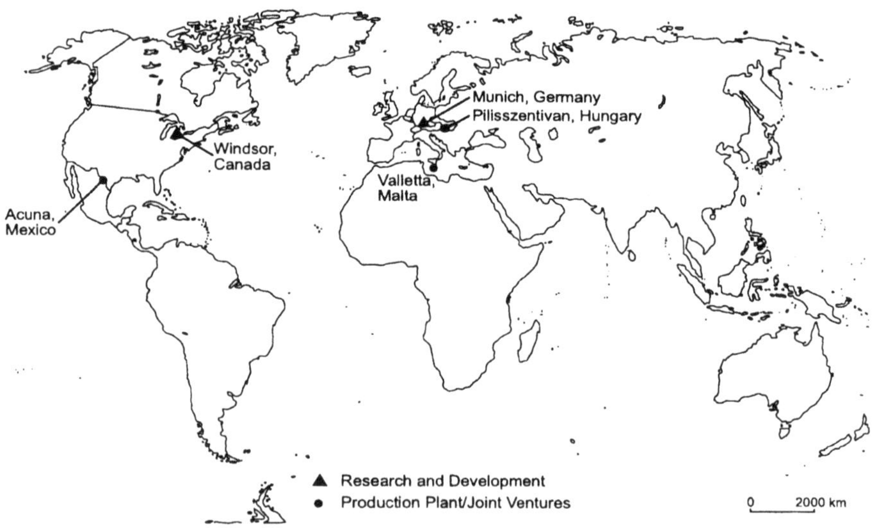

Figure 6.5 Locational structure of W.E.T., 2002

Compiled by: M. Fuchs, *Cartography*: T. Eckhardt.

The importance of labour costs in the shared vision of central management can be illustrated by the example of the plant in Malta. This demonstrates the importance of labour costs in the managements' shared vision, which ultimately also leads to the internationalisation of R&D. This plant in Valletta was founded in

1987 as the company's first foreign location. Labour costs in Malta then rose, and since the mid-1990s, although the Maltese workers were highly qualified, the Hungarian plant became more important than the Maltese one. In 2002, about 200 employees were working in Valletta (Malta), while there were 700 to 800 in Pilisszentivan (Hungary). When W.E.T. opened a new production plant in Langfang (China) in 2003, top management decided that production in Malta should be phased out.

However, labour costs are not the only factor in international competition; good quality is another, which according to the shared vision of the company's management, can be developed only by employees who are highly motivated. Thus, the management of this company is trying to decentralise skills and responsibilities, including some aspects of the organisation of communication processes, and is developing a new R&D centre in Pilisszentivan. This more offensive strategy of R&D internationalisation is a product of the company's history, which was shaped by the necessity to internationalise in order to obtain cheap labour.

Initially, this strategy to internationalise R&D had two counter-productive effects: the engineers at headquarters, in particular, feared they would lose their jobs in the long term. Management had to face these fears and tried to reassure the engineers, because if these highly qualified employees were to leave the company would undoubtedly run into difficulties. The second counter-productive effect was that the engineers at the Hungarian location were not accustomed to longer-term planning and steering processes. But, the management assumed that the decentralisation of engineering skills close to production would be of great advantage to the whole company, because it would minimize some of the detailed work at headquarters. The shared vision at headquarters was that plant engineers, trained in a centrally planned socialist economy, had to learn how to direct the process of communication and consensus-finding. Thus, central management wanted to decentralise some steering skills and responsibilities, but from the point of view of the German management this strategy had its limitations in that the labour force was not very highly qualified, relatively speaking. Therefore, engineers from headquarters accompanied these learning processes by staying in Hungary for the period necessary to complete them.

Thus this is an example of the deliberate decentralisation of some steering skills which are closely related to the engineering of the production process. Communication between the parent company and its foreign plants is sometimes difficult, especially if the plants have been in existence for many years. Often, the employees in these old plants are very confident and will only accept suggestions from top management if they are convincing enough. Managers have to find their own way of decentralisation; there is no structural necessity for any one specific path.

Conclusion

The three case studies have shown the shared visions and the decisions made in three companies which have chosen different paths within the constraints of *proximity to the client's R&D* and *experienced employees to guarantee high quality* as well as *labour costs*. Figure 6.6 illustrates the position of the case studies with regard to this heuristic model of skills and responsibilities in R&D: The first case study shows the locational stability of R&D for product and process design as a strategy to maintain standards required by the clients. The case study makes clear that such a strategy of continuity in R&D can be combined with an internationalisation strategy for production plants, and that international production requires some decentralised design activities in the foreign plants. The second example stands for a planned but cautious strategy to combine labour costs with the factor of skilled labour. The third case study is an example of a company where labour costs are the most important factor. This company has transferred process design to the foreign plant in Hungary. The last two case studies bear out the assumption that, in case of internationalisation, instrumental skills internationalise earlier than the steering skills and responsibilities. Management can treat instrumental skills in the R&D-process as a type of 'commodity'. Hitherto, the decentralisation of steering skills has usually led to problems in the implementation of internationalisation, at least in medium-sized automobile components supply companies. As a result, most steering skills remain and grow in the centres. Furthermore, the examples illustrate that in the rare case of steering skills growing on an international level, this usually involves the internationalisation of the ability to direct the process of communication and consensus-finding in a social system. Innovative skills, such as the ability to create structures or institutions (in the sense of codes of conduct) or organisations (such as new departments in the company) are rarely found in the foreign plants, and seem to be a domain of headquarters.

Although steering skills have remained and grown in the centres, in future they may perhaps expand more intensively in the foreign plants too. The internationalisation of knowledge which was formerly to a large extent tacit can set in, because usually each process, in R&D or elsewhere in the production process, needs some know-how and responsibility to organise the business effectively, as case study 1 has illustrated. The selected medium-sized companies possibly still have to learn how to decentralise steering skills. Some other case studies of global players among automobile supplier companies, such as Valeo, show that headquarters have specific programmes for management in foreign plants in order to decentralise instrumental and steering skills in R&D, because this reinforces the individual advantages of the different plants in global competition. Thus it should be borne in mind that this is a sector which is in a specific phase of a dynamic globalisation process. Up to now, it is not clear to scientists and to the actors in the companies precisely which trend could be the 'best way'.

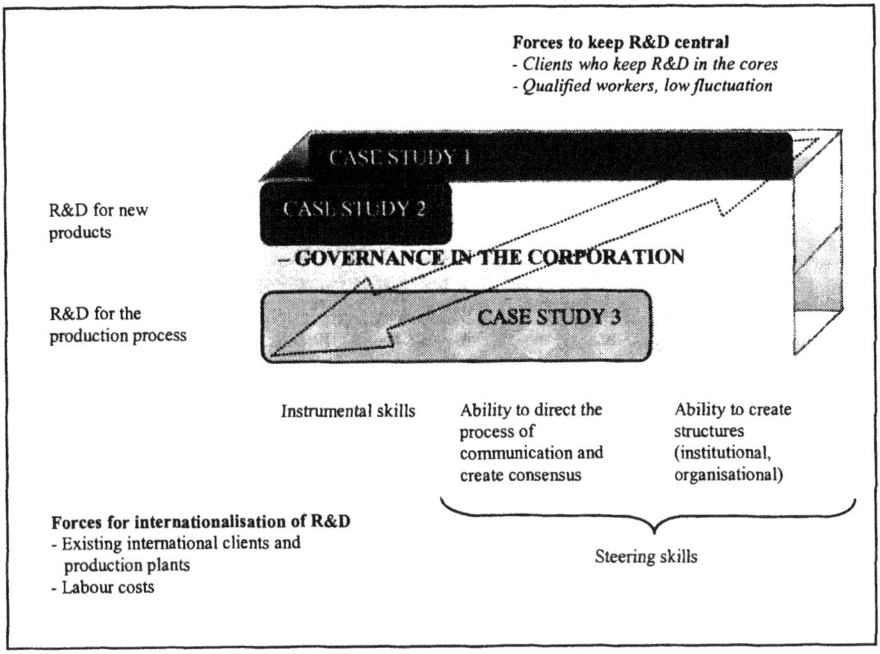

Figure 6.6 Skills in R&D in the selected case studies of automobile components supply companies

One factor which is also important, but which is held invariable in this case study, is the specific nature of the production process. This strongly influences the conditions for the international division of labour and skills. For example, a pressing shop in a foreign country has a greater requirement for decentralised process engineering than for other production processes, because every product needs the development of an adequate tool. On the other hand, the highly automated assembly of mechanical parts makes it reasonable for maintenance of the expensive machines to be carried out by the firms that sold them to the assembler. The sub-sector of this analysis, electronics, may be a section in an intermediate position: it does not make decentralisation of R&D either very important or very unimportant.

As processes have been discussed in an explorative manner and on the micro-level, the influences of the macro-system of regulations that result from specific European and national conditions (Dicken/Hudson/Schamp 1995, pp. 10-14) have not been taken into account here. The differences between the international car manufacturers and their various corporate cultures on a global level must be noted, but could not be discussed at any length in this chapter. As major US automotive supply companies such as Delphi show, the globalisation of R&D can go very far. Delphi has built up more than 30 technology centres for research, development, design, and testing all over the world. Thus, the German automobile supply industry with its many medium-sized companies may be less globalised than other companies of this sector.

The explorative insights gained here cannot lead to representative conclusions, but do highlight the important aspect of the complexity of steering R&D. It shows the openness of processes resulting from different shared visions of the management; there is no absolute need for internationalisation, but it is a possibility. And not every possibility of internationalising R&D means a loss of skills in the parent companies, because they can receive more steering competences and can thus improve on their 'central' role. At the same time, the foreign plants can learn. The analysis illustrates that the specific history of the company can affect the managements' construction of reality. Particularly the second and third case studies suggest that the firm's history influences the managers' shared visions. As the experience acquired in the two companies is different, this leads to specific visions and strategies. The second case study shows a cautious strategy of internationalising instrumental skills in product design to India, and in the third case stronger efforts to internationalise process design, including some steering skills, can be seen.

Finally, at the regional level, this insight illuminates some problems of regional policy that tries to make regions attractive for R&D (Markusen 1996). Although there are convincing examples of 'dynamic technological integration' (Ivarsson / Vahlne 2002, p. 1) as possibilities for productive networks, it should be borne in mind that the regional setting is only an opportunity, and there is no absolute need for such advantages to attract R&D (see Ivarsson 2002, p. 226ff, Larsson 1999). The shared visions in the companies lead to the internationalisation or stability of R&D, and often they are bound up with the specific company history.

References

Berger, S., Sturgeon, T., Kurz, C., Voskamp, U., Wittke, V. (1999) *Globalization, Value Networks, and National Models.* Cambridge (USA) (= MIT IPC, Globalisation Working Paper, 99-000).

Boyer, R., Freyssenet, M. (2002) *The Productive Models. The Conditions of Profitability.* Paris.

Cantwell, J. (1995) The Globalisation of Technology. What Remains of the Product Cycle Model? In: *Cambridge Journal of Economics* 19 (1), pp. 155-174.

Dicken, P., Hudson, R., Schamp, E.W. (1995) New Challenges to the Automobile Production Systems in Europe. In: Hudson, R., Schamp, E.W. (Eds.), *Towards a New Map of Automobile Manufacturing in Europe? New Production Concepts and Spatial Restructuring.* Berlin etc., pp. 1-20.

Fromhold-Eisebith, M. (1999) Bangalore. A Network Model for Innovation-Oriented Regional Development in NICs? In: Malecki, E.J., Oinas, P. (Eds.), *Making Connections. Technological Learning and Regional Economic Change*. Series of the IGU-Commission on the Organisation of Industrial Space. Aldershot, pp. 231-260.

Fuchs, M. (2003) 'Learning' in Automobile Components Supply Companies. The Maquiladora of Ciudad Juárez, Mexico. In: Lo, V., Schamp, E.W. (Eds.), *Knowledge, Learning, and Regional Development*. Münster, Hamburg, London, pp. 107-130.

Humphrey, J., Lecler, Y., Salerno, M.S. (2000) Introduction. In: Humphrey, J., Lecler, Y., Salerno, M.S. (Eds.), *Global Strategies and Local Realities. The Auto Industry in Emerging Markets*. Houndsmills, Basingstoke, Hampshire, pp. 1-15.

Humphrey, J., Schmitz, H. (2000) *Governance and Upgrading: Linking Industrial Cluster and Global Value Chain Research*. Brighton (= IDS Working Paper, 120).

Ivarsson, I. (2002) Transnational Corporations and the Geographical Transfer of Localised Technology. A Multi-Industry Study of Foreign Affiliates in Sweden. In: *Journal of Economic Geography* 2 (2), pp. 221-247.

Ivarsson, I., Vahlne, J.-E. (2002) Technology Integration through International Acquisitions. The Case of Foreign Manufacturing TNCs in Sweden. In: *Scandinavian Journal of Management* 18 (1), pp. 1-27.

Jürgens, U. (2000) Toward New Product and Process Development Networks. The Case of the German Car Industry. In: Jürgens, J. (Ed.), New Product Development and Production Networks. *Global Industrial Experience*, Berlin etc., pp. 259-287.

Jürgens, U. (2001) Approaches towards Integrating Suppliers in Simultaneous Engineering Activities. The Case of Two German Automakers. In: *International Journal of Automotive Technology and Management* 1 (1), pp. 61-77.

Kaplinski, R. (2000) Spreading the Gains from Globalisation: *What can be Learned from Value Chain Analysis?* Sussex (= IDS Working Paper, 110).

Lacertera, N. (2001) Corporate Governance and the Governance of Innovation. The Case of the Pharmaceutical Industry. In: *Journal of Management and Governance* 5 (1), pp. 29-59.

Larsson, A. (1999) *Proximity Matters? Geographical Aspects of Changing Strategies in Automotive Subcontracting Relationships. The Case of Domestic Suppliers to Volvo Torslanda Assembly Plant*. Göteborg (= Department of Human and Economic Geography, School of Economics and Commercial Law, University of Göteborg, Series B, No. 97).

Malecki, E. J. (1997) *Technology and Economic Development. The Dynamics of Local, Regional, and National Change*. Harlow, London.

Malmberg, A., Maskell, P. (1999a) Localized Learning and Regional Development. In: *European Urban and Regional Studies* 6 (1), pp. 5-8.

Malmberg, A., Maskell, P. (1999b) The Competitiveness of Firms and Regions. 'Ubiquitification' and the Importance of Localized Learning. In: *European Urban and Regional Studies* 6 (1), pp. 9-25.

Markusen, A. (1996) Sticky Places in a Slippery Space. A Typology of Industrial Districts. In: *Economic Geography* 72 (3), pp. 293-313.

Messner, D. (1997) *The Network Society. Economic Development and International Competitiveness as Problems of Social Governance*. London.

Messner, D. (2002) *Nationalstaaten in der Global Governance-Architektur. Wie kann das deutsche politische System Global Governance-tauglich werden?* Duisburg (=INEF-Report, 66).

De Meyer, A. (1992) Management of International R&D Operations. In: Granstrand, O., Håkanson, L., Sjölander, S. (Eds.), *Technology Management and International Business. Internationalization of R&D and Technology*. Chichester etc., pp. 163-179.

Pries, L., Schweer, O. (2004) The Product Development Process as a Measuring Tool for Company Internationalization? The Case Studies of Daimler-Chrysler and Volkswagen. In: *International Journal of Automotive Technology and Management*, (forthcoming).

Schoenberger, E. (1999) The firm in the region and the region in the firm. In: Barnes, T.J., Gertler, M.S. (Eds.): *The New Industrial Geography. Regions, Regulations and Institutions*. London, New York, pp. 205-224.

Segrestin, B., Lefebvre, P., Weil, B. (2002) The Role of Design Regimes in the Coordination of Competencies and the Conditions for Inter-firm Cooperation. In: *International Journal Automotive Technology and Management* 2 (1), pp. 63-84.

Serapio, M.G., Dalton, D. (1993) Foreign R&D-Facilities in the United States. *Research – Technology Management* 36 (6), pp. 33-39.

Spohr, M.O. (2002) *Inter- und intraorganisationale Beziehungen in der europäischen Autoelektronikbranche. Eine ego-zentrierte Analyse*. Frankfurt/Main.

Sugiyama, Y., Fujimoto, T. (2000) Product Development Strategy in Indonesia. A Dynamic View on Global Strategy. In: Humphrey, J., Lecler, Y., Salerno, M.S. (Eds.), *Global Strategies and Local Realities, The Auto Industry in Emerging Markets*. Houndsmills, Basingstoke, Hampshire, New York, pp. 176-206.

VDA (2001a) *Auto Jahresbericht 2001*. Frankfurt.

VDA (2002) *Auto Jahresbericht 2002*. Frankfurt.

Chapter 7

Transnational Corporations and the Post-Socialist Economy: Learning the Ropes and Forging New Relationships in Contemporary Poland

Boleslaw Domański

Introduction

Transnational corporations are perceived as leading components of contemporary capitalism and a major vehicle of globalisation. Their embeddedness in economies and societies where they carry out their activities is a popular concept in studies of the relationship between the global and the local (see Dicken et al. 1994; Ettlinger 1999; Phelps 2000). The concept implies that the transnational corporation (TNC) is implanted in local networks of production, institutional and social relations. Many studies approach the embeddedness of TNCs investing in less developed areas from the point of view of their effects on the host economy. The latter may be treated as a passive recipient, not to say a 'desert' (Grabher 1994; Hardy 1998). A different perspective on the impact of local networks on the activity of the foreign company is adopted here. Schoenberger (1999, p. 211) points out that firms:

> live in different places and must confront and respond to the particularities of these places across the whole range of practices and issues'. Phelps & Fuller (2000, 241) show 'the highly politicised nature of relations in which individual affiliates are embedded both internally, within their parent companies, and externally, with governments and other supporting organisations. ... The sorts of locational hierarchies produced from corporate spatial divisions of labour are not, if they ever were, determined purely by an economic rationality but are shaped, albeit at the margin, by the political behaviour of the MNEs and their affiliates in an increasingly murky firm-state nexus at the national and local scales.

Business strategies are both path- and place-dependent at different spatial scales (Cox 1996).

Since 1990 large foreign companies have increasingly penetrated Central and Eastern Europe. Their power may seem tremendous in countries that have just embarked on the dismantling of overwhelming state systems and are characterised by a low level of economic development and a lack of well-established institutions of civic society. On the other hand, TNCs had very limited knowledge about and experience of doing business in these territories, which were previously closed to foreign capital, and hence their investment required intensive learning of national and regional economic environments, cultures, institutional and social networks. This entails the development of extensive relations with various actors and adjustment of the TNCs' strategy and everyday practices.

The fundamental question addressed in this chapter is how the embeddedness of large foreign manufacturers within the network of economic, social and political relations affects their behaviour in post-socialist Poland. In other words, the author aims to provide an insight into how strategies and practices of transnational manufacturing corporations unfold as a result of their relationships with a multitude of local actors. Poland is an interesting case here for at least three reasons: first, it represents the second largest post-socialist economy after Russia, second, it has recently experienced fast industrial growth (86 per cent 1991-2001, following a 30 per cent decline 1989-1991), and third, trade unions played a unique role, having been instrumental in bringing down the communist regime and part of all government coalitions throughout the 1990s (Domański 2003).

The chapter begins with a discussion of various theoretical approaches to the firm and foreign investment from the point of view of the role of local relations. It is followed by an analysis of a variety of TNCs' relationships with nation-state institutions, labour, other firms, and local governments.[1] This leads to some generalisations about the dynamics of investors' strategies in Poland and concludes with the interpretation of foreign investment in terms of learning as a source of intangible assets which give rise to localised capabilities of the TNC and sunk costs.

Theoretical Approaches to the TNC and its Relationships with the Host Economy

In the long prevailing neo-classical paradigm in mainstream economics, foreign investment has been accounted for by differentials in rates of return underlain by the relative labour and capital endowment of countries. New trade theory attempts to incorporate the TNC into a general equilibrium framework. Increasing returns – both internal to the firm (Chamberlinian) and linked to the industry (Marshallian) – are used to explain trade, international location of economic activity and agglomeration (Krugman 1991, 1998).

History matters; as companies may gradually achieve internal economies, regions are affected by historical accidents (e.g. location of certain industries) and

1 The analysis is based on empirical research which covered a few hundred industrial investors and included about 130 interviews with TNC managers, trade union leaders and local governments (Domański 2001a, 2001b).

a virtuous circle of agglomeration may develop. Ietto-Gillies (2002) argues that this approach does not actually distinguish the reasons behind inter-regional and international production. It ignores TNCs' relations with rivals, labour and government. Space is reduced to geometrical relations (Lee 2002).

International involvement may be seen as part of the decision to internalise market transactions within a firm to avoid costs and risks in the allocation of resources in situations of market imperfections. The transaction costs theory takes an internalisation perspective, where foreign direct investment may prove superior to intermediate contractual arrangements, e.g. licenses and alliances (Buckley and Casson 1998). It is a microeconomic approach related to the governance theory of the firm developed by Williamson (1975), where TNCs are studied in terms of efficient organisation of production. Internalisation theory refers mainly to vertically integrated production (Cantwell 2000). A firm's ownership advantages are perceived in a comparative, static (historical) perspective. External relations that do not find expression in a firm's costs are ignored.

A broader dynamic interpretation of international production and trade, with explicit locational dimension, is found in Vernon's (1996) international product life-cycle concept. It identifies different stages in the life cycle of a product, each with different consequences for the internationalisation of the company. The implication for the post-socialist economies of Central and Eastern Europe is that they could expect TNCs to be involved in standardised, low-cost production. The fundamental reasons for change are of a technological character; place-specific factors matter as long as labour and market attributes are concerned.

The (Uppsala) internationalisation process model is another sequential perspective, in which business moves from irregular exporting to eventually full production in foreign markets, driven by experiential knowledge accumulation (Johanson and Vahlne 1990). It is based on behavioural theories, with assumptions about lack of information and the importance of perceived risk and uncertainty. Internationalisation is seen not as a sequence of deliberately planned steps founded on rational analysis, but an incremental, successive learning through stages of increasing foreign commitment. Relationships with local actors can be analysed here from the point of view of knowledge building by the investor. The Uppsala model is criticised as a poor explanation of the causes of changes. Likewise, Vernon's theory proposes a unitary progression from one stage to another, whereas it is unrealistic to assume a simple evolutionary sequence for firms that are already global (Edwards 2002).

A dynamic (organisational) capability approach shares with internalisation (transaction cost) theories the view that markets fail to define property rights in non-imitable and non-tradable firm-specific resources (Teece et al. 1997). 'Dynamic capabilities reflect the TNCs' skill in learning, integrating, building and augmenting existing organisational assets as a platform for future growth' (Nicholas and Maitland 2002, p. 15). It borrows from evolutionary theories of the firm (Nelson and Winter 1982). 'The principal feature of the modern business enterprise is that it is an organisational entity possessing know-how' (Teece 1980, p. 226).

The competitive advantage of the firm is built on its access to, and control over, tangible or intangible assets and its ability to combine them in building distinctive competences. The firm's structure and decisions are understood as an inherently dynamic process, commonly of path dependent nature. The evolutionary competence theory of the firm has become the leading perspective in managerial science and industrial organisation. Maskell (2001) argues that it suits economic geography, because it can work on different levels of spatial aggregation and is based on the assumption of bounded rationality. He argues that firms' competences can be built on 'created localised capabilities', which make territorial entities differ from one another. This approach allows the exploration of the role of learning and relations with various local agents in the dynamics of foreign investment and is adopted here.

Dunning's (1977, 2000) eclectic paradigm proposes three inter-related principles which are fundamental to the understanding of international production: ownership-specific advantages (not possessed by competitors), internalisation (advantages are exploited by the firm itself rather than by exports or licensing) and location-specific factors. They are derived from the theory of the firm, organisation theory, trade theory and location theory. Dicken (2003) argues that it is an extremely useful conceptual framework within which specific cases can be examined. The impact of local networks can be considered here. Both eclectic and internalisation approaches treat FDI decisions as if made in isolation, not as a long-term sequential process as in product life-cycle, internationalisation or dynamic capability views (Edwards 2002).

Several studies show the sequential pattern of FDI where re-investment is based on learning from prior experience (Kogut 1993; Kogut and Zander 1993; Chang 1995). TNCs' firm-specific advantages may originate not only in the parent; subsidiaries can also be important in creation of capabilities. Ietto-Gillies (2002) stresses the importance of the strategic behaviour of TNCs towards governments and labour and argues that learning about a location may enhance the ability to operate in additional locations. Safarian (1993) identifies various stances adopted by TNCs in their dealings with national and sub-national institutions from 'coping', to 'avoidance', 'insulation' and 'exit'. Many studies show that TNCs become sensitive to the opportunities for negotiation created by the local systems of governance (Peck 1996; Phelps et al. 1998). Dicken and Malmberg (2001, p. 358) write about multiscalar territories – from the local state through the national to the supranational – as bounded systems of governance. 'Such governed spaces are one of major ways in which the resources on which firms depend are "packaged"'. The nation-state remains a principal system of governance.

Networks are seen as breaking the market/firm dualism. Yeung (2000, p. 302) says that networks are 'both the governance structure and a process of socialisation through which disparate actors and organisations are connected in a coherent manner for mutual benefits and synergies'. Social bonds and trust intensify through repeated contacts. Business networks, supplier commodity chains, production networks and innovative networks are identified. It is the author's belief that political and social networks, which involve trade unions, state agents and local administration, should be considered too.

All in all, the evolutionary view of the firm and the dynamic capability approach seem like an appropriate perspective to explore a variety of networks in which global corporations are embedded in the specific transition economy of Poland. This advocates a longitudinal approach in empirical analysis, but the comprehensive paradigm of Dunning can also be useful.

Types of Relationships

The relationships of TNCs in the host economy and society can be analysed from the point of view of various criteria. First, they involve numerous external actors including state agents, local governments, other TNCs, domestic firms, labour and trade unions as well as other units of the corporation. Second, they concern all sorts of domains, especially ownership of assets, market, product quality, supplies, technology, organisation (restructuring) of production, employment, public subsidies, environment and location. Relationships among these main groups of actors in the various spheres of contact take the form of competition, conflict, cooperation, and control (Figure 7.1). Selected relationships are examined in some detail.

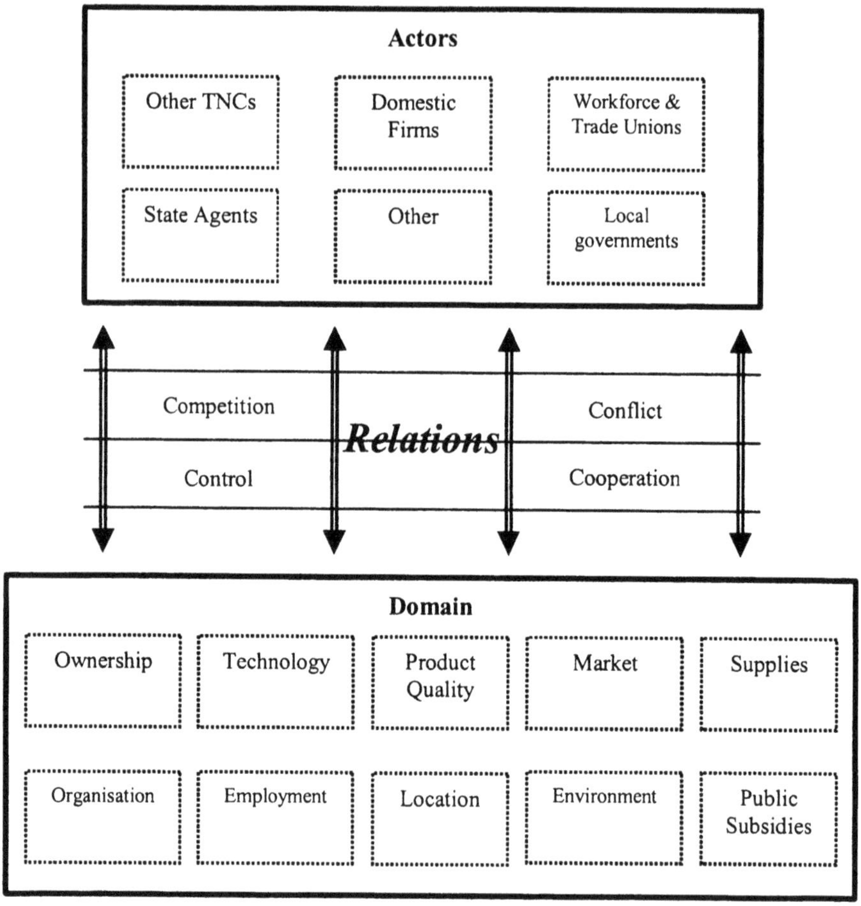

Figure 7.1 Main relationships of the transnational corporation in the host economy

TNCs and the Nation-State: General Regulation and Ownership Changes in Industry

Contrary to the 'hollowing out' thesis (Ohmae 1990), state regulations have had a considerable impact on foreign investors. The bargaining power of the Polish government vis-à-vis TNCs stemmed, first, from almost full control of production facilities when communism fell. Second, it was the state's role to overhaul the legal system regulating various domains of economic activity, i.e. to devise and implement a new framework for market mechanisms. In a comparative perspective, the predictability and stability of the general political and economic situation are pointed to as preconditions for FDI in Poland vis-à-vis many transition countries. Since the legal system was shaped at the same time as TNCs began to invest in Poland, they could get involved in formal or informal bargaining for favourable solutions.

The state controls structures and resources, which are of critical importance to TNCs, but this control is exerted by many state institutions (functional division of competences), which have specific interests and practices.[2] Moreover, these practices change as a result of shifts in policies and personnel, not least due to government shake-ups. Privatisation, domestic market protection and taxes are areas with national policy meanders, while environmental and labour regulations reveal more stable patterns. This lack of coherence and stability inevitably undermines long-term strategies on the part of the government.[3] The question is how this affects foreign investors. On the one hand, this may create extra opportunities for negotiation with individual institutions of the post-socialist state. On the other, the multitude of state bodies involved, changing rules and practices as well as the discretionary power of particular agents make the system of governance less comprehensible and cause delays. This can be illustrated by the problems of ownership changes.

The key asset controlled by the national government is factories themselves, their privatisation is inevitably a political decision. Conflicts over privatisation have two major sources: its heterogeneous objectives and the number of actors involved. Privatisation is at the same time expected to bring high revenue for the state budget (selling a factory), improve the long-term competitiveness of the privatised firm, and ensure good remuneration and employment prospects. These aims tend to be contradictory. For example, investment commitments of the investor, which affect company prospects, may be valued less than maximisation of the state's revenues, i.e. a short-term government goal. The effects of bargaining in the form of final conditions of privatisation by and large depend on how valuable the 'product' (enterprise) is for the investor. If it controls an attractive market, the state may be able to achieve generous investment and social packages, a good price and occasionally other commitments, e.g. guaranteed purchase and support for tobacco growers by cigarette companies. One should note several actors involved in the privatisation process: ministry officials, politicians associated with a particular branch and/or region, company managers and trade union leaders. In addition, there are claims of pre-war owners, e.g. lawsuits about company name or trade mark. Legal uncertainties concerning land ownership sometimes delay or entirely frustrate investment.

---

2 Bargaining among state agents is rooted in the gatekeeping role they performed under socialism (Domański 1997).

3 As a result greater success of national policy is observed in the domains where consistent and coherent regulation has been introduced and a stable institutional framework exists, e.g. in environmental policy.

Although there is no formal preference for indigenous investors in Poland, we can find many cases in which they are chosen, even though a transnational corporation submits a more generous bid together with a comprehensive investment plan. Concern over a hostile takeover and/or disappearance of a national brand name could be a reason behind practical discrimination against foreign bidders, but more often this can be attributed to the interests of various lobbies. There has been practical preference for insiders' privatisation in Poland, i.e. management and employee buyout (Williams, Baláž 1999), especially of small and medium-sized firms. Managers and trade union leaders in big state-owned enterprises may be afraid of losing their position and influence as a result of privatisation; political parties would no longer appoint their people to company boards. Powerful lobbies make privatisation of some enterprises and industries more politicised than others. The joint privatisation of Huta Katowice and Huta Sendzimira, the largest Polish steelmakers, was opposed by management and union leaders in the 1990s. They pursued independent investment plans and ran into heavy debts, which ultimately led to their consolidation by the state. This shows evidence of poor ownership control in state-owned firms and institutional lock-in. The sugar industry is one of the most strongly regulated and politicised. The privatisation contract for 16 plants signed by Saint Louis Sucre in 2000 was blocked by the Ministry of Internal Affairs, which refused permission for land purchase under the influence of lobbying by supporters of the national sugar holding creation.[4] The delay in privatisation may lead to economic deterioration, even bankruptcy of state-owned industrial enterprises, although the collapse of firms privatised with TNC participation has been uncommon, e.g. Daewoo.

Obstacles and delays in privatisation could push TNCs to greenfield investment or the purchase of private firms. Privatisation by a domestic investor could precede the take-over by a TNC. A shift of competences on a larger scale took place in the mass privatisation programme, where National Investment Funds proved effective in selling more than a hundred enterprises fast and often at a better price than the ministry, e.g. cement mills.

There has been no stable national policy on financial incentives offered to foreign investors in Poland. The Act of December 1988 granted all companies with foreign participation universal three- or six-year tax holidays. In June 1991 they were replaced with individual tax exemptions awarded by the Ministry of Finance. A study by Błuszkowski and Garlicki's (1996) shows that these were enjoyed by 8 per cent of foreign firms. Since 1996 foreign and domestic companies are eligible for the same corporate tax relief and exemptions, e.g. based on investment, exports and location in special economic zones. There is little room for bargaining now, except for the location of economic zones. What may be negotiated is the property tax imposed by local governments, and especially infrastructure provision.

4 The investor has recently won the legal battle at the Polish court.

TNC and Trade Unions: From Suspicion to Bargains

Poland has a reputation for having strong trade unions. In fact, TNCs often find the influence of the unions greater than in many other European countries. The generally strong political position of trade unions since 1989 curbed deregulation and flexibility in labour relations and pushed labour costs above Czech and Hungarian levels, even though Polish GDP per capita is lower. At the microeconomic level trade unions can effectively lobby for or against a particular investor during privatisation negotiations. The content of a social pact signed with the unions is instrumental here, in particular employment guarantees and remuneration conditions. The failure to negotiate a satisfactory social pact would obstruct privatisation; there were investors who managed to outbid competitors through a generous deal with the unions. The expectations of the unions gradually grew; a three-year employment guarantee became a standard in the late 1990s, even 5- or 6-year ones existed.[5] These could constrain rapid restructuring, but TNCs are able to circumvent them by offering high severance pay to people who voluntarily quit their jobs, e.g. Siemens in Wrocław and Warsaw, and ABB in Warsaw.

A special challenge for many investors is learning how to function in multi-union plants, as their experience at home could be that of bargaining at the national level or single-union agreements. In Poland workplace collective bargaining is central, in contrast to sectoral or national social dialogue. There are between two and more than ten union organisations in a single firm. This gives the employer opportunities to play one union against another, but on the whole frustrates collective bargaining, as unions may compete among themselves. In such instances smaller ones are prone to put forward new radical demands in order to gain popularity. The bargaining between management and trade unions sometimes leads to collective labour disputes.

Wage level is the most common concern of the employees, who encounter an intensified labour process and greater worker responsibility. Several interviewed managers argue that a regular pay rise above the inflation rate accepted in Polish affiliates in the 1990s would be an unusual phenomenon in their home country, e.g. Sweden or the UK. The rise resulted not only from the position of the unions in foreign-owned companies, but also from growing earnings in indigenous firms, as TNCs always tend to pay above the local average. A steep increase in unemployment checked wage growth in 2001-2002.

Enhanced productivity often leads to redundancies; their size, timetable and financial compensation are toughly negotiated. Unions in foreign-owned companies normally win higher severance pay than the three months provided for by the labour code. Outsourcing, which undermines employment certainty, is contested with limited success; flexible working hours are rejected. New production organisation is generally accepted. The labour experience of various corporate

5 Some TNCs argue that long employment guarantees discourage their investment in ailing Polish companies.

experiments in car factories is interesting here. Lean production at a greenfield GM plant at Gliwice produced a largely satisfied, newly-recruited workforce, which benefited from high wages, skill upgrading, good working conditions and employment stability. On the other hand, the preservation of elements of a Taylorist regime with gradual outsourcing of some activities was contested at older Fiat plants and Daewoo's petrification of the production model delayed restructuring, which brought only short-lived positive effects. Strategic development issues are rarely part of the bargaining agenda. Conflicts with the unions in the home country over the relocation of production to Poland may be deliberately publicised, e.g. a transfer of engine manufacturing by Fiat.[6] The participation of Polish representatives in European Works Councils adds another dimension to these relationships.

The prevailing experience among the interviewed unionists is that early fears about the short-term commitment of TNCs proved groundless, similar to high expectations concerning wages. There is a widespread opinion that the foreign investor makes the Polish firm competitive and hence brings brighter prospects for the workplace, which represents the upper segment of the labour market;[7] this is reinforced by comparisons with state-owned enterprises that are plagued by sluggish production, irregular pay and endless layoffs. Consequently, unions in foreign-owned plants are not prone to strike even in cases of prolonged disputes. It seems that the necessity to negotiate and reach agreement with the investor on the social pact at the very beginning abolishes some barriers and provides a basis for further contacts. Neither union representatives nor managers expect serious changes in employment conditions when Poland joins the EU.

The majority of foreign employers evaluate union activity in their factories positively. Their current views on industrial relations cannot be separated from the general role of the workforce in the success of foreign manufacturers in Poland. Undoubtedly, a vital advantage they exploit here is the opportunity to select the best employees due to the prominent position of TNCs on the labour market. The relatively skilled, adaptable and low-cost (on a European scale) labour force constitutes a crucial resource captured by TNCs, which underlies productivity growth, good product quality and hence overall profitability of the venture. The effects of union bargaining – wage increase, employment guarantees, severance pay, etc. – have to be understood in this context. They may be regarded as costs of achieving the investor's strategic goals and are accompanied by substantial effort and expenditure on training. The TNC's behaviour in labour bargaining also reflects its

---

6 This contributes to the propagation of an 'all-in-one-boat' philosophy, a belief in labour and managers' affinity of interests in the face of competition, both inside and outside the corporation (see discussion in Berndt 1998).

7 Still, there may be an internal segmentation within the corporation in terms of job stability, which reflects a division of company production in Poland into core and peripheral activities, e.g. in the case of Philips, from a long-term commitment to the manufacturing of lighting and tuners, through large-scale assembly of TV sets to entirely peripheral assembly of domestic appliances subcontracted to an indigenous producer.

individual strategy, as can be illustrated by Daewoo's activity, when it chose Poland as its export platform for the European Union. The Korean investor accepted a three-year employment guarantee for about 20,000 people in the Warsaw-based FSO in 1996, which had been rejected by General Motors negotiators earlier. It began investment by raising wages considerably, improving locker-rooms, showers, etc., and distributing 5 per cent of the shares among the employees, as part of its strategy of creating an image of a new 'good citizen' of Europe, contrasting with Daewoo's practices in Korea.

It is to be remembered that at the beginning there was much uncertainty among investors about labour attributes and industrial relations. Generally low cost was the only tangible benefit of the Polish labour force and some managers openly admitted to having being afraid of the reputedly militant trade unions when they came to Poland. Investors' relationships with labour and the unions have been a dynamic process of learning through bargaining. In most of the companies the search for workforce loyalty, low turnover and good work has produced considerable progress in working conditions and wages as well as some sort of stable relationship with major unions. The exact forms and how they are achieved differ; there are also firms with poor industrial relations.[8] The interviewed top executives of major corporations operating in Poland, e.g. Philips and ABB, emphasise the role of local managers in this process. Local managers were a vital component in the TNCs' strategy of learning the ropes in a generally unknown business and social environment – taking advantage of local knowledge and contacts to restructure the enterprise and transfer corporate culture, new organisation and management techniques. Labour bargaining is a domain which in many corporations is left to be decided locally.[9] Dunn's (1999) study confirms that US management practices in a Polish firm are negotiated and transformed by the workers.

Finally, we can put forward the question as to how labour issues and bargaining influence the strategies and practices of TNCs in other spheres. First of all, positive experience with labour quality and industrial relations is a prerequisite for decisions on further expansion. There are no examples of large investors who left Poland for labour reasons.[10] Secondly, corporations have learnt that trade unions may be a crucial ally in lobbying for governmental concessions, e.g. protectionist measures, or special economic zones. Moreover, labour factors are important for the organisation of production both at individual plants and on a broader scale. The former finds expression, for example, in negotiations over automation levels. The

---

8 Industrial relations in large foreign firms vary from a sort of partnership between unions and management, through various forms of contestation and bargaining over certain issues, to strained relations, which may occasionally be underlain by attempts to marginalise the unions (Domański 2001b).

9 Some union leaders attribute unsatisfactory industrial relations to local, often young Polish managers too.

10 However, there is no information on potential investors who might give up investment plans in Poland on the ground of supposed labour or industrial relations problems.

GM factory opened at Gliwice in 1998 – based on the experience of Japanese plants, such as the Californian NUMMI joint venture with Toyota, and the East German plant at Eisenach – is regarded as the most advanced example of lean manufacturing within the corporation. Staff from GM plants in other countries is brought there for learning. Decisions on the mode of entry (acquisition, joint venture or greenfield) and geographical location of production are also contingent on labour conditions. Some large greenfield investments were preceded by investor encounters with 'old' work practices and unrealistic, in the investor's opinion, expectations as to the size and length of employment guarantees in existing factories, e.g. GM's joint venture with FSO in Warsaw, or Volvo assembly at Jelcz. Donnelley's difficulties in improving the performance of a printing company purchased earlier in the U.K. lay at the root of the choice of greenfield rather than acquisition in Poland.

TNCs and Other Companies: Markets and Supplies

TNCs get involved in a network of relations with numerous domestic and foreign firms, which inevitably affects their activities in Poland. The fundamental reason behind the bulk of manufacturing FDI in Poland is the size of the market and its faster growth than in the European Union. Early investors who acquire leading local producers may easily gain a dominant position due to the heritage of socialist monopolies, customer habits or protectionism. Still, a monopolistic situation is less common than in the smaller Czech or Hungarian economies. The involvement of several foreign firms in the same production area brings competition to industries that were monopolised under state socialism, e.g. TV sets, tyres. There are several new factories built by joint ventures of two TNCs, e.g. Philips and Matsushita, VW and Siemens. Global alliances find their expression in Poland too, a recent example being Powertrain (General Motors and Fiat) which was created to manufacture engines, e.g. in Bielsko-Biała.

Market position may be reinforced by non-market means sought in bargaining with the government. For example, tariffs imposed on imported cars were an incentive for Fiat in 1992.[11] The exclusive rights to supply the national telecom were granted to AT&T (Lucent Technologies), Alcatel and Siemens, which invested in Polish communication equipment factories. TNCs may directly or indirectly (e.g. through EU institutions) put pressure on imposing 'minimum standards' and other requirements to fight local competitors. There are no fiscal incentives available to investors from abroad alone, but they may take advantage of new

11 Concessions given to Fiat stimulated the government to encourage other investors and were followed by silent approval of Daewoo's simple assembly of cars, which were knocked down in a South European port, circumventing tariff barriers. Fiat and Daewoo's gains in individual bargaining over national imports regulation were an important, though temporary, success and were eroded within a few years. This is part of broader meanders in car imports policy, including fluctuating regulations on second-hand cars.

plants located in special economic zones vs. existing producers of similar products. In many branches, foreign corporations increase competition for indigenous producers and may push them out of their traditional markets. Still, there are industries where major corporations have been unable to outshine domestic brand-name rivals, not only historically dominant manufacturers, but also new private firms established in the 1990s, e.g. Tele-Fonika (cables), Atlas (adhesive mortars), Koral (ice-cream). The success of the latter is built on local knowledge and capabilities.

Various types of imitative behaviour of TNCs should not be ignored. The very decision to undertake investment in Poland is often underlain by earlier decisions of competitors or customers, it may also be an attempt to run ahead or prevent acquisition by another TNC. These strategies are evident in many Polish industries. They also work on a broader Central European scale; for instance, Philips first investment in Poland (Piła in 1991) could be a response to General Electric's purchase of Tungsram in Hungary. Foreign investors carefully watch the behaviour of other firms in the same industry, to some extent of all large TNCs, and try to learn from their success and failure, especially in marketing, technology and location. One should also note a considerable learning process on the part of domestic producers under the impact of foreign companies, which is visible in general progress in technology, organisation and management (Durka 1999; Błuszkowski and Garlicki, 2000).

Joint venture with a domestic partner is a mode of entry, which lowers uncertainty and allows learning through the contacts of the partner. After a few years, the latter may no longer be needed and the foreign investor often seeks full control. The knowledge and capabilities of the domestic partner have an impact on various features of the investment, including its location. For example, National Forests seemed to be an ideal partner to manufacture furniture in the late 1980s as a monopoly controlling the supply of timber. However, after an initial success the joint venture (Furnel) collapsed. The passive power of a huge state organisation together with generic advantages of the TNC (ICL) were insufficient to adjust to a changing environment and compete with other producers, which had or acquired expertise in the furniture business faster. This also illustrates the TNCs' attempt to build local competence on short-lived economic and political conditions (weak competition, control of raw materials, favourable exchange rate).

The Foreign Investors Chamber of Industry and Commerce (IPHIZ) was founded in the 1980s to represent small foreign firms and was joined later by some large investors. Foreign enterprises are also members of a few other nation-wide business organisations, e.g. Polish Confederation of Private Employers (PKPP). Social dialogue in the Tripartite Committee is obstructed by conflicts between major trade unions (Solidarność and OPZZ), lack of representation of some groups of employers and inconsistent government policy. On the whole, employers' organisations are not considered to be influential in Poland at the moment. TNCs' lobbying tends to take place through their individual channels rather than as a collaborative effort in the host economy, whereas political networks in the home country may also be important.

The scope of domestic supply linkages is very important from the point of view of the host economy, as they generate direct and indirect multiplier effects. TNCs' purchase of component and raw materials differs considerably depending on the industry in question, and partly on export-orientation and Greenfield / acquisition type of investment (less domestic supplies in export-oriented and new plants). Companies that operate longer in Poland rely on domestic suppliers to a greater extent, as is most evident in the car industry. The expansion of supplier networks reflects TNCs' local learning and allows them to lower costs even further vis-à-vis Western Europe. This helps embed the investor if long-term relations develop rather than temporary subcontracting. In addition, foreign manufacturers create a demand for various producer services as well as consumer goods and services through income effects (Domański 2004).

TNCs and the Locality: Regional Policy, Local Government and Environment

In 1995 the Polish government introduced a new measure of regional policy – special economic zones. They were a form of localised opportunities for large and medium-sized investors, which could gain up to 10 years' tax exemption plus another 10 years of 50 per cent tax relief. Their geographical location became an element of bargaining between numerous local authorities and the government. Seventeen economic zones were originally created comprising more than 40 towns, and some localities were added later. Most of the zones have achieved little success in attracting large investors – most of the capital went to a few places. One of them is Gliwice, which did not actually meet the criteria for governmental support (high unemployment). The economic zone was established here under pressure from General Motors, which chose the town as a site for its factory in 1996. Bielsko-Biała and Kwidzyn provide even more striking cases of investors bargaining for the location of the zones. Areas next to the *existing* factories of Fiat and Philips, respectively, were incorporated into the zones in 2000 as a result of effective lobbying of temporary coalitions of company officials, trade union representatives and local politicians. The TNCs presented this as a requisite for a major expansion.[12]

Special economic zones arouse conflicts. Protests from domestic manufacturers of similar goods occasionally resulted in investors being refused permission to locate in the zone, e.g. Unifi (fibres) and Merloni (cookers) in Łódź. Economic zones became one of the most contentious issues in Poland's negotiations with the European Union. There was a strong pressure on Brussels from many European companies to abolish Polish zones on the grounds that they violate EU regulations. It is interesting to note in this context that only 8 per cent of the European greenfield investment in Poland was located in the zones, against 21 per cent of US greenfield investment. Poland eventually agreed to change the incentives so that

---

12 One could see this as an element of after-care policies aimed at developing and embedding foreign affiliates, e.g. as discussed by Amin et al. (1994) and Young and Hood (1998).

they comply with the EU limits on public assistance, but the rights obtained by investors before 2000 have not been abolished.

Local governments (*gmina*) are other significant players that can influence locational choice. They are part of infrastructural deals and may act as a guide in the local institutional network that includes utility companies and various state agents. More important, they can speed up the investment process, e.g. by prompt issuing of building permits. The interviews reveal that local governments have greater impact on location decisions outside major cities and there is better co-operation between investors and small communities. At the same time, foreign firms participate in the investment in local infrastructure for greenfield plants in 46 per cent of such plants located in rural communes, 40 per cent of those in small towns (below 25,000 people) and only in 15 per cent of those in large and medium-sized urban places. This may lead to the conclusion that foreign companies find some intangible benefits in small communities, such as faster decision-making and limited bureaucracy, and hence prefer to undertake their bargaining here (Domański 2004).

Pilkington at Sandomierz provides an example of the transfer of a firm's experience from the home country in actively coping with redundancies in a community. The investor maintains a foundation that provides financing to firms set up by people made redundant or enterprises that employ them. In the first stage more people found jobs than lost them as a result of new technology, so access to the funds was broadened to cover all local residents and firms.

Polish environmental regulations have been gradually brought in line with the EU legislation, so TNCs had little difficulty to accommodate to them. Some foreign investors encountered local protests, e.g. they made Elf abandon its plans to open a polystyrene factory in Gdańsk. A serious conflict was ignited by Goodyear's decision to open a logistic centre next to its factory at Dębica in 1998, which required the clearance of 4 hectares of forest. The investor put enormous pressure on the local council, which eventually approved the location, as did the Ministry of Environment. The deal was to plant the same number of trees elsewhere and pay a compensation of $1.2 million. The company ultimately decided to build the centre at Tarnów to avoid a delay caused by potential protests and lawsuits from environmental groups.

The enforcement of regulations, which were ignored under socialism, sometimes appears as a barrier to environmental abatement. Pilkington's investment in a new float glass plant, which was to replace an old noxious factory, was nearly blocked because the new facility was located close to the residential housing built next to the factory grounds a few decades earlier.

The Dynamics of TNCs' Strategies, the Position of Polish Affiliates in Broader Production Networks and Changes on Locational Behaviour

Many theoretical approaches and empirical studies demonstrate that company strategies are modified in line with changing conditions. As a matter of fact, they also evolve because companies learned from these conditions. Managers of many firms that expand their activity in Poland, e.g. ABB, Philips, car manufacturers, admit that when they came here, they had little knowledge of the situation (except for generally low wages), and consequently started out with limited investment and production plans. It was only when they learnt that company quality standards could be met and healthy growth in productivity achieved, that they assigned local factories to undertake broader functions within the TNC production network. This could mean widening of a range of products and/or more advanced technology. Moreover, Polish plants are moving from serving post-socialist markets to EU-oriented production; entirely new export-oriented plants are also being built. This trend is most evident in the automobile, electronics and electrical industries, and increasingly in machinery, paper, rubber and plastic products. For example, Piła and Kwidzyn became the largest European Philips factories for lighting and TV equipment, respectively. Volvo, which started with SKD assembly of trucks and buses, decided to make the Wrocław plant its biggest bus factory in Europe. Volkswagen built a diesel engine factory at Polkowice after a 6-year involvement in more simple production at Gorzów/Wielkopolski and Poznań. Most investors emphasise that they did not buy factories in Poland with the intention of relocating production from Western Europe. The relocation and consequent closure of some plants in EU countries was decided later as part of a broader spatial reorganisation of production within the company, where economies of scale and the supra-national specialisation of factories were important. The decisions were underlain by comparisons of costs, quality, productivity, market dynamics and geographical location. This means that the strategies of TNCs in post-socialist Poland tend to be *emergent* rather than *deliberate* (see Mintzberg 1987; McGrath-Champ 1999), that is they do not simply follow a grand scheme devised at the beginning of their investment in Poland, but develop in response to opportunities perceived later.

Gradually increasing foreign commitment based on accumulation of knowledge and reduction in uncertainty through the investor's experience is in accordance with the internationalisation process model. However, the important point is that much of FDI in Poland was not preceded by prolonged exports through sales outlets or licensing. The perception of a unique historic opportunity caused many TNCs to act fast. There are also companies, in particular the latecomers, which take the plunge on the basis of observation of other investors and begin with new factories geared to exports of advanced products, e.g. Toyota gear-boxes and engines at Wałbrzych. The trajectory of certain firms reveals clear path dependency. Remote causes – decisions on license agreements and contracts between Fiat and the communist government – triggered sequences of events leading to increasing involvement of the Italian concern in Bielsko-Biała and Tychy: the purchase of the Polish car-maker in 1992 for $180 million and total investment of $1.7 billion meanwhile with an extensive network of suppliers.

The strategies discussed manifest a long-term TNC involvement entailing costly investment and offensive restructuring of an acquired enterprise or expansion of a greenfield plant. The investors seek comparative advantage based on enduring conditions, which may bring advantages to both the company and the community. This contrasts with another strategy, where investors use local opportunities or resources without significant sunk costs pursuing short-term profits. Central and Eastern Europe may be treated similar to unstable Third World countries, where the rule 'make profits fast and escape' is applied. The success of this strategy relies to a large extent on extraordinary conditions and/or political concessions, e.g. Furnel. It rests on narrow learning of the local environment, in contrast to broader knowledge necessary in the case of long-term strategies. As a matter of fact, the 'long-term' and 'fast-profit' strategies are ends of a spectrum, between which actual firm strategies are placed.

The company strategy finds its expression in the position of a Polish subsidiary within the entire corporation and hence in relations between the former, the mother company and its affiliates in other countries. This reflects the character of manufacturing activity carried out in Poland, e.g. high/low-value added, specialised/standardised, final product/component, type of customers, the extent of intra-firm supplies and imports. The majority of foreign-owned factories manufacture standard products for the Polish or Central-East European markets, similar to those produced by the company in other countries. They are what Hamill (1993) calls 'miniature replicas'. There are also Polish plants that have become principal producers of more advanced components or final goods for the European market, e.g. certain car makes and engines or ABB distribution transformers in Łódź. Such specialised factories are integrated into the global production network and take over some manufacturing activity from Western Europe. Their products differ in value added, technological sophistication, and the share of non-routinised, skill-based operations. Finally, there is assembly activity and the manufacture of simple labour-intensive goods for exports or domestic market, which primarily rely on the low cost of labour and may have short-term prospects.

The position of Polish manufacturing affiliates can be entrenched or gradually upgraded. This can be viewed from the perspective of value chains. Success in intra-corporate competition within successive rounds of investment depends on various factors including economic performance, labour skills and sunk costs (Clark 1994; Young and Hood 1998; Phelps and Fuller 2000; Melachroinos and Spence 2001). The evolution of TNCs' strategies indicates that their learning and relationships developed in Poland lead to upgrading of the role of many affiliates. If the augmentation of production competences is accompanied by non-production functions (R&D, design and marketing), the subsidiary may move towards a 'product specialist' role within the corporation, which reflects enhanced intra-firm trust and power, e.g. Philips Lighting at Piła. Withdrawals and plant closures have been rare so far in Poland.

We can identify a few fundamental types of TNC development in Polish space on the basis of two criteria: the firm's behaviour after initial investment is completed (further investment, stabilisation, withdrawal) and the number of locations (Table 7.1). Spatial reorganisation of company activity in Poland includes transfer of production to other foreign or domestic manufacturers and/or closure of some branch plants as a result of the concentration of production in fewer factories. In the case of multi-plant expansion, relations between the factories should also be taken into account. Thus, the transnational corporation in Poland may comprise:

1. Several plants manufacturing a similar product, e.g. cement, soft drinks, animal fodder, industrial gases;
2. Different standardised or specialised products manufactured by various plants with limited linkages among them, e.g. food products, furniture, machinery;
3. Integrated network of factories manufacturing different specialised products with extensive linkages among the plants, e.g. automobiles, electrical equipment.

The vast majority of TNCs with multiple plants in Poland belong to the two former groups, i.e. they do not develop an intra-corporate production network in the host territory.

Table 7.1 Types of TNC development in Poland's economic space

Company development in Poland	**Number of locations**	
	single location	**multilocational**
single investment	cardboard plants	British Oxygen, Air Liquide, Cargill
further expansion	Goodyear, Michelin, Procter & Gamble, Mars, Mondi, BAT, Lucchini	Philips, Alstom, Volkswagen, Nestle, Unilever, SAB, Ikea, Saint Gobain, Lafarge
expansion with spatial reorganisation including relocation and/or withdrawal	United Technologies, Danfoss, GKN, Reemtsma	ABB, Siemens, Fiat, Heineken, Philip Morris, Coca Cola
withdrawal or collapse	Ford, TV assemblers	Daewoo, ICL

From a geographical perspective it is interesting to see how TNCs alter their locational behaviour and why. The purchase of a privatised local producer may represent a historic opportunity. It is a faster and cheaper way of establishing an industrial base in a new market than greenfield investment, and frequently makes it possible to capture a segment of a market with limited initial competition. On the other hand, a greenfield plant allows recruitment of new employees, the forging of new work practices, and avoidance of bargaining with the unions,[13] redundancy costs as well as technical constraints. It has already been pointed out that some companies decide to build a new factory in the wake of their failure in bargaining on acquisition and/or difficulties encountered in existing plants. Greenfield investment constitutes more than one quarter of the total manufacturing FDI in Poland. It is interesting that American and Japanese TNCs are far more inclined to a greenfield mode of entry than their European counterparts.

Changing locational trends observed among the investors indicate their obvious learning of Polish economic space. The location of new factories in large cities and metropolitan areas was far more common till 1995 than in the later period. The region of Warsaw attracted as much as 40 per cent of greenfield FDI in the first half of the 1990s and only 17 per cent later on. The study reveals that the geographical scope of locational search is limited: in half of the new factories studied, all the locations considered were situated within a single region. The narrowest search is typical of plants built in the Warsaw metropolitan area – only a quarter of these firms took locations in other regions into account. The availability of information on the capital region and other major cities, including earlier contacts, had a significant impact on location choices in the early 'pioneer' years. It is American managers who put the strongest emphasis on access to information about a locality and the positive experience of other investors; at the same time US investors show the highest level of spatial concentration in the Warsaw region. This contrasts with a far greater dispersal of new factories built by companies from neighbouring Germany, which may be expected to have broader knowledge of Poland. A general tendency to invest in metropolitan areas, characterised by relatively high wages and strong unions, proves that the opportunity to recruit skilled employees in a large labour market is the most important variable among labour-related location factors. Access to the market and road connections are further salient factors behind the preference for metropolitan areas.

Another process that contributes to the spatial concentration of new plants in some regions and localities is imitation of locational decisions of well-known companies. For example, GM's location at Gliwice in the old industrial region of Upper Silesia was followed by a few dozen investors. An important finding is that TNCs often choose places that have already shown some sort of economic success, which may mean foreign investors, but also endogenous growth and/or visible infrastructural improvement, e.g. Niepołomice near Cracow and Tarnowo

13 It usually takes 1-3 years for trade unions to appear in new factories; some lack a union even later.

Podgórne near Poznań. Thus, TNCs' locational learning includes both imitation of other firms and 'picking winners'. There is evidence that investors present in Poland for a long time have much broader knowledge about economic conditions in various regions and hence may carry out a wider locational search; Polish managerial staff is an important factor here. Growing awareness of the unusually high costs in Warsaw vis-à-vis other parts of the country also influences locational decisions nowadays.

Conclusion: Learning and Localised Capability of the TNC

A firm's geographic embeddedness is not just an outcome of its actions, but also a mediating factor that shapes the trajectory of the firm in a host territory. The analysis of the behaviour of large corporations in Poland since 1990 indicates that the strategies of a vast number of TNCs are more emergent (evolutionary) than deliberate. They evolve gradually based on the experience achieved at earlier stages, rather than following some 'grand plan' devised beforehand. Many companies began production in Poland with modest plans, testing quality standards, productivity, reliability and adaptability of the labour force of the newly acquired or built Polish factories, before they decided to enhance the role of these establishments. The involvement of the majority of TNCs in Poland appears to be part of their long-term strategy now and entails increasing sunk costs. This often means a change from assembly to broader manufacturing, from simple to more advanced products and sometimes R&D functions, from the initial orientation towards the domestic market to exports to the European Union. The learning about the reliability and quality of local suppliers underlies a successive increase in local purchases. They include both indigenous producers and other foreign firms, which may be encouraged to follow their customers or the leaders. There are also manufacturers who carry on their original strategy of progressive expansion in a new market or export-oriented production (e.g. furniture). More temporary-oriented strategies, withdrawals and disinvestments may be an attribute of assembly activities and the production of simple, labour-intensive goods for exports.

It is interesting to note the path dependency in the development of companies. Contingent initial events, e.g. previous decisions on the type of investment, mode of entry or location of a factory, set in motion reactive or reproducing mechanisms that limit the future scope for action. TNCs' pursuit of their interests in Poland causes them to become embedded in a dense network of economic, social and political relationships. While most of these relations can be asymmetrical in terms of power, they influence the redefining of investors' objectives and changes in their practices. Broad networks of local relations are a mode of learning. Learning is an indispensable part of the investment process. It reduces the TNC's uncertainty and creates distinct dynamic capabilities by securing assets of a rather non-tangible nature: knowledge, stability of relations, trust, alliances, and control.

The investor gradually builds up his localised knowledge about the market, competitors, suppliers, labour, union bargaining practices, enterprise restructuring, regulation and policies of state institutions, privatisation practices, infrastructure provision, location opportunities, etc. Some of this knowledge can be transferred and imitated elsewhere within the corporation or out of its boundaries, e.g. organisation of production, technology, market culture. There are investors who believe they can benefit from the experience gained in Poland in future investment in Russia and Ukraine. Still, much of this is non-imitable and non-tradable know-how, especially concerning specific immobile assets such as labour.

The activity of various agents in a company environment is a source of uncertainty. It can be reduced by establishing a network of broad and stable relations with various stakeholders. This allows them to acquire knowledge about changes in the environment faster and helps them to respond more effectively. Non-conflicting industrial relations are considered a requisite of radical restructuring of acquired firms – the price to be paid in the form of wage increases and other provisions is worthwhile. Stability in the networks of suppliers is associated with their reliability. Maintaining contacts with the local government may prove useful in the case of further investment. Subsidiaries seek a greater role within the TNC, which can change their internal and external relations and require extended learning.

Stable and close relations reinforce trust. It is a salient ingredient of the relationships with suppliers, local governments, and trade unions. Besides the trust that TNCs bestow on specific actors and enjoy on their part, there may be a more general trust that TNCs strive for among top politicians and the media. The author's research shows that major investors go to great lengths to build and maintain an image of a 'good investor' in the country. This general trust in a stable investor with long-term commitment is something the company may fall back on in case of unpopular decisions, e.g. closure of an ABB branch plant at Lębork and relocation of production to other factories in Poland.

Alliances are a special form of TNCs' relationships. Bargaining with the national government and EU institutions over quality standards or protectionist measures often depends on industrial lobbies composed of competing firms. Personal relations with top political figures may prove more influential here than contacts at the ministry level, which are undermined by frequent personnel changes. Alliances forged with trade unions may be crucial for privatisation, protection against imports, and occasionally, inclusion into an economic zone. Employers' organisations seem to be less important in Poland.

Last, but not least, foreign investors learn how to exercise control over labour, suppliers, local government and other actors. Networks of relations are constituted and transformed by TNCs, which are able to 'enrol, convince, and enlist others into networks on terms which allow initial actors to "represent" the others' (Murdoch 1995, 748). The power of TNCs in various relations is either legitimised or contested. The legitimisation may involve mobilisation and enrolment of others (labour, local authorities) around 'mutual objectives' (Smith 2003). Control can be accompanied or replaced by trust.

Poland provides little support for general opinions about the de-regulated post-socialist state: a vast new system of regulation has been devised and implemented since 1990. Some impact of the legacy of regulations and practices from the socialist era can be observed. The state can be viewed as a network of institutions with different goals and practices, albeit linked by certain rules and dependences. There is some divergence between formal and actual power. This complexity, lack of coherence and stability weakens the bargaining position of the state as a whole far more than the apparent 'de-regulation', though it does not necessarily extend the power of business enterprises. In fact, the activity of foreign corporations is made more difficult. The question is to what extent the overall economic stability is jeopardised, which could discourage existing and new investors. General TNCs dependence on governmental decisions decreases as privatisation in many industries is nearly complete, protectionist measures wane and there are fewer legal changes in the system. Still, investor's sunk costs are influenced by bargaining with the government: increased by investment commitments in privatisation contracts or reduced by financial incentives. Moreover, one should not forget the sub-national public authorities – communes represent governed spaces where important resources are 'packaged' (land, infrastructure, utilities, labour force), to use Dicken and Malmberg's (2001) expression. There are other actors whose actual power goes far beyond their power in terms of formal authority, e.g. industrial lobbies and some trade unions.

In the discussion of embeddedness of foreign investors it is not often realised that extensive local networks of stable relationships and non-tradable know-how constitute a vital element of sunk costs. Accumulated sunk costs include 'initial set up costs, the costs of learning over time about particular environment, and the costs of building reputation, gaining acceptance among government, employees, and other firms regarding their reliability as producers, employers and suppliers in each market' (Wade 1996, p. 81). It is also emphasised that sunk costs include the skills of the local labour force. In fact, the labour force with its specific mix of attributes (cost, skills, adaptability, and geographical proximity to Western Europe) is a critical asset exploited by transnational corporations in Poland and other Central European countries. Investor's effective learning of how to use this asset and convert it into desired product quality and satisfactory productivity constitutes a TNC's localised capability,[14] which determines the success of investment in the post-communist land. This know-how is both company and location-specific, that is it can not easily be transferred to another firm, even less to another national territory. Proper and stable industrial relations require both learning and expenses and constitute another sort of sunk cost, which may influence future re-investment (expansion) or exit decisions (e.g. relocation to a cheaper East European country).

---

14 Maskell (2001) uses the term 'capabilities' when referring to territorial entities and 'competences' when referring to firms.

If TNCs are vehicles of globalisation that want to move effectively and efficiently in this part of European territory, they have to learn and observe local traffic rules as well as depend on reliable local drivers and other team. In Poland, foreign manufacturing corporations have by and large learnt how to forge a non-conflicting relationship with trade unions at the shop-floor level as well as how to use their political clout for their own interest at the national level. On the whole TNCs have acquired broad knowledge on how to operate here and have become a powerful force lobbying for Poland's swift integration into the European Union.

All things considered, this is not meant to claim that the behaviour of transnational corporations in Central and Eastern Europe is solely or predominantly affected by conditions they find locally, but that the relationships they forge here and lessons they learn matter. The TNCs' investment is immanently a learning process, which takes place through their embeddedness in networks of local relationships. Learning and bargaining involve acquiring and combining various tangible resources (including skilled and inexpensive labour) and intangible, non-tradable assets (knowledge, stable relations, trust, alliances and control) in order to build enhanced localised competence. Dicken and Malmberg (2001) argue that the relative bargaining power of national and sub-national governance units is primarily a function of the size of the country/region and its locational specific assets. The analysis of TNCs' activity in Poland reveals that this is true of *various agents* operating in the national territory, including state institutions, local authorities, trade unions, suppliers, and competitors. Their bargaining power is underlain by three crucial national factors: the size of the Polish market, which is larger than in Hungary, the Czech Republic and other countries of the region, labour qualities and the geographical location close to the EU and Russian markets.

References

Amin, A., Bradley, D., Howells, J., Tomaney, J. and Gentle, C. (1994) 'Regional Incentives and the Quality of Mobile Investment in the Less Favoured Regions of the EC', *Progress in Planning*, vol. 41, pp. 1-112.

Berndt, C. (1998) 'Ruhr Firms Between Dynamic Change and Structural Persistence Globalisation, the 'German Model' and Regional Place-Dependence', *Transactions of the Institute of British Geographers*, vol. 23, pp. 331-352.

Błuszkowski, J. and Garlicki, J. (1996) *Companies with Foreign Participation in Their Local Environment*, PAIZ/Indicator, Warsaw.

Błuszkowski, J. and Garlicki, J. (2000), *Opinia inwestorów zagranicznych o społecznych I ekonomicznych warunkach działalności w Polsce*, PAIZ/Indicator, Warsaw.

Buckley, P.J. and Casson, M.C. (1998) 'Analyzing Foreign Market Entry Strategies: Extending the Internalization Approach', *Journal of International Business Studies*, vol. 29, pp. 539-562.

Cantwell, J. (2000) 'A Survey of Theories of International Production', in C.N. Pitelis and R. Sugden (eds), *The Nature of the Transnational Firm*, Routledge, London, pp. 10-56.

Chang, S.J. (1995) 'International Expansion Strategy of Japanese Firms: Capability Building Through Sequential Entry', *Academy of Management Journal*, vol. 38, pp. 383-407.

Clark, G.L. (1994) 'Strategy and Structure: Corporate Restructuring and the Scope and Characteristics of Sunk Costs', *Environment and Planning A*, vol. 26, pp. 9-32.

Cox, K. (1996), 'Period and Place, Capitalist Development, and the Flexible Specialisation Debate', in D.C. Knudsen (ed.), *Transition to Flexibility*, Kluwer, Boston, pp. 155 -177.

Dicken, P. (2003) *Global Shift: Reshaping the Global Economic Map in the 21st Century*, Sage Publications, London.

Dicken, P., Forsgren, M. and Malmberg, A. (1994) 'The Local Embeddedness of Transnational Corporations', in A. Amin and N. Thrift (eds.), *Globalisation, Institutions, and Regional Development in Europe*, Oxford University Press, Oxford, pp. 23-45.

Dicken, P. and Malmberg, A. (2001) 'Firms in Territories: A Relational Perspective', *Economic Geography*, vol. 77, pp. 345-363.

Domański, B. (1997) *Industrial Control over the Socialist Town: Benevolence or Exploitation?* Praeger, Westport, CT.

Domański, B. (2001a) *Kapitał zagraniczny w przemyśle Polski*, Institute of Geography and Spatial Management, Jagiellonian University, Cracow.

Domański, B. (2001b) 'Poland: Labour and the Relocation of Manufacturing from the EU', in G. Gradev (ed.), *CEE Countries in the EU Companies' Strategies of Industrial Restructuring and Relocation*, European Trade Union Institute, Brussels, pp. 21 - 49.

Domański, B. (2003) 'Industrial Change and Foreign Direct Investment in the Postsocialist Economy: The Case of Poland', *European Urban and Regional Studies*, vol. 10, pp. 99-118.

Domański, B. (2004) 'Local and Regional Embeddedness of Foreign Industrial Investors in Poland', *Prace Geograficzne, Institute of Geography and Spatial Management, Jagiellonian University* (forthcoming).

Dunn, E. (1999) 'Slick Salesmen and Simple People: Negotiated Capitalism in a Privatised Polish Firm', in M. Burawoy and K. Vedery (eds), *Uncertain Transition: Ethnographies of Change in the Postsocialist World*, Rowman and Littlefield, Lanham, pp. 125-150.

Dunning, J.H. (1977) 'Trade Location of Economic Activity and the MNE: A Search for an Eclectic Approach', in B. Ohlin, P.D. Hesselborn and P.M. Wijkman (eds), *The International Allocation of Economic Activity*, Macmillan, London, pp. 395-431.

Dunning, J.H. (2000) 'The Eclectic Paradigm as an Envelope for Economic and Business Theories of MNE Activity', *International Business Review*, vol. 9, pp. 163-190.

Durka, B. (ed.) (1999) *Foreign Investments in Poland*, Foreign Trade Research Institute, Warsaw.

Edwards, R. (2002) 'FDI: Strategic Issues', in B. Bora (ed.), *Foreign Direct Investment: Research Issues*, Routledge, London, pp. 28-45.

Ettlinger, N. (1999) 'Local Trajectories in the Global Economy', *Progress in Human Geography*, vol. 23, pp. 335-357.

Grabher, G. (1994) 'The Disembedded Regional Economy: The Transformation of East German Industrial Complexes into Western Enclaves', in A. Amin and N. Thrift (eds), *Globalisation, Institutions, and Regional Development in Europe*, Oxford University Press, Oxford, pp. 177-195.

Hamill, J. (1993) 'Employment Effects of the Changing Strategies of Multinational Enterprises', in P. Bailey, A. Parisotto and G. Renshaw (eds), *Multinationals and Employment: The Global Economy of the 1990s*, ILO, Geneva.

Hardy, J. (1998) 'Cathedrals in the Desert? Transnationals, Corporate Strategy and Locality in Wrocław', *Regional Studies*, vol. 32, pp. 639-652.

Ietto-Gillies, G. (2002) *Transnational Corporations: Fragmentation amidst Integration*, Routledge, London.

Johanson, J. and Vahlne, J. (1990) 'The Mechanism of Internationalisation', *Internationalisation Marketing Review*, vol. 7, pp.11-24.

Kogut, B. (1993) 'Learning, or the Importance of Being Intert: Country Imprinting and International Competition', in S.G. a. D.E. Westney (eds), *Organisation Theory and the Multinational Corporation*, pp. 136-154.

Kogut, B. and Zander, U. (1993) 'Knowledge of the Firm and the Evolutionary Theory of the Multinational Corporation', *Journal of International Business Studies*, vol. 24, pp. 625-646.

Krugman, P. (1991) *Geography and Trade*, The MIT Press, Cambridge, MA.

Krugman, P. (1998), 'What's New about the New Economic Geography?', *Oxford Review of Economic Policy*, vol. 14, pp. 7-17.

Lee, R. (2002), 'Nice Maps, Shame about the Theory? Thinking Geographically about the Economic', *Progress in Human Geography*, vol. 26, pp. 333-355.

Maskell, P. (2001), 'The Firm in Economic Geography', *Economic Geography*, vol. 77, pp. 329-344.

McGrath-Champ, S. (1999) 'Strategy and Industrial Restructuring', *Progress in Human Geography*, vol. 23, pp. 236-252.

Melachroinos, K.A. and Spence, N. (2001) 'Conceptualizing Sunk Costs in Economic Geography: Cost Recovery and the Fluctuating Value of Fixed Capital', *Progress in Human Geography*, vol. 25, pp. 347-364.

Mintzberg, H. (1987) 'Crafting Strategy', *Harvard Business Review*, vol. 65, pp. 66-77.

Murdoch, J. (1995), 'Actor-Networks and the Evolution of Economic Forms: Combining Description and Explanation in the Theories of Regulation, Flexible Specialization, and Networks', *Environment and Planning A*, vol. 27, pp. 731-757.

Nelson, R. and Winter, S. (1982) *An Evolutionary Theory of Economic Change*, Harvard University Press, Cambridge, Mass.

Nicholas, S. and Maitland, E. (2002) 'International Business Research: Steady States, Dynamics and Globalisation', in B. Bora (ed.), *Foreign Direct Investment: Research Issues*, Routledge, London, pp. 7-27.

Ohmae, K. (1990) *The Borderless World: Power and Strategy in the Interlinked Economy*, Free Press, New York.

Peck, F. (1996) 'Regional Development and the Production of Space: The Role of Infrastructure in the Attraction of New Inward Investment', *Environment and Planning A*, vol. 28, pp. 327-339.

Phelps, N.A. (2000) 'The Locally Embedded Multinational and Institutional Culture', *Area*, 322, pp. 169-178.

Phelps, N.A., Lovering, J. and Morgan, K. (1998) 'Tying the Firm to the Region or Tying the Region to the Firm? Early Observations on the case of LG in South Wales', *European Urban and Regional Studies*, vol. 5, pp. 119-138.

Phelps, N.A. and Fuller, C. (2000) 'Multinationals, Intracorporate Competition, and Regional Development', *Economic Geography*, vol. 76, pp. 224-243.

Safarian, A.E. (1993) *Multinational Enterprise and Public Policy: A Study of the Industrial Countries*, Edward Elgar, Aldershot.

Schoenberger, E. (1999) 'The Firm in the Region and the Region in the Firm', in T. Barnes and M. Gertler (eds), *The New Industrial Geography: Regions, Regulation and Institutions*, Routledge, London, pp. 205-224.

Smith, A. (2003) 'Power Relations, Industrial Clusters and Regional Transformations: Pan-European Integration and Outward Processing in the Slovak Clothing Industry', *Economic Geography*, vol. 79, pp. 17-40.

Teece, D.J. (1980) 'Economies of Scope and the Scope of the Enterprise', *Journal of Economic Behaviour and Organisation*, vol. 1, pp. 223-247.

Teece, D.J., Pisano, G. et al. (1997) *Dynamic Capabilities and Strategic Management*, Strategic Management Journal, vol. 18, pp. 509-533.

Vernon, R. (1996) 'International Investment and International Trade in the Product Cycle', *Quarterly Journal of Economics*, May, pp. 190-207.

Wade, R. (1996) 'Globalisation and Its Limits: Reports of the Death of the National Economy are Greatly Exaggerated', in S. Berger and R. Dore (eds), *National Diversity and Global Capitalism*, Cornell University Press, Ithaca, pp. 60-88.

Williams, A.M. and Baláž, V. (1999) 'Privatisation in Central Europe: Different Legacies, Methods, and Outcomes', *Environment and Planning C: Government and Policy*, vol. 17, pp. 731-751.

Williamson, O.E. (1975) *Markets and Hierarchies: Analysis of Antitrust Implications*, Free Press, New York.

Yeung, H.W.C. (2000) 'Organizing 'the Firm' in Industrial Geography I: Networks, Institutions and Regional Development', *Progress in Human Geography*, vol. 24, pp. 301-315.

Young, S. and Hood, N. (1998) 'Attracting, Managing and Developing Inward Investment in the Single Market', in A. Amin and J. Tomaney (eds), *Behind the Myth of European Union*, Routledge, London, pp. 282-304.

Chapter 8

From Transfer to Hybridisation? The Changing Organisation of Taiwanese PC Investments in China

Jinn-yuh Hsu

Introduction: Transfer versus Hybridisation?

In the globalising economy, cross-border business networks are increasingly important and constitute the channels of capital, technology, people, and information flow. Thus they usually have social and economic impacts on both home and host countries and regions. As the production networks grow transnational, divergent national business systems meet and organisational tension and contradiction arise. A key question is whether, in the process of transplanting, the practices of transnational production change or are changed in response to host country industrial environments and regulatory systems. The practices include those both internal and external to the firm. The former refers to the production systems, employment relations, and inter-divisional relationships, while the latter covers inter-firm and firm-government relationships.

Some scholars (Kenney & Florida 1993, Florida et al. 1998, Womack et al. 1990, Rutherford 2000) predict that common institutional configurations and ways of organising the economy will emerge as globalisation and direct competition between different production systems lead to the survival of the most efficient economies. Under these circumstances, inward FDI serves as a transmission vehicle for the best practices of the investing regions. Thus their organisational configurations and business strategies will remain as intact as possible so that they can insulate themselves from local disturbances, and even subdue the local industrial complex. In contrast, scholars such as Hollingsworth (1997, 1998), Abo (1994, 2000), Boyer (1998) and Zeitlin (2000) have argued that as investing firms get embedded into host environments in complex ways, constraints, amongst other things, are placed on their behaviour. Hybridisation occurred where forms became separated from existing practices and recombined with new forms and new practices.

Both theoretical threads agreed on the existence of divergent national business systems, but disagreed on the transformation of organisational configurations in the process of cross-border investment. While those who believe that the transfer thesis points out the deliberate dis-embeddedness in the new institutional environments of the incoming firms, the proponents of the hybridisation thesis stress that the imperative re-embedding of inward FDI in the host social system is a critical process. On the one hand, although Kenney & Florida (1993) observe that the Japanese transplants are capable of isolating themselves from the American 'bad' environment and imposing better practices on the existing systems, it is still an open question as to how these transplants can change the industrial conventions that have long been influential in host regions. Based on the concept of social embeddedness, Hollingsworth persuasively demonstrates the possibility of firms harnessing the links with their surrounding institutions, thus creating hybrid forms of business systems. However, it is not clear why the foreign firms have to infuse their better practices, in most instances, with the lousy ones in the alien social spaces.

Rather than engaging in the sterile debate on whether hybridisation occurs or not, we would take the transplanting in the globalisation process as one of firm-territory de-coupling and re-coupling (Jessop 2000). In other words, the attention should focus on how the firms respond strategically to the divergent local institutional environment and the corresponding organisational reshuffling. We argue that the interplay of four elements at the regional level constituted the divergent local institutional systems, which led to different profit strategies, business models, and organisational configurations (Storper 1997, Whitley 1992). The four elements include: industrial system, government policy, market operation, and labour regime. First, industrial systems vary in accordance to the standing conventions, which legitimate the exploitation modes of resources and constraints (Saxenian 1994, Storper 1997). Second, government policies and subsidies carry divergent levels of capabilities to shape the context for the market mechanism to operate, while the constituted market further restructures the state autonomies in different localities (Evans 1995). Third, the market is the socially and culturally embedded field where firms and institutions interact to allocate resources and realise profit (Block 1990). Finally, the different regulations and activities of the local labour markets shape the different industrial trajectories in different regions, and consequently create the limit and the potential for firms to exploit and transform it (Peck 2000). Investigating the difference in the interaction patterns in the divergent regions will illustrate the rich varieties of production worlds in the geographical industrialisation process of the Taiwanese capital–China localities connection. We argue that geographical strategies always constitute part of the key competitive advantage for the multinationals (McKendrick et al. 2000).

As a matter of fact, although the transfer thesis and hybridisation argument stand respectively on the rival extremes of the theoretical spectrum of organisational transformation, the real situation would be much more complicated than their predictions. As Boyer (1998) argued, a multiplicity of transformations occur when the transplants attempt to apply the same principles to spaces which are divergent and heterogeneous in terms of natural resources, economic organisation, skill competence, governing regime, and market formation. The process of transplanting is full of learning by trial and error in the construction of supply chains, the development of labour regimes and the engagement of market probing. After investigating the literature on the convergence of production systems in the global economy, Gertler

(2001) concluded that the final consequence will depend on the degree of similarity in the institutional environment between the transplants and their host regions. However, and as a matter of fact, the power relations between them will be more influential in the development of cross-border diffusion of industrial systems. The various levels of bargaining power possessed by the transnational corporations and their host regions would lead to divergent forms of transplanting. Those transplants with superior performance will diffuse widely and, in contrast, those with inferior or equal operation will assimilate themselves with the local standard. The geographical spread of hybridisation will accordingly vary depending on the struggle between the transplanting profit strategies and the hosting societal constraints. As a result, multiple forms of hybridisation and transfer will occur, from the complete diffusion of the same productive and organisational routines and the adaptation of part of the old principles in combination with the host ones, to the innovation of new industrial systems in the host regions, as Boyer (1998) demonstrated.

This research aims to explore the changing organisation of Taiwanese PC (Personal Computer) investors in China in the light of the debate between transfer and hybridisation theses. A number of vital concerns will be attacked: how exactly did the organisational reshuffling process happen as the Taiwanese investors met the Chinese contexts? How do Taiwanese investors exploit the diversity between the regional institutional environments in order to handle the uncertainties and opportunities in their Chinese operations? In what sense does the transplanting industrial system take advantage of geographical strategies to transform itself? How do the divergent local institutional mechanisms work for the transformation of the cross-border production system, and how should we evaluate them appropriately? Finally, as most of the inhabitants of both Taiwan and China share the same culture and language, but simultaneously live under contrasting socio-economic systems, how does the institutional similarity and difference affect the developmental patterns of cross-border investments? Among these issues, the role played by *guanxi* (interpersonal ties) in Chinese business practice will be tackled first in the following section.

The Role of Guanxi: Organisational Buffer versus Cultural Advantage

As for the role of cultural affinity in the governance of cross-border investment, two contrasting perspectives are at stake. Hsing (1998) demonstrated that Taiwanese small and medium-sized enterprises (SMEs) took advantage of the cultural affinity with Chinese local cadres and built up alliances with them by exploiting ethnic Chinese connections. He argued that Taiwanese SMEs, being familiar with the norms and practices of *guanxi* in China's social context, and mostly in combination with bribery or 'gift economy', could soon became insiders, building up friendships and providing an efficient way for the Taiwanese capital to obtain information, raise capital, reduce labour costs, and enforce contracts.

In contrast, Wu (1997) preferred the concept of interest conflict and power bargaining over cultural similarity in explaining the interactions between Chinese local authorities and Taiwanese investors. Rooted in new institutionalism, he argued that cross-border Taiwanese investments should be taken as part of the reorganisation of global production networks and that they had to solve the problems of governance in the divergent institutional environments. The Chinese local cadres, particularly those in the Pearl River Delta (PRD) Region, were painted as the rent-seekers, who exploited the autonomous powers granted by the central government through its reform policies to prefer some investors to others. Under these circumstances, Taiwanese investors could not but establish *guanxi* with the local government and pay extra fees above the nominal tax to avoid trouble from the local authorities.

In fact, the wisdom of new institutionalism cast light on the solution: it argued that the outcome of the interactions between the state, market and social connections, such as *guanxi*, could vary dramatically, from unproductive rent-seeking to advantageous relational assets. For Taiwanese investors, the socio-economic space of China is both a land of cheap labour and a potential market, and a place of turbulent institutions and dull practices. As a consequence, it makes sense for the Taiwanese small-medium-sized investors who lack capital to do nothing but utilise the informal institutions such as *guanxi* manipulation to keep their supposedly superior intra-firm management and inter-firm production system intact in the initial stage (cf. Yeung (1998), and Wong et al. (1998) on Hong Kong transnational corporations in Southeast Asia and China).

This chapter investigates these issues using a case study of the cross-strait investments made by Taiwanese informatics industrial firms in the Shanghai region in the Yangzi River Delta (YRD). The choice of the informatics industries, including mainly PC peripherals and components, is based on the observation that since the early 1990s, it has become one of the most strategically important cross-strait investment sectors. Moreover, most of the research on cross-strait investments, such as Hsing (1998) and Wu (1997), have so far concentrated on the traditional sectors, such as shoes, garments, and toys. This study will update the theoretical understanding and empirical findings. Finally, Taiwanese electronics firms have clustered geographically in two regions, mainly the PRD and the YRD respectively (Figure 8.1). Diverging from the geographical patterns demonstrated by the traditional sectors, which concentrated in the PRD region, Taiwanese PC investors chose towns in Southern China in the first wave of investments in the late 1980s, and after the mid-1990s, the investors gradually moved to Eastern China (see Figure 8.2). While most accounts of cross-strait investments are based on the case studies conducted in the PRD (Hsing 1998, Wu 1997, Cheng 1999), this chapter hopes to provide a preliminary comparison of firm behaviour in a divergent institutional context.

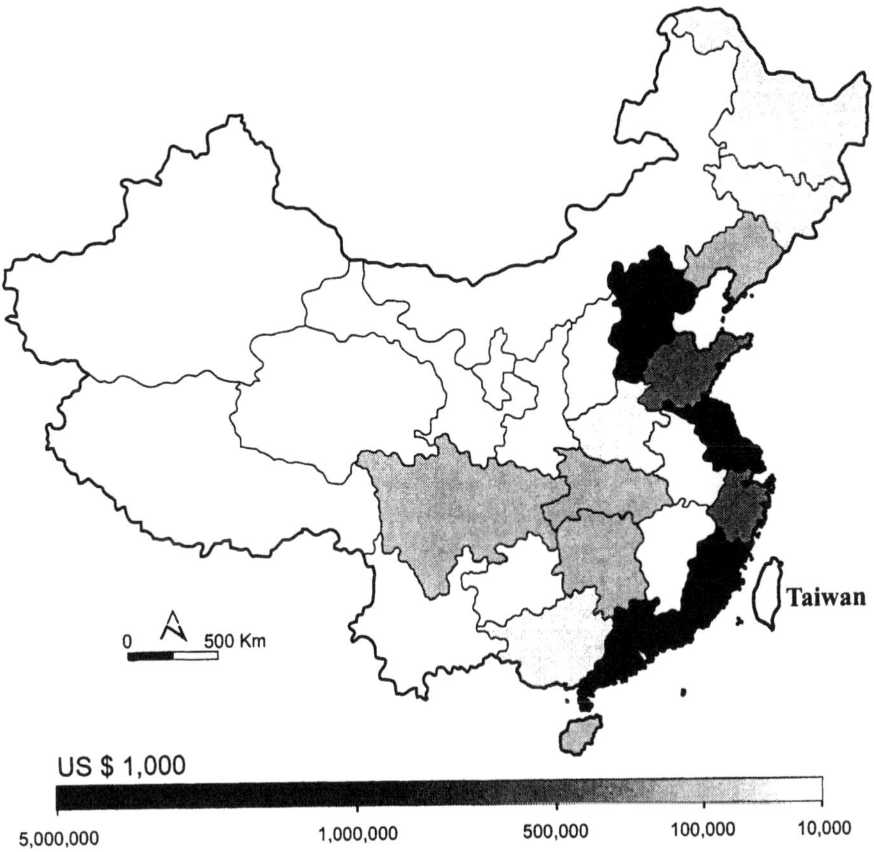

Figure 8.1 The locations of Taiwanese electronic investors in China, in investment value volume

Source: China's Third Industry Survey, 1997.

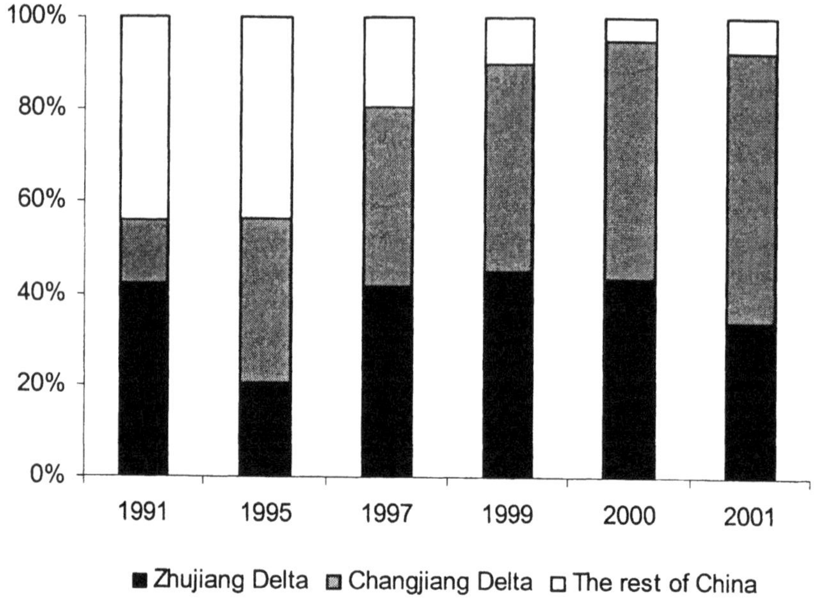

Figure 8.2 The geographical patterns of Taiwanese electronics cross-strait investments in the 1990s

Source: Statistical Reports of Cross-strait Investment, the Investment Commission of Ministry of Economic Affairs, various years.

In the next section, the methodology will be described before the empirical details. Next, we will portray the geographical patterns of the cross-border investments. The remainder of this chapter will tackle the changing organisation of cross-strait investment. The discussion will evolve around the changes in the multiple governance mechanisms at various levels, including labour regimes, material input control, inter-firm subcontracting and the reorganisation of cross-border production chains. Finally, a theoretical dialogue with the hegemonic discourses of institutionalism, predominantly the thesis on social embeddedness or soft institutionalism (MacLeod 2001), will be considered. A deliberation on cultural turns in economic geography and the study of informal institutionalism (Wang 2000) and *guanxi* (Smart 1999, Yeung 1998) in Chinese business will conclude the chapter.

Methodology

This chapter is based on the findings of an ongoing research project that started in January, 2001. In the research process, the researchers spent about two months on four field trips and conducted about 40 interviews with Taiwanese PC sector (including PC component, peripherals and system house) investors, managers and two local government agencies. The selection of the sampled firms was based on the database published by key investment consulting companies such as the Capital Group (2001) and a special issue by the noted Taiwanese technology business journal publisher, Business Next (2000), which was widely believed to keep more accurate data on the 'real' investments made by Taiwanese high-technology firms'.[1] In the database, the items and locations of the major investments in the PC sector were listed (Table 8.1). Accordingly, six key PC system firms, including Acer Peripherals (PC peripherals maker), Foxconn Electronics Group (precision machinery), Inventec (notebook computer maker), Delta Electronics (switch power supply), Quanta (notebook computer maker), and ASUS (motherboard maker) were identified because of their recognition as relatively large in investment amount and ability to attract subcontractors to follow in cross-strait activities. All of the targeted firms were on the list of the top 50 Taiwanese electronics firms (including mainly semiconductor and PC firms).[2] After contacting them, the latter two declined to be interviewed. Among the other four key firms, two (Acer Peripherals and Inventec) chose the YRD as their first location in their Chinese investment, and the other two firms had set up plants in the PRD before they moved their core operations northward.[3] In the course of the interviews, the subcontractors for leading firms were added to the interview list, making the final number of firms 39. Company profiles of the firms interviewed are shown in Table 8.2.

1 Due to the 'don't rush, be patient' policy stipulated by former president Lee Teng-hui, many Taiwanese investors, particularly high-technology firms, choose to register in third countries such as the Cayman Islands as foreign companies prior to engaging in their Chinese investments, rather than report to the government. Thus it is generally alleged that the official cross-strait investment records are less accurate than those kept by the investment consulting companies.

2 Among these 50 firms, 26 belong to the category PC industry, and 15 of them have engaged in FDI in China.

3 Delta Electronics entered Dongguan in the PRD in 1992 to start producing power supply and expanded to five plants there. Since acquiring 400 acres of land in Shanghai in 1999, it has set up its research center there and expects to invest in 12-15 plants to produce high value-added product in the YRD in the next 5 years. Similarly, Foxconn started its Chinese operation in Shenzhen and Kunshan in 1993 and expanded to 23 plants after 8 years. It has established its R&D unit in Shanghai in cooperation with a local Shanghai laser research center.

Table 8.1 Key PC sector cross-strait investments

Region	Location	motherboard	PC systemhouse	Notebook PC	PC component	monitor	Disc player	Other PC peripherals
PRD	Shatoujia	Elite						
	Shenzhen	Elite-Huandian-MSI	Acer-Huashen		Honyun-System-Xieyi-Foxconn-Enlightcop			Zhimau-Hanping
	Guangzhou	Advantech	FIC		Guangbau-Enlightcop			Xuli
	Longhua	AOpen						
	Huiyang	Lunghua						
	Huizhou				Lixin			
	Shijie				Juxiang			
	Zhongshan	AOpen	Acer-Xiangfeng		Zhenfa			
	Shunde		Mitec			Guofeng		
	Dongguan	Elite-Gigabyte MSI-Youtung-Yintai-Chengchi			Guangbau-Delta-Honyun-Xinju-Xieyi-Shuttle-Fuyie Xuande-Enlightcop-Juxiang	Liteon-ADI	Liteon-Yinchun	Xuli-Mustek-Genius-Yuncheng-Lianchang
YRD	Shanghai	ASUS-Advantech	Mitec, Huanda-Tatung	Inventec-CLEVO-Quanta	Delta-System-Jinhau-Foxconn-Juxiang	Tatung		
	Wuxi				Guangbau			Guangqun
	Changzhou				Zhenfa			
	Suzhou	ASUS-Abit			Lixin-Enlightcop-Juxiang	AcerPeripherals,Meichi	Acer Peripherals	Umax-Lianchang
	Wuxian							
	Hangzhou				Foxconn			
	Kushan	Advantech-Yuangang	Mitec, FIC	Compal-Inventec Twinhead	Jinhau-Foxconn-Xuande-Tongxie, Logitech			Zhongshan
	Ningpo	Gigabyte-EPOX						
	Nanjing			Inventec-CLEVO				
	Wujiang	Lunghua	Tatung	Huayu	Delta	ADI-Tatung		Microteck-
The rest of China	Beijing	Advantech	Acer	CLEVO	Xuande			Global View
	Dandong				Xuande			
	Tianjin				Guangbau-Delta			
	Shenyang		Heda					
	Xiamen	Huandian						
	Wuhan			Twinhead				
	Xian			Inventec				
	Chengdu			CLEVO				

Source: adapted from The Capital Group (2001) and The Business Next (2000)

Table 8.2 The company profiles of interviewed PC firms

	Main business	No. of employees	location of operation
Delta	power management, video display, component, networking, electro-mechanic	>10,000 (group)	Dongguan (P)[4] Kunshan (Y), Shanghai (Y)
D1[5]	screw nuts	about 200	Dongguan (P), Wujiang (Y)
D2	plastics	N.A.	Dongguan (P), Kunshan (Y)
D3	plastics	<500	Shunde (P), Wujiang (Y)
Foxconn	Computer components, cable, electro-mechanic parts	>10,000 (group)	Shenzhen (P), Kunshan (Y)
F1	electronic parts	N.A.	Dongguan (P), Kunshan (Y)
F2	electronic parts	N.A.	Shijie (P), Kunshan (Y)
F3	plastics	350	Kunshan (Y)
F4	molding development	150	Kunshan (Y)
Acer Peripherals	PC peripherals, monitor, mobile phone	8,000	Suzhou (Y)
A1	machinery, molding	2,000	Dongguan (P), Shenzhen (P), Shanghai (Y)
A2	molded cable, adapter	N.A.	Shanghai (Y)
A3	precision machinery	800	Shenzhen (P), Shanghai (Y)
A4	plastics	N.A.	Shanghai (Y)
A5	plastics	N.A.	Wujiang (Y)
A6	screw nuts	about 500	Guangzhou (P), Wujiang (Y)
A7	wire harness	N.A.	Wujiang (Y)
A8	plastics	about 800	Dongguan (P), Wujiang (Y)
A9	Silicon anode cap, focus pack	N.A.	Wujiang (Y)
A10	plastics	about 700	Shunde (P), Wujiang (Y)
A11	electronic parts	N.A.	Huizhou (P), Wujiang (Y)
A12	electronic parts	about 500	Dongguan (P), Wujiang (Y)
A13	cables	N.A.	Wujiang (Y)
A14	electronic parts	about 1,000	Dongguan (P), Kunshan (Y)
A15	plastics, electronics	N.A.	Kunshan (Y)
A16	plastics	N.A.	Suzhou (Y)
Inventec	notebook PC, sever system, wireless internet appliance, translator machine	>5,000	Shanghai, Kunshan (Y)
I1	electronic parts	N.A.	Kunshan (Y)
I2	electronic parts	N.A.	Kunshan (Y)
I3	wire harness	about 700	Dongguan (P), Kunshan (Y)
I4	precision machinery	N.A.	Shanghai (Y)
I5	molding	about 200	Shanghai (Y)
I6	plastics	N.A.	Shenzhen (P), Shanghai (Y)
Microtek	scanner	>1,000	Wujiang (Y)
Logitech	mice	>1,000	Suzhou (Y)
WUS	PCB products	about1,500	Kunshan (Y)
FIC	motherboard	>5,000 (group)	Guangzhou (P), Kunshan (Y)
Compal Electronics	notebook PC	>3,000	Kunshan (Y)
ADI	monitor, LCD screen	>1,000	Dongguan (P), Wujiang (Y)

4 P connotes PRD. Y indicates YRD. A number of firms have plants outside of both regions, such as Xian, which are not included in the table.
5 For anonymity purposes, the first character "D" means subcontractor or supplier of Delta group. The rest can be reasoned by analogy.

In addition to in-depth interviews, government publications, business surveys and journal reports provided valuable material for this research. However, in order to avoid turning the research into a collection of business anecdotes, the researcher has been particularly careful about drawing conclusions from these reports. The researcher double-checked with the relevant people or agencies before making the final judgments.

The General Location Patterns: From PRD to YRD

By the mid-1990s, a decade after Taiwanese SMEs first went to China, the majority of the cross-strait investments shifted from traditional sectors, such as garments and footwear, to informatics industries, particularly PC components and peripherals.[6] An acute labour shortage and relatively high costs triggered the emigration of Taiwanese PC industries, starting with the most labour-intensive and price-sensitive keyboards and PC mice to power supply units, and then to motherboards and monitors.[7] In undertaking cross-strait investments, a series of governance and coordination issues posed themselves for Taiwanese PC firms, the first question coming to mind being the location of the new plant. As Figure 8.2 demonstrates, most Taiwanese PC firms chose the PRD and the YRD as their destination. In fact, the locale of Taiwanese investments has shifted from the former to the latter since 1997, when China's membership in the WTO was assured.

The acquisition of overseas Chinese FDI in China has been extensively explained (Hsing 1998, Wang 2000, Lin 1997, Smart 1998). According to the Ministry of Foreign Trade and Economic Cooperation (1999), from 1979-98, the sources of FDI in China concentrated highly in the Great China Circle, including Hong Kong, Taiwan and Singapore. In total they accounted for over 70 per cent of the total FDI in China. At the same time, most of these ethnic investments went to the culturally and geographically proximate open-door regions, such as Guandong and Fujian.

After 1997, following the negotiation of membership in the WTO, the opening of China's market to foreign companies was assured. The YRD, particularly Shanghai City and its neighbouring region, was assessed as a better location as a market centre than its counterpart in South China. From a strategic perspective, it makes sense to make one's presence felt in the area. In addition, the role of the international procurement offices (IPOs) of key PC buyers, such as Dell, IBM, and

---

6 The production value of Taiwan's PC sector reached 3,380 billion US dollars in 1998, and over 29 per cent of the value was created by Taiwanese investors in China (Wall Street Journal, Oct. 21, 1998). In addition, the ratio of made-in-Taiwan PC products declined from 72 per cent in 1995 to 52.7 per cent in 1999, and the portion of made-in-China PC products rose from 14 per cent to 33.2 per cent in the same period (MIC, 2001).

7 According to Chung (1997), the estimated cost savings (including material cost, direct labour and indirect labour) to Taiwanese PC companies ranged from 22 per cent in mouse production, to 8 per cent in monitor making, in comparison with offshore manufacturing in China in 1993. The range was between 16 per cent and 4 per cent, compared with Malaysia.

Compaq, is critical in organising the spatial division of labour for the PC production chains.[8] It is recounted that major Taiwanese PC motherboard and notebook producers rush to cluster in the YRD (Zhuang et al. 2001).

According to a recent survey conducted by Taiwan's major electronics industrial association, Taiwan Electricity and Electronic Manufacturer Association (TEEMA, 2001), the YRD ranked as the top priority, but the PRD region the lowest, even lower than the South Eastern Region (including Fujian Province) and North Eastern Region (including Liaoning Province), on the recommendation list for Taiwanese investment. In terms of investment environment, the survey classified the towns and cities in the YRD, such as Wujiang, Kunshan and Shanghai as category A cities (the most highly recommended), and put most PRD cities, even Shenzhen and Guangzhou, in the category D (the least recommended). In conclusion, the survey attributed the results to the lack of a stable legal system and social environment in these PRD cities. Furthermore the frequent disputes on tax and customs issues in the region were deemed most annoying for Taiwanese investors.

The survey was echoed by Mr. Lee of Goda Membrane Switch Co., a PCB maker who possessed plants in Dongguan and Suzhou. He expressed concern about the capriciousness of the local policies in the PRD region.

> Sometimes the local government people came to us and required us to pay extra fees for something strange like urban beautification, or to sponsor their travel. We did not worry about bribery or gift sending as we might count them as costs, and we could make decisions if we could cover them in the business operation plan. But unexpected fees or requirements were beyond our calculation, and we could not invest under such uncertainty. Not complying with their requirements could mean penalties, even imprisonment. We could not afford it.[9]

In fact, as Hsing (1998) demonstrated, the lax legal system can grant extra incentives for cross-border investment. Thus, how to take advantage of the system and at the same time keep their core operation intact became the primary consideration for Taiwanese investors. An 'export enclave' strategy was adopted.

Transfer and Institutional Dis-embeddedness: The Export Enclaves

Encountering the new institutional environments, Taiwanese investors had to re-establish their own production networks and effective governance systems. To work these out, they assessed the possibility and limitation of local sourcing and

8 In a questionnaire conducted by Taiwan's Ministry of Economic Affairs, 81.9 per cent of the interviewed Informatics firms agreed that the huge potential market in China was the key pull factor for Taiwanese investors. In addition, 73.5 per cent of the respondents thought requests by the International Procurement Offices (IPO) of the key buyers were one of the major forces pushing them to move operations across the Strait (MOEA 1999).

9 Interview with Mr. Lee, L-P, Vice General Manager, Goda Membrane Switch Co. June 26, 2001.

adjustment. More often than not, Taiwanese investors came across the following concerns as they transplanted part of their operations to China.

First of all, the local institutional regulations and enforcements are critical for Taiwanese investors to set up and operate their business. Among the institutional regulations, the tax and custom rules raise most anxieties for the inward investors.[10] According to China's state laws, foreign investors can be exempted from between 15 and 24 per cent of income tax, depending on whether they locate their plants in the special zones or high and new technology zones (Gu & Zhau 1998). Even the incentives to attract foreign capital were well defined in the law (tax free for the first two years and half tax reduction for the following three years, *liang-mian-san-jian-ban*). The final deals between the foreign investors and local governments vary depending on the bargaining powers of both sides (Li & Yeung 1999). Most Taiwanese investors relocate their plants to China for the tax and land subsidies at an initial stage. According to a survey, over 58 per cent of Taiwanese investors are engaging in OEM business and mainly export to the US market, although the share that targets the local market is rising as China gradually opens up its market (MOEA 2001). They usually import raw and intermediate inputs from overseas, mainly Taiwan, then process and assemble the products, and finally export them to the final markets.[11] As these investors enjoy preferential dealings, they are not allowed to sell products locally without special permission, otherwise they will be charged with smuggling, which is severely punished in China, even with the death sentence. More to the point, when the business firms are suspected of smuggling, the owners are arrested first, before a series of frightening investigations take place. As a result, the foreign investors have to carefully manage the material inputs and pay the additional tax if they wish to sell the products locally.[12]

To handle the tax and customs issues, Taiwanese investors usually take some measures to avoid getting into trouble. In the early stage, Taiwanese investors mostly chose to adopt joint ventures with local Village and Township Enterprises (VTEs) and let the latter take charge of dealing with the administrative matters, as they are usually part of the administration. In spite of a local counterpart who

---

10 According to a survey, complaints about customs and tax administrations as the most bothering offices in China rank number one (76.3 per cent) and number two (60.8 per cent) among Taiwanese investors, much higher than number three (15.2 per cent), which is the labor bureau (MOEA 2001). Mostly the investors applied customs based on the contracted prices, but they were paid the market prices, which were typically higher than the contracted ones. This led to disparities between the custom tax accounts and company financial ones, and frequently incurred criminal charge of tax evasion.

11 In the late 1980s, most Taiwanese imported inputs from Taiwan to China for processing. As more materials were gradually sourced locally, the ratio of imports from Taiwan declined to about 30 per cent in 1998. However, most critical and high value intermediates were still being imported from Taiwan. See discussion later.

12 In accordance with the laws on special economic zones, all final products have to be exported in order to qualify for business tax breaks. They are also only allowed to be sold locally with the special permission of the governments at the same time being levied the 17 per cent business tax.

occupies a number of positions such as vice president and manager, these are in most cases powerless at running the joint venture. In other words, it is a kind of 'fake joint venture, real foreign direct investment' (Wu 1997). These VTEs were paid in dividends or other benefits in returns, and thus meant extra costs to Taiwanese investors.

Besides the hassles related to the tax and customs systems, Taiwanese PC firms have to attack the problems of reconstructing the logistic supports and production system. Generally speaking, they had two options to choose between: either outside sourcing or inside making. For Taiwanese SMEs, the conventional wisdom led them to conceive of vertical integration, which might not be the best policy in the volatile new economy. Accordingly, they included a number of key parts in their own operations and subcontracted most of the components to makers whom they trusted. The case of Acer Peripherals exemplifies this point.

Acer Peripherals expanded its overseas operation to Suzhou in 1994.[13] In the beginning, they tried to find qualified local subcontractors for production purposes, but soon they gave up the idea and contacted their original partners in Taiwan. As indicated by Mr. Pan, the former vice president of Acer Peripherals (Suzhou), who was in charge of the early plant construction and operation:

> I was instructed to reconstruct the 'industrial infrastructure' in Suzhou as soon as possible. The infrastructure meant the collaborative manufacturing system. To get the job done, I travelled to Suzhou and the neighbouring cities, even to Ningpo in Zhejiang Province which was noted for good moulders, to find if any qualified industrialists could become parts of our subcontracting system, but I failed to get qualified persons. Their moulding technical level could not match our precision requiremenst. In spite of a number of local component makers that could offer lower prices than our original ones, the key criteria of a successful subcontractor included not only cost down, but also the capability of engaging engineering change to match our demands. I estimated that the competent subcontractors would not be available within the next 3-5 years.[14]

Local sourcing was often taken as the best policy for multinationals to optimise resource allocation and save costs (Dunning 1993). In fact, Taiwanese investors like Acer Peripherals tried to source locally at first, but they had to resort to other options, as the local environment could not make the necessary skills and quality available.

The alternative was to move the whole subcontracting system altogether. This strategy was dubbed 'the hen brought little chickens together'.[15] Acer

13 Before the China subsidiary, Acer Peripherals set up another plant in Malaysia in 1989. The product groups included CRT monitors, LCD monitors, LCD projectors, DVD players, telecommunication system, and cellular phone products.

14 Interview with Mr. Pan, J-S, Deputy Managing Director, Darfon Electronics (Suzhou) Co. May 23, 2001.

15 In fact, this strategy was adopted by other core PC makers, such as Inventec Group, a top notebook computer maker in Taiwan, who brought more than 40 subcontractors to visit sites before settling in Shanghai in 1997. The subcontracting firms regularly meet to discuss issues concerning law interpretation, tax matters, labour control and other management problems. They dubbed the meeting the 'Inventec Association'.

Peripherals asked more than 20 subcontracting partners to visit the potential sites in 1993, and finally 14 decided to follow along. These subcontracting firms consisted of specialists in precision machinery, plastic ejection, cable assembly, and even varieties of screws. All of them clustered in the Wujiang Economic Development Zone, within one-hour driving distance of Suzhou. Most of the subcontractors had more than 5 years' working relationship with the core firm, Acer Peripherals. The long-term stable relationship helped Acer Peripherals convince the collaborators to move together, though there were not any written documents to guarantee the success. As Mr. Hsu of KuaoLeng Electronics Co., a screw maker partner of Acer Peripherals, pointed out,

> Acer Peripherals approached us and asked if we would like to follow them to China in 1993. More than 80 per cent of our capacity was ordered by them, and if they moved, and we did not move together, I was not sure if we could find the replacement orders....They did not have to give us any written promise, and we knew each other well. They trusted our capabilities, and we were not easily replaced.[16]

The previous social relations persuaded the subcontracting firms to make the move, and helped the core firms to copy the supply chains quickly in the alien land. In this sense, the social embeddedness in Taiwan was transferred across the Strait and worked well in sustaining production networks in China without outside disturbance.[17]

On the whole, the subcontracting relationship was not exclusive, but could be extended to other PC makers. Under such an arrangement, over 70 per cent of the components for Acer Peripherals still came from its original Taiwanese subcontractors in 2001, according to Mr. Pan. This verified the copying of the decentralised industrial system in Taiwan. In fact, more component subcontractors were gradually drawn to the core firm's geographical orbits than the latter mandated; otherwise, the former would lose its business. As a result, almost the whole Taiwanese 'chicken nest' moved to the YRD. One exception, which is still left in Taiwan, was certain R&D jobs such as the moulding process and products with high precision requirements. This was because local Chinese technical skills could not satisfy the requirements. The trend of attracting more subcontractors across the Strait led the share of material and intermediate input procurements of Taiwanese investors from donor country to decline from 27.04 per cent in 1993 to 18.2 per cent in 1998 (Kao 2000).

The efforts to transfer the production system also appeared in the aspect of labour regulation. Most Taiwanese PC investors, particularly those in the PRD, hire people from rural areas (*ming-gong*), mostly young women in their twenties or less,

---

16 Interview with Mr. Hsu J, General Manager, KuaoLeng Electronics Co., June 25, 2001.

17 Strategies of lead firms to 'convince' their home suppliers to follow are very common in the automobile industry (Japanese carmakers going to the US or Europe, German carmakers going to the US all took the same strategy). This seems to be a general strategy in sectors where a dominant lead firm has sufficient demand power over its suppliers.

and pay them in the range of RMB 400 to 700 per month (Zhu 2001). Even those located in the YRD, which was noted for its previous industrial bases of State Owned Enterprises (SOEs), purposely avoid recruiting experienced workers, and choose to employ labour from other rural regions, such as Anhui Province, Hunan Province, and even the North of Jiangsu Province (*Subei*). By doing so, Taiwanese investors could 'produce' labour at their will, and circumvent the constraints of the existing industrial practices in the collectivist SOEs. A number of interviewed managers listed the 'bad habits' of the veteran workers, including irresponsibility (*chi-da-guo-fan*), opportunism, laziness, and gaining extra advantage at the publics' expense.

> We would like to hire a novice, rather than an old hand with 'bad habits'. As far as an old hand was concerned, you had firstly to let him/her forget the traits that he/she had lived with for a long time. It was much more difficult than teaching an employee from nothing.[18]

To control and train labour effectively, each Taiwanese PC investor dispatched numerous cadres (*Tai-Gan*) from the mother company to the subsidiaries in China.[19] These expatriates generally played the roles of both overseers and taskmasters in the factories. They were the pillars of the Taiwanese inward firms, and alleviated the frictions accrued in the acclimatisation process. They fundamentally transferred their technical skills and industrial practices to the new plants. Besides labour control in the shop floor, most Taiwanese investors constructed bulky dormitories to house their workers from other towns, and imposed strict rules, such as curfew, to discipline their off-duty lives.

By and large, Taiwanese investors chose to respond to the tensions caused by the process of transition in the alien institutional environments in China in the late 1980s to the mid-1990s strategically. They played *guanxi* games with local governments to form a buffer as insulation from external turbulence. They moved core business to 'sound' institutionalised regions, adopted fake joint ventures to attack the annoying tax and customs issues, organised collective associations to voice their discontent, implanted entire subcontracting chains, and sent out cadres to manage the labour training and regulation issues. Since most Taiwanese PC investors came to China to take advantage of cheap labour for export, compliance with the mandates from the key buyers (IPOs) of the global production chains may be the best way to govern the cross-strait investments, so that the competitive production systems can be kept as intact as possible. In other words, they adopted dis-embedded strategies to prevent local interruption, and in so doing, they could enjoy the benefit of substantial

---

18 Interview with Mr. Lee M, Vice President, WUS Printed Circuit Group, December 5, 2001.
19 The number of expatriates varied in different scales and sectors of firms. Basically, the larger the firms, the more overseers were needed. Also the more sophisticated the sector, the more taskmasters required. According to the report, 72.1 per cent Taiwanese inward companies had less than 5 Taiwanese managers, 19.1 per cent less than 10, and only 0.3 per cent over 20 cadres from Taiwan (MOEA 2001).

tax subsidies and low labour costs, and remained spontaneously competitive in the production systems.

Hybridisation and Institutional Re-embeddedness: The Emergent Local Market

The situation changed as China's market emerged confidently and gradually opened to foreign investors after the mid-1990s. Exploring the huge internal China market became the mandate for Taiwanese investors to survive and to prosper. Accordingly, their management and operation systems evolved around different principles than those previous 'enclave' strategies. At the same time, China's institutional environments had been under reform since the National State Council (*Guo-Wu-Yuan*) acted on a series of corruption scandals in the coastal cities and promised to attack the issue of barriers to trade under the new Premier Zhu in 1998. Tax and customs administrations were reshuffled and there was some evidence of a gradual shift from personal particularism to impersonal universalism in the most developed areas in China, such as the YRD (Guthrie 1998).[20] At the same time, the local governments in the YRD have learnt from the previous bargaining experiences with foreign investors to recognise that improving their efficiency in administrative procedures and the spirit of 'ruling by law' is more critical than hardware infrastructure construction for attracting foreign investments (Chang & Chiu 2000). In the new situation, Taiwanese investors started to look seriously at the growth potential of the largest market in the world.

A number of issues obstructed the road to China's market for Taiwanese PC SMEs. First of all, they had to solve the tax and customs matters as they shifted from export orientation to local targeting. In accordance with the rules, the foreign investors should pay back the tax subsidies, which were exclusively granted to export processors, and a clear distinction should be made between materials for exports and those for the local market; otherwise, they would be considered to be smuggling. In practice, it was more than likely that the investors mixed together the materials, machinery, people and the like, and got into serious troubles. Consequently, some Taiwanese investors prefer to set up new autonomous companies to be responsible for local marketing[21] or at least split off the local marketing division from other departments in the firm organisation.

In order to build up good relations with local key customers, it is vital that the primary mission for the inward investors be to capture the opportunities to

20 Yang (2002) cast doubts on the argument about the decline of *guanxi* in China, and insisted that the practice of *guanxi* might be replaced by a legal system in some economic areas, but it would arise in others. Nevertheless, she agreed that the dynamic evolution of the institutional environment, particularly the establishment of a stable legal system, would increasingly gain influence in the business practices in China.

21 For example, the Sky Hawk Computer Group established a new company to take charge of its business with key local customers, such as the Legend Computer and Haier Group.

transform themselves. The transformation appeared in two aspects: on the one hand, subcontracting to the major electronics groups in China, such as Legend Computer, might help the Taiwanese SMEs to penetrate the local market, which was seemingly open but really carried mountains of barriers. On the other hand, the collaboration could also extend to new differentiated product development, which might meet the local demand. Let us take the collaborative project between Taiwan's First International Computer (FIC) Group and China's Legend Computer Group as an example. To facilitate the collaborative project, FIC sent out a team of several key engineers to Legend's new plant in Shanghai to develop new products such as notebook computers. In return, FIC established a stable subcontracting relation with the largest local computer group in China (*China Times*, Jan. 9, 2002).

Another famous instance was the joint ventures made by Foxconn Group with promising start-ups founded by local engineers to access the emerging product areas in wireless telecommunication, as China's innovation base was believed to be more competitive than its counterpart in Taiwan. These investments were to target new product development and to enhance the innovative activities for the Foxconn Group in China (*China Times*, Jan. 29, 2002). In a sense, they were different from the fake joint ventures observed in the previous stage. The Foxconn Group also expanded its business operations in the YRD to cope with the new opportunities and challenges in China, and created good relations with China's government by setting up communist party office in its plants.

The local market imperative is not just confined to Taiwanese investors' usual practices as subcontracting partners for key PC buyers, but also extended to envision the possibility of metamorphosing as own brand-name (OBN) makers with a high profit margin. Becoming an OBN producer was a totally new and knotty practice for most Taiwanese PC investors, who were long well-noted for their hidden OEM factories in the global production networks (Lin 2000). It required not only production capabilities, which most Taiwanese PC investors knew quite well, but also marketing competence, which consisted of the tacit knowledge of local tastes, specific standards, and exact regulations, to make headway into the long-fortified Chinese market. Besides collaboration with local customers like the FIC-Legend case, a number of Taiwanese investors had to recruit local talents who had better sales experience to run the newly-added marketing division. Acer Peripherals hired a group of local managers, including some who spent several years in State Owned Enterprises, to strengthen its task force in carving out a Chinese market, according to Mr. Lee, the CEO (Zhuang 2001). Inventec Group was also noted for its early establishment of local innovation teams to develop products (e.g. translator appliance) and sold the products locally. As Mr. Huang, vice president of Inventec (Shanghai) indicated,

> We had been OEM partners for key international electronics buyers for a long time. We could never dream of creating our own brand products until we arrived in Shanghai. But selling products in China was not a piece of cake, even though we Chinese shared the same culture. So, we hired qualified local people with product design capabilities and marketing experiences to lead the division. They usually

> needed guidance in the beginning, but they could adjust themselves very quickly.[22]

After localisation efforts, a number of Taiwanese investors started their own brand of products, such as Acer Peripherals switching from a member of Acer group to BenQ brand maker. Quanta Computer, the largest notebook maker in the world, sold its products under the name of Getac in China, and Inventec Group promoted its Okwap cellar phones in the domestic market in 2001. Adding new divisions such as R&D and marketing resulted in the reshuffling of the organisational fields of the inward investors. Taiwanese investors perceived the local institutional environments more as resources than as constraints in the previous export-orientation stage. They had to tap these resources by opening up firm boundaries and they absorbed the nutrition by behaving like local companies. In other words, this involved the process of organisational de-coupling and re-coupling, and engendered governance tensions within the organisation and with outside institutional arrangements. It is still too early to tell the trend of organisation-institution interaction between Taiwanese investors and China's emerging market. In fact, other possible trajectories taken by Taiwanese investors still exist, if the promised opening of China's market is not realised in the future. It might force Taiwanese firms to withdraw from the hybridisation attempt, and return to the OEM model. Boyer (1998) called this 'aborted hybridisation'. The result will depend on the power relations and negotiations between the transplanting firms and local institutional environments, which include the industrial system, government policy, market operation, and labour regime.

Conclusion: Reflections on Institutionalism and Territory-Firm Nexus

It is widely recognised that firms should not just be conceived as legally bounded entities and owners of property assets, but also as institutions with permeable and highly blurred boundaries, or in Dicken & Malmberg's term (2001), 'networks within networks'. In the networking process, the firms opened up their boundaries and tapped into the surrounding networks, and created certain realms of firm-territory interaction. Dicken (2000) even dubbed the nexus as the process of 'placing firms and firming places', and Schoenberger (1999) called it 'firm in region and region in firm'. All of the literature assumed that the hybridisation of the firm and geographical organisations would occur naturally once the intra-firm networks were grafted onto the inter-firm and firm-space ones. However, such an assumption could not hold, as firms should be regarded as economic units with power (Maskell 2001, Prahalad & Hamel 1990, Teece & Pisano 1994), which could mobilise their resources to colonise their threatened environment without transforming themselves. In this regard the most appropriate theory of the firm for economic geography is one that takes the firm seriously by identifying its internal competences and its clear boundaries in the process of institutional (re)embeddedness, rather than just viewing the firm as social networks with blurred limits. By doing so, it is possible to bring the

22 Interview with Mr. Huang P., Vice President, Inventec Group (Shanghai), August 16, 2001.

firm back in as an agent with a different proactive capacity or power structure and then further decode the peculiarity of the firm-place nexus in different times and geographies (Hudson 2001, Yeung 2002).

This paper also validated the flexibility of small firms in relocating their investments. Most of the foreign investments, other than overseas Chinese, took advantage of their scale advantage to utilise cheap labour and explore the market in mainland China (Wang 2000). In contrast, the small Taiwanese investors had to build up interpersonal relations with local cadres in the host PRD region to avoid the turbulence from the environing institutions, and gain a foothold in the risky but profitable land. However, while market opportunities opened for the small Taiwanese PC investors, they adopted new organisational configurations to meet the new situation. At the same time, these small firms were alert to PDR regional lock-in as the onetime institutional flexibility became an institutional barrier, and could not adjust itself to meet the firms' demands. It seemed the change in environment affected these small firms severely, and the litheness of the latter allowed them to escape from the turbulence caused by the former. While Romo et al. (1988) predicted that large firms with monopoly power were better candidates than small ones to migrate elsewhere, as they could easily embed themselves in each regional economy, the Taiwanese small PC investors demonstrated their fluidity by not letting themselves be detained by sunk cost.

From the perspective of regional development, this also demonstrated the limit of a number of endogenous theories that prioritised social capital and institutional thickness over power relationship and political economy (Lovering 1999, Fine 2001). This research demonstrated that different modes of institutional embeddedness could create various effects in divergent times and geographical scales: Taiwanese PC SMEs took advantage of *guanxi* or flexible citizenship (Ong 1999) as relational assets in the beginning of the cross-border investment at the international level and buffered themselves from the disorders in the state regulation in the PRD, and they moved to comply with the formal regulations and made use of the YRD's regional advantage on the domestic level as China's internal market emerged. The key issues for the regions here were not to establish much thicker institutions, but to learn to monitor the firm-region interaction by experimenting with institutional adjustment (Sabel 1996).

In brief, this research resonated with the arguments asserted by Gertler (2001:20) that 'any revised theory of firm practice must take adequate room for individual, collective and corporate agency within the firm'. Based on the understanding that the firm is a competence-creating agent, this research has shown that cross-border PC investors have adopted divergent strategies to disconnect and reconnect with hosting regions with the goal of strengthening the firm's core competence. As a result, the latter changed the regional development of the firm's investments (firming place) and imbued the firm with the regional stamp (placing firm). Transfer and hybridisation, in this analytic vein, could be geographically sensitive strategies for firms, as reflexive agents, to protect and to reinforce their competence in the evolution of the firm-territory nexus.

References

Abo, T. (1994) (ed.) *The Hybrid Factory: The Japanese Production System in the United States*. Oxford: Oxford University Press.

Abo, T. (2000) Spontaneous integration in Japan and East Asia: development, crisis, and beyond. In *The Oxford Handbook of Economic Geography*, eds. G. Clark, M. Feldman, and M. Gertler, 625-648. Oxford: Oxford University Press.

Block, F. (1990) *Postindustrial Possibilities: A Critique of Economic Discourse*. Berkeley: University of California Press.

Boyer, R. (1998) *Hybridisation* and models of production: geography, history, and theory. In *Between Imitation and Innovation: the Transfer and Hybridisation of Productive Models in the International Automobile Industry*, eds. R.Boyer, E. Charron, U. Jurgens and S. Tilliday, pp. 23-56. Oxford: Oxford University Press.

Business Next (2000) *Special Issue: Top 100 Taiwanese High Technology Enterprises*. Taipei: Business Next Journal (in Chinese).

Capital Group (2001) *March on China: the Special Issue on Taiwanese Investments in China*. Taipei: Capital Group (in Chinese).

Chang, J-M. and Chiu, S-L. (2000) *Globalisation and Suzhou's outward orientation economic development: a comparative study of four technology development parks*. Paper presented in *Symposium of Globalisation, Sunan Economic Development and Taiwanese Investment*, October 31, 2000, Taipei (in Chinese).

Cheng, L-L. (1999) The invisible elbow: semiperiphery and the restructuring of international footwear market. *Taiwan: A Radical Quarterly In Social Studies* 35: 1-46 (in Chinese).

Chung, C. (1997) Division of labor across the Taiwan Strait: macro overview and analysis of the electronics industry. In *The China Circle*, ed. B. Naughton, 164-209. Washington, D.C.: Brookings Institution Press.

Dicken, P. (2000) Places and flows: situating international investment. In *The Oxford Handbook of Economic Geography*, eds. G. Clark, M. Feldman, and M. Gertler, pp. 275-291. Oxford: Oxford University Press.

Dicken, P. and Malmberg, A. (2001) Firms in territories: a relational perspective. *Economic Geography* 77(4): pp. 345-363.

Dunning, J. (1993) *Multinational Enterprises and the Global Economy*. Wokingham: Addison-Wesley.

Evans, P. (1995) *Embedded Autonomy: States & Industrial Transformation*. Princeton: Princeton University Press.

Fine, B. (2001) *Social Capital versus Social Theory: Political Economy and Social Science at the Turn of the Millennium*. London: Routledge.

Florida, R., Jenkins, D. and Smith, D. (1998) The Japanese transplants in North America: production organisation, location, and research and development. In *Between Imitation and Innovation: the Transfer and Hybridisation of Productive Models in the International Automobile Industry*, eds. R.Boyer, E. Charron, U. Jurgens and S. Tilliday, pp. 189-215. Oxford: Oxford University Press.

Gertler, M. (2001) Best practice? Geography, learning and institutional limits to strong convergence. *Journal of Economic Geography* 1(1), pp. 5-26.

Gu, C-L. and Zhau, L-X. (1998) *China's High Technology Industries and Parks*. Beijing: Zhongxin Press (in Chinese).

Guthrie, D. (1998) The declining significance of guanxi in China's economic transition. *The China Quarterly* 154: pp. 254-282.

Hollingsworth, R. (1997) Continuities and changes in social systems of production: the cases of Japan, Germany, and the United States. In *Contemporary Capitalism: the Embeddedness of Institutions*, eds. J. Hollingsworth and R. Boyer, 265-310. Cambridge: Cambridge University Press.
Hollingsworth, R. (1998) New perspectives on the spatial dimensions of economic coordination: tensions between globalisation and social systems of production. *Review of International Political Economy* 5(3): pp. 482-507.
Hsing, Y-T. (1998) *Making Capitalism in China: The Taiwan Connection*. New York: Oxford University Press.
Hudson, R. (2001) *Producing Places*. New York: The Guilford Press.
Jessop, B. (2000) The crisis of the national spatio-temporal fix and the tendential ecological dominance of globalising capitalism. *International Journal of Urban and Regional Research* 24, pp. 323-360.
Jessop, B. (2001) Institutional re(turns) and the strategic – relational approach. *Environment and Planning A* 33(7): pp. 1213-1235.
Kao, C. (2000) *The localisation of Taiwanese investment in China and its effects on Taiwan's economy*. CIER Working Paper (in Chinese).
Kenney, M. and Florida, R. (1993) *Beyond Mass Production: the Japanese System and Its Transfer to the U.S.* Oxford: Oxford University Press.
Li, X. and Yeung, Y-M. (1999) Bargaining with Transnational Corporations: The Case of Shanghai. *International Journal of Urban and Regional Research* 23(3), pp. 513-533.
Lin, G. (1997) *Red Capitalism in South China: Growth and Development of the Pearl River Delta*. Vancouver: UBC Press.
Lin, T-L. (2000) *The Social and Economic Origins of Technological Capacity: A Case Study of Taiwan's Personal Computer Industry*. Doctorial Dissertation, Sociology, Temple University.
Lovering, J. (1999) Theory led by policy: the inadequacies of the 'new regionalism' (illustrated from the case of Wales). *International Journal of Urban and Regional Research* 23, pp. 379-395.
MacLeod, G. (2001) Beyond soft institutionalism: accumulation, regulation, and their geographical fixes. *Environment and Planning A* 33(7): pp. 1145-1167.
Maskell, P. (2001) *The firm in economic geography. Economic Geography* 77(4): 329-344.
McKendrick, D., Doner, D., and & Haggard, S. (2000) *From Silicon Valley to Singapore: Location and Competitive Advantage in the Hard Disk Drive Industry*. Stanford: Stanford University Press.
MIC (Marketing Information Center) (2001) *Yearbook of Taiwan's Informatics Industry*. Taipei: MIC (in Chinese).
Ministry of Foreign Trade and Economic Cooperation (1999) *Almanac of China's Foreign Economic Relations and Trade*. Beijing: China Economic Publishing House.
MOEA (Ministry of Economic Affairs) (1999) *Survey on Taiwanese Manufacturing Firms' Outward Investments*. Taipei: MOEA (in Chinese).
MOEA (2001) *The Effect of the Changing China's Investment Environment on Taiwanese Investors Decision*. Taipei: MOEA (in Chinese).
Ong, A. (1999) *Flexible Citizenship: the Cultural Logics of Transnationality*. Durham: Duke University Press.
Peck, J. (2000) Places of work. In *Companion to Economic Geography*, eds. T. Barnes and E. Sheppard, 77-94. Oxford: Blackwell.
Prahalad, C. and Hamel, G. (1990) The core competence of the corporation. *Harvard Business Review* 68: pp. 79-91.

Romo, F., Korman, H., Brantly, P. & Schwartz, M. (1988) The rise and fall of regional political economies. *Research in Politics and Society* 3: pp. 37-64.
Rutherford, T. (2000) Re-embedding, Japanese investment and the restructuring buyer-supplier relations in the Canadian automotive components industry during the 1990s. *Regional Studies* 34(8): pp. 739-751.
Sabel, C. (1996) *Experimental regionalism and the dilemmas of regional economic policy.* Paper presented to the conference on socio-economic systems of Japan, the US, the UK, Germany, and France. Institute of Fiscal and Monetary Policy, Tokyo, Japan.
Saxenian, A. (1994) *Regional Advantage: Culture and Competition in Silicon Valley and Route 128.* Cambridge: Harvard University Press.
Schoenberger, E. (1999) The firm in the region and the region in the firm. In *The New Industrial Geography: Regions, Regulation and Institutions*, eds. T. Barnes & M. Gertler, pp. 205-224. London: Routledge.
Smart, A. (1998) Economic transformation in China: property regimes and social relations. In *Theorising Transition*, eds. J. Pickles and A. Smith, pp. 428-449. London: Routledge.
Smart, A. (1999) Expressions of interest: friendship and guanxi in Chinese societies. In *The Anthropology of Friendship*, eds. S. Bell and S. Coleman, pp. 119-136. Oxford: Berg Press.
Smart, A. (2000) *The emergence of local capitalisms in China: overseas Chinese investment and patterns of development.* In *China's Regions, Polity, and Economy: A Study of Spatial Transformation in the Post-Reform Era*, eds. S-M. Li and W-S. Tang, pp. 65-96. Hong Kong: The Chinese University Press.
Storper, M. (1997) *The Regional World: Territorial Development in a Global Economy.* New York: The Guilford Press.
Teece, D. and Pisano, G. (1994) *The dynamic capabilities of firms: an introduction. Industrial and Corporate Change* 3(3): pp. 537-556.
TEEMA (Taiwan Electricity and Electronic Manufacturer Association) (2001). *A Survey Report on the Evalution of Investment Environments and Risks in China.* Taipei: TEEMA (in Chinese).
Wang, H. (2000) Informal institutions and foreign investment in China. *The Pacific Review* 13(4): pp. 525-556.
Whitley, R. (1992) *Business Systems in East Asia: Firms, Markets, and Societies.* Newbury Park: Sage.
Womack, J., Jones, D. and Roots, D. (1990) *The Machine that Changed the World.* New York: Rawson Associates.
Wong, S-L. and Salaff, J. (1998) Network capital: emigration from Hong Kong. *British Journal of Sociology* 49(3), pp. 358-374.
Wu, J-M. (1997) Strange bed-fellows: dynamics of government-business relations between Chinese local authorities and Taiwanese investors. *Journal of Contemporary China* 6, pp. 315-346.
Yang, M. (2002) The resilience of guanxi and its new developments: a critique of some new guanxi scholarship. *The China Quarterly* 170: pp. 459-476.
Yeung, H. (1998) *Transnational Corporations and Business Networks: Hong Kong Firms in the ASEAN Region.* London: Routledge.
Yeung, H. (2002) *Towards a relational economic geography: old wine in new bottles?* Paper presented at the 98th Annual Meeting of the AAG, Los Angles, March 19-23, 2002.
Zeitlin, J. (2000) Americanization and its limits: reworking US technology and management in post-war Europe and Japan. In *Americanization and Its Limits*, eds. J. Zeitlin and G. Herrigel, 1-52. Oxford: Oxford University Press.

Zhu, Y. (2001) The management patterns of Taiwanese informatics investors in China. *Taiwan Research Quarterly* 74(4): pp. 95-99 (in Chinese).

Zhuang, S-Y., Chen, Z-J., & Chen, B-H. (2001) *Taiwanese High Technology Firms Cluster in YRD*. Taipei: Global View Publisher (in Chinese).

Chapter 9

Contrasting Experiences with Business Networking in a Transition Economy: The Case of Poland

Tadeusz Stryjakiewicz

Introduction

The transformation of the economies of Central and Eastern Europe that started in 1990 has entailed, on the one hand, their ever broader inclusion in global industrial networks, and on the other, the creation of several types of new local networks (which are also affected by the process of globalisation). Geographers and economists tend to view globalisation as a crucial driving force in regional development. However, for the countries in transition, a radical change in their political and socio-economic systems is another, equally important, driving force. Naturally, there is feedback between these two forces.

The aim of this chapter is to depict the heterogeneity of business networking (driven by both global and local forces) using Poland as an example. The task is not easy because the pre-1990 command economy was characterised by a rigidly hierarchical system of links controlled by the state, and the lack of networking structures was one of the factors responsible for the poor flexibility, innovativeness and performance of this type of economy. No wonder, then, that in the present post-Communist economies one can hardly speak of fully matured, well-developed forms of industrial networks such as, e.g., strategic alliances or innovation networks.

Another reason making the analysis difficult is the fact that the very concept of industrial (or business) networks and networking is fuzzy and is applied in a variety of ways (cf. Mønsted 1995: 193). The attempts undertaken so far in spatial-economic studies to make it systematic and operational (e.g., Håkansson 1987; Axelsson, Easton 1992; Conti 1993; Grotz, Braun 1993; Grabher 1993; Conti, Malecki, Oinas 1995; Park 1996; Schamp 2000; Stryjakiewicz 2001) have mainly drawn from the experiences of highly advanced countries. Hence, for the purposes of this chapter, industrial networks are defined in very general terms, after Håkansson and Johanson (1993: 40), as 'sets of connected exchange relations among actors performing industrial activities'. This definition allows the inclusion in the analysis of various 'non-standard' cases typical of economies in transition. Care was taken that the forms covered showed at least four basic features characteristic of networks, viz. interdependence of actors, reciprocity, loose coupling, and power (cf. Grabher 1993: pp. 8-12).

Forms of Business Networks in a Transition Economy

The present stage of transition of most post-Communist economies is what Grabher and Stark (1998) call compartmentalisation. It consists in the separation, but also co-existence (owing, among other things, to selective state intervention), of effective enterprises, sectors and regions that are adjusting well to the market system, and those that still more or less follow the rules of the old system in their performance (cf. Stryjakiewicz 1999, 2002). The above statement is also true of networking in these countries. This issue has been tackled, among others, by Grabher and Stark (1996, 1998) as well as Pickles and Smith (1998). In the opinion of Grabher and Stark (1998: p. 16):

> a fundamental concern of our understanding of social and economic change in transition must be the way in which networks (...) are transformed and institutionalised in new forms.

In turn, Smith and Pickles (1998), who try to theorise post-Communist transformation (including the transformation of networking) as an evolutionary and path-dependent process, emphasise in unison 'the diversity of forms of transition' (p. 2), 'the multiple and differentiated strategies for transformation that are at work' (p. 6), and 'struggles over pathways that emerge out of the intersection of old and new' (p. 15).

The result of this intersection of old and new is a diversity of network types in the transition economy. In terms of their contribution to the transition process, they range from disastrous to beneficial. In general, the reach of business networking as well as the number and interdependence of the actors involved are on the increase. What is growing in particular is the 'network-forming' role of foreign enterprises (cf. the chapter by Domański in this volume). If we take into consideration power relations, domestic firms seem to have too little importance in this matter (with the exception of some of those controlled by the state).

At present two opposing tendencies can be observed in networking. The first is to conserve some mechanisms typical of the former command economy, or to make superficial, cosmetic changes in them by way of adjustment to the new conditions. The aim of this type of network creation (or conservation) is to exclude entire sectors or some fields of economic activity from the rules of the market economy, to preserve the dominance of politics over the economy, and to perpetuate the *nomenklatura* system and the vague, uncontrolled capital flow between state or local budgets, enterprises, and private bank accounts of politicians or political parties. This peculiar type of networking has much in common with the Chinese system called 'a socialist market economy'.

The other tendency in the process of the emergence and development of networks is closely connected with globalisation and the transition to a real market economy. The target is to obtain or make use of a competitive advantage in those fields where this has been impossible so far owing to the closed character of the economy and other limitations of the former command system. Taking into consideration the above contrasting aims of networking and its different driving forces, the following typology of the various network forms in a transition economy can be made:

1. 'Old' networks (surviving from before 1990):
    1.1. modified command-economy networks;
    1.2. traditional local networks.
2. Networks emerging as an effect of the transformation process:
    2.1. Externally driven networking through transnational corporations (TNCs) and EU institutions;
    2.2. Informal 'transition' networks;
    2.3. Grass-roots networks of the 'new economy' (in an initial stage).

These forms, presented in Figure 9.1, can be found primarily in the Polish economy, but it seems that they are also identifiable in the other post-Communist economies of East-Central Europe, though to varying degrees. The diagram reflects the intersection of the old and new types of networks as well as the main driving forces (external-internal) behind their creation and/or operation. Selected case studies illustrating the particular forms of networking in Poland are presented in the next section.

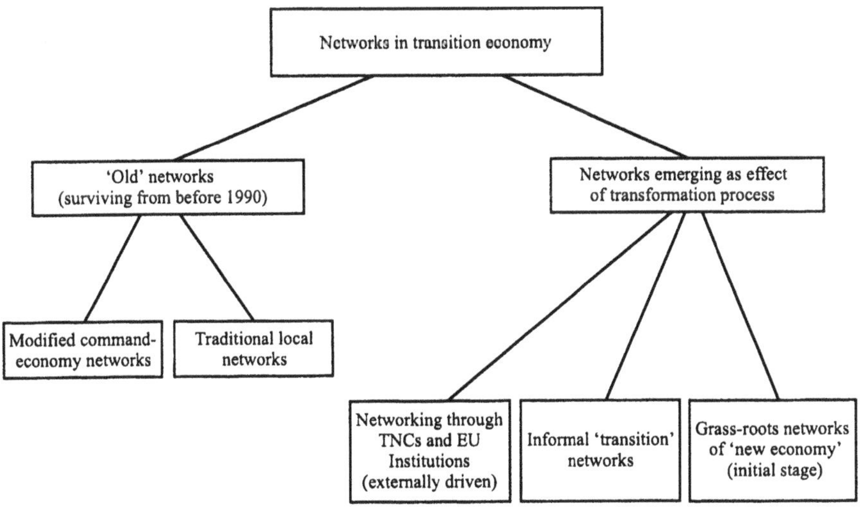

Figure 9.1 Forms of business networking in an economy in transition

Case Studies of 'Networking in Transition', the Example of Poland

'Old' Networks Surviving from Before 1990

Command-Economy Network – The Case of the Polish Copper Mining-Metallurgical Combine The central planning (or command) system generated a 'hub and spokes' type of network between the political centre and outlying actors. As a result, enterprises, formal regional institutions and local authorities were merely 'transmission belts' for decisions taken elsewhere (cf. van Zon 1992; Pickles 1995; Kosonen, Salmi 1999). In this regime some large industrial plants/combines (e.g., mines, steelworks, shipyards, oil refineries) achieved a privileged position and power relations were based on the *nomenklatura* system, i.e., the appointment to key managerial posts in enterprises of people politically affiliated with the Communist Party.

As a result of the transformation of the economy – its marketisation and the emergence of a private sector (including foreign companies) – the position of large, state-owned firms has generally been weakened. On the other hand, in many cases the old power relations based on the predominance of political over economic targets and the *nomenklatura* system have survived in a modified, more complex form. What is more, the patterns of interdependence and reciprocity have become even more extensive in personal, economic and spatial terms. An example of this type of network, which I call a modified command-economy network, is one involving the Polish Copper Mining-Metallurgical Combine with headquarters at Lubin (see Figure 9.2).

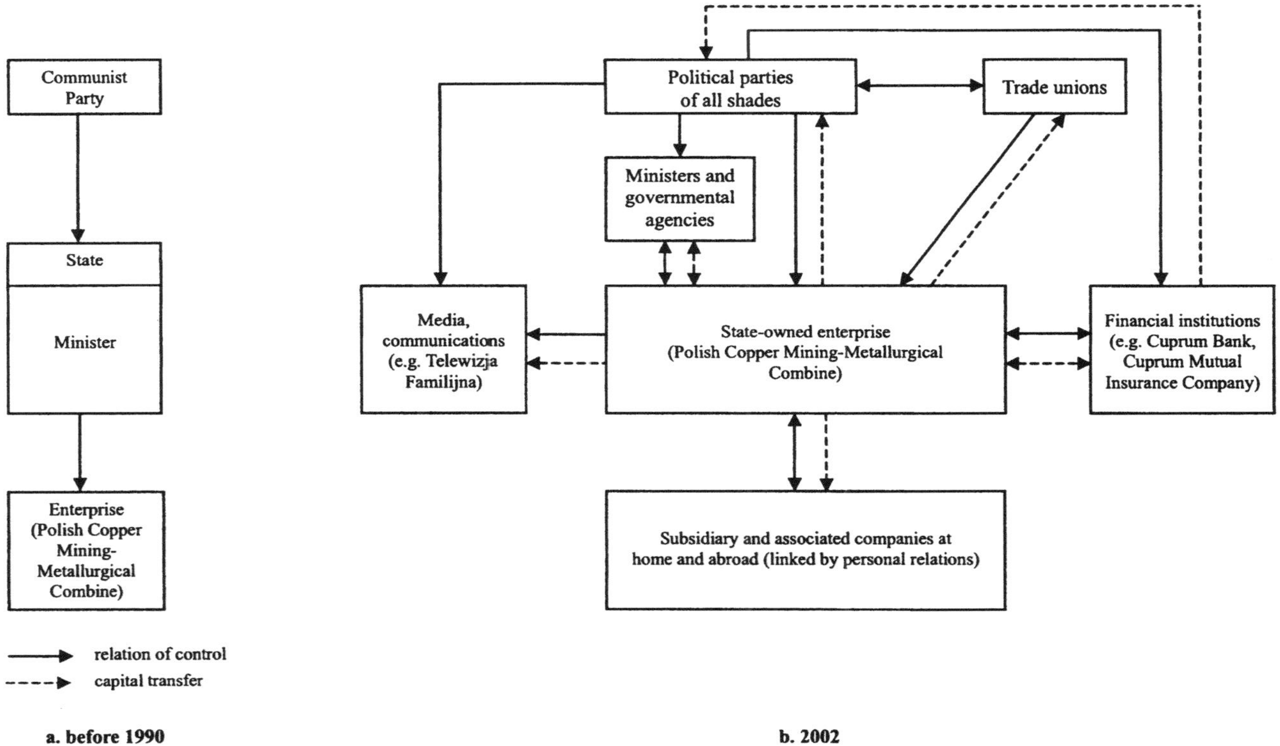

Figure 9.2 Evolution of a command-economy network on the example of the Polish Copper Mining-Metallurgical Combine

The Combine is one of the biggest Polish enterprises, with 18,500 employees and another 10 thousand in 28 subsidiaries, including those located abroad, e.g. in London and Vienna.[1] Although formally the Combine is a joint stock company, the largest block of shares (in various forms) is in the possession of the State Treasury, and it is the Treasury that is its de facto owner. As a consequence, for more than a decade now the firm's management has changed every time there has been a takeover of top political power in Poland. This is the pattern exhibited also by other Treasury-controlled companies, such as the Orlen Polish Oil Concern, Polish Telecommunications, Polish Power Networks, or Polish National Insurance. There are personal links between most of the board members and leading politicians, both at the central and local levels. The reason politicians are so greatly interested in acquiring influence over the Combine and other such companies is the transfer of capital from which the activities of the parties, especially their election campaigns, are financed. What also serves a political goal is the capital involvement of the Combine and the other Treasury-controlled companies in the media, e.g. in the loss-making Family Television. The Combine also sponsors the activity of trade unions representing different political orientations. This system of relations has ensured the Combine a long life without significant structural changes as well as subsidies from the state budget whenever it shows a debit balance (as happened in 2001). A separate problem which smacks of a swindle is transfers of capital to private accounts (e.g., for fictitious consulting services) and to subsidiary and associate companies. One such company, located in Lumumbashi to mine copper and cobalt in the Democratic Republic of Congo (Congo Kinshasa), was set up in 1997 through another company registered in the Virgin Islands. According to Urbanek (2002), this attempt at 'internationalisation' of relations incurred losses for the Combine to the amount of $32 million.

The case of the Polish Copper Combine shows that while the transformation process has caused a transition from a strongly hierarchical to a network-like structure, such a network is not a *tertium datur* between a hierarchy and the market. On the contrary, it petrifies some old relationships and induces the development of a system called 'political clientism'. The system tries to preserve the communist principle of the dominance of politics (and politicians) over the economy and to change one-party *nomenklatura* into a multi-party one (or more precisely, into a *nomenklatura* of the parties which have won elections). The best solution to this problem would be speedy and honest privatisation. However, the persistence of command-economy networks produces a typical lock-in, because their aim is to hinder privatisation and the introduction of full-blown market rules to the economy (including real competition and the appointment of managers based on the criteria of professional qualifications and competence). One must agree at this point with the statement made by Buravoy and Krotov (1993: p. 58):

---

1 The emergence of subsidiaries from a once homogeneous structure of the Combine and the establishment of a capital group were made possible by the systemic transformation; the location of a few companies abroad has been the result of global pressure.

> Commentaries which focus on the political conditions of the transition to a market economy underestimate the capacity of the [Soviet-style] economy to reproduce itself and resist transformation.

Traditional Local Networks – The Case of the 'Furniture Town' of Swarzędz Luckily, the above-presented dark side of networking in the conditions of transition is neither the only nor even the predominant one. Under the command system Poland was unique in the Communist bloc in that it had allowed the survival in some local milieux of highly specialised handicraft networks with very old traditions, resembling the 'Third Italy' model, though naturally at a considerably smaller scale.

The best example of such a traditional local network is Swarzędz near Poznań, widely known throughout Poland as 'the furniture town'. At the beginning of the 1990s some 80 per cent of its residents earned their livelihoods in the crafts and wood industry; besides carpenters, they were upholsterers, glaziers, mechanics designing and servicing woodworking machines, etc. With a population of 25,000, the town features more than a thousand workshops (usually family-based), and besides, about 1,200 people are employed in the two large plants of the once-state-owned Swarzędz Furniture Works.

The carpentry traditions of the town date back to the 17th century and flourished at the close of the 19th. Since then, the local crafts have developed the following characteristics ensuring them adaptability to any conditions:

1. far-reaching specialisation within the trade: the workshops specialise not only in particular kinds of furniture (e.g., room, kitchen, office or representative furniture), but also in the manufacture of, say, a specific kind of sideboard or drink trolley;
2. a high quality of products;
3. flexibility and customisation of production to meet the wishes of each individual buyer;
4. versatility of the craftsmen – their qualifications cover not only all stages in the manufacture of furniture, but also designing and marketing;
5. reduction in transaction costs through an extensive web of informal links embracing supplies, production, and distribution, thus cutting out all middlemen;
6. the development of a local 'corporate culture' manifesting itself, for instance, in the organisation of joint furniture exhibitions and promotional events, and in supporting vocational training.

The vitality of the local crafts, though diminished in the Communist times, did not vanish even then. Swarzędz, together with a similar town of Kalwaria Zebrzydowska near Cracow, were then 'islands of capitalism' and produced furniture for Party committees, among other customers. One might even claim that it was then that 'solidarity for survival' was reinforced, becoming an integral component of the mentioned local milieu.

In the period of transition this local milieu has been faced with a new challenge. The Swarzędz furniture establishments are now undergoing an accelerated process of adaptation to the new economic conditions, especially the ever stiffer competition at home and abroad. The result has been a growing diversification of the once fairly homogeneous local milieu. What is changing are both the organisation of production within the town and its system of external links (they were recently analysed in detail by Stryjakiewicz 1999 and Muent 2000). Some craftsmen have decided to gear their production to the regional and national markets, and to preserve the old design of Swarzędz furniture in the hope that after a period of a depressed market their traditional goods will be in vogue again. Others are trying to use the rapidly changing market situation and introduce post-Fordist methods of labour organisation,[2] change the line (e.g., by undertaking new tasks involving interior designing and furnishing of small pensions, hotels, and offices), and above all, to expand the spatial range of market links. An example of the great adaptability of this group of craftsmen is their immediate response to the emergence of a market niche after the collapse of export by large state-owned firms to the former Soviet Union in the first half of the 1990s. With no capital to set up foreign sales offices of their own and no help from business-environment institutions (such as, e.g., obtaining low-interest credit), they took the advantage of the location of Swarzędz on the A-2, the international route Berlin-Moscow, and drew customers from the East to the local exposition pavilion. Busloads of 'tourists' from Belarus, Russia and Ukraine ordered and collected furniture here (which often necessitated the dismantling of a portion of the bus seats). According to the Carpenters' Guild data, it was due to this informal export that about 60 per cent of Swarzędz furniture appeared on the eastern markets in 1993. The exports declined after Poland's eastern neighbours had introduced new customs tariffs which allowed their nationals to bring only 20 kg of goods tax-free, and after small furniture-making businesses had been set up in Polish frontier voivodeships in the east, where labour is much cheaper, and hence manufacturing costs lower. The decline, however, was largely offset by the exports of a new private Swarzędz-based enterprise, the Wiza Furniture Works, which employs 175 people in three manufacturing plants and exports half of its goods to the eastern markets, this time on regular principles (Boruc 1997: p. 41). The firm specialises in the manufacture of exclusive furniture (lounge sets)

2 Muent (2000) has shown that the owners of thriving businesses set great store by expanding the autonomy of their highly skilled workers in the process of production, while those of stagnating firms tend to give their employees precise and detailed instructions as to how each task should be done, and to keep strict control of their performance. They are also more wary of subcontracting part of their work to outside firms.

meeting the needs of the richest Russians, and its leading development strategies focus on original design and quality. Other, less well-off Swarzędz businesses are increasingly operating as subcontractors (also for German firms). Grabher's loose coupling seems a typical feature of this kind of network.

It is hard to tell the direction the local response to the pressure of globalisation will take: whether it is going to be erosion of the local system of production or its constructive reconfiguration along the lines of the 'Third Italy' model. The latter would involve a change in the present institutional embeddedness (cf. Dei Ottai 1996, Muent 2000). For example, some craftsmen complain that the institutional representation of the Swarzędz crafts, viz. the Carpenters' Guild, is unable to keep pace with the progress of the transformation. In particular, they feel too little is being done in the field of marketing, and the provision of knowledge and information is insufficient. Taking into consideration the financial contribution of craftsmen to the upkeep of the Guild, the reciprocity relation is clearly asymmetrical here.

The local government has not developed any specific policy to support furniture firms, either. It focuses rather on attracting inward investment in order to diversify the town's economic structure. One of the results has been a steady increase in the number of private economic entities. In turn, the number of woodworking and upholstery businesses has been on the decline since the mid-1990s (see Figure 9.3), even though it still greatly exceeds the pre-1990 figure. What is more, there has been an increase in the number of establishments providing specialised business services (including services to furniture-makers). One can, therefore, put forward the thesis that in the economy of Swarzędz an adaptive process of diversification is taking place whose effect has been a drop in the proportion of the crafts and furniture industry in the economic structure of the town. However, it need not mean the atrophy of the traditional function of this carpenters' town, but its enrichment and augmentation with related and supporting industries (e.g., marketing, financial brokerage, etc.). Paradoxically, one of the biggest threats to the survival of traditional local networks like the one at Swarzędz may be the success of the transformation and the lack of skilled labour (especially young) that it may engender. At Swarzędz, where the unemployment rate equals 4 per cent (as against the national average of 18 per cent in 2002), work in carpentry is becoming less and less attractive in comparison with jobs offered by other enterprises, such as Eurotranscenter (ETC) or a nearby Volkswagen plant. Thus, the effect of globalisation manifests itself not only through product competition, but also through competition on the local labour market.

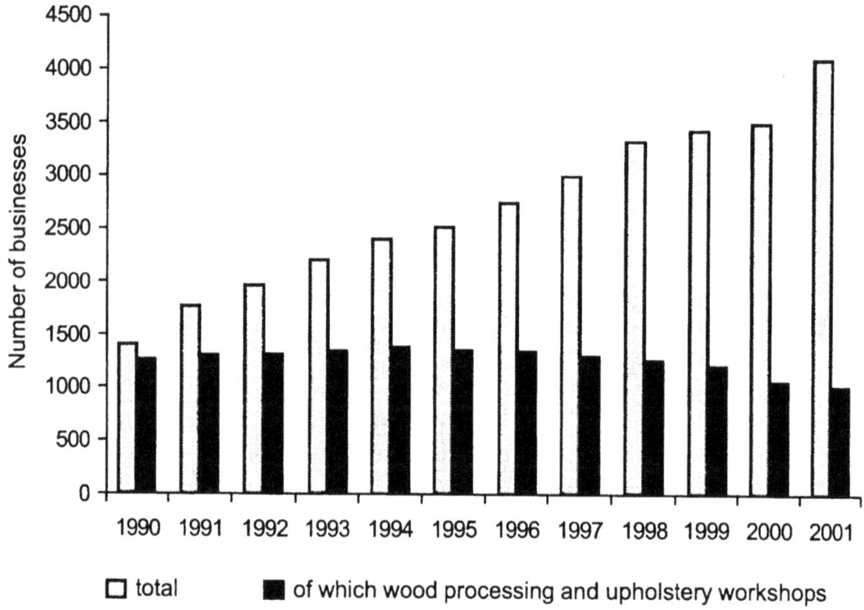

Figure 9.3 Dynamics of the number of wood processing and upholstery workshops against all economic entities in the town and commune of Swarzedz in 1990-2001

Source: Based on the data of the Town and Commune Office at Swarzedz.

Emerging Networks – An Effect of Transition

Networking Through International Firms and Institutions – The Case of the EU PHARE-Struder Programme The main actors representing 'global forces' in transition economies are transnational corporations (TNCs), international financial institutions, and European Union institutions. As far as Poland is concerned, there are several studies analysing the 'networking role' of TNCs (see, e.g., Matykowski, Stryjakiewicz 1993; Bąk, Kulawczuk 1996; Dziemianowicz 1997; Zorska 1998; Hardy 1998; Olesiński 1998; Stryjakiewicz 1999, 2000; Durka 1999; Domański 1999, 2001; this issue is also discussed by Domański in this book). Therefore, in order to illustrate the role of external driving forces in the creation of new network structures, this chapter focuses on the European Union's PHARE-Struder programme. The experiences its implementation has afforded seem especially worth noting on the eve of EU enlargement, because it was the first regional programme in East-Central Europe to be financed by the European Union. One of its principal targets was to mobilise local resources in order to attract new kinds of economic activity and increase employment. The programme, which amounted to 76.7 million ecus in the years 1994-1997, embraced six voivodeships (provinces) suffering a structural crisis of industry or agriculture (see Figure 9.4). It was addressed primarily to small and medium-sized businesses (employing up to 100 people) in the private sector, and was to support their investment activity. Subsidies amounted to 25 per cent of investment outlays estimated in a firm's project, with the firm's own minimum input to be 20 per cent. A condition for apportioning the means was the firm's obligation to carry on business in the region for at least three years. Details of the programme's assumptions and general assessments of its effects are presented in Kozak (1996), *Implementation* ... (1997), and Konopielko and Bell (1998). The following conclusions can be drawn from the above studies:

1. The majority of accepted projects concerned manufacturing activities with highly diversified product ranges and technological levels (from computer software to the manufacture of building materials, beer, furniture, coaches, and even coffins).
2. Unlike some other EU programmes (especially those from the initial stage of the transformation), most of the money spent was earmarked for direct investments (e.g., the expenditure for foreign advisers and experts was less than 10 per cent of the total expenses).

The regional effect of the implementation of the PHARE-Struder programme, apart from creating new jobs, manifested itself in the generation of a new network of links embracing various actors at a number of levels (see Figure 9.4).[3] At the central level, the Polish Agency for Regional Development (PARR) was set up in 1993 to co-ordinate the programme as a whole. Also, its regional counterparts were established. The network embraced banks servicing the Struder programme. Its small infrastructure projects fund made it possible for towns and communes to join the programme (in the town and commune seats information desks were also set up for entrepreneurs interested in it). The network embraced not only money flows, but also the relations of opinion-giving, co-ordination, assessment, and control. EU organs were also involved, e.g. through their experts and advisers. In the light of Grabher's characteristics of networking, the interdependence of actors is of primary importance in this case.

3 For graphic reasons, the figure only includes major elements of the network connected with the principal fund – investment grants for small and medium-sized firms. Equally extensive is the network of links connected with, e.g., the training and advisory fund.

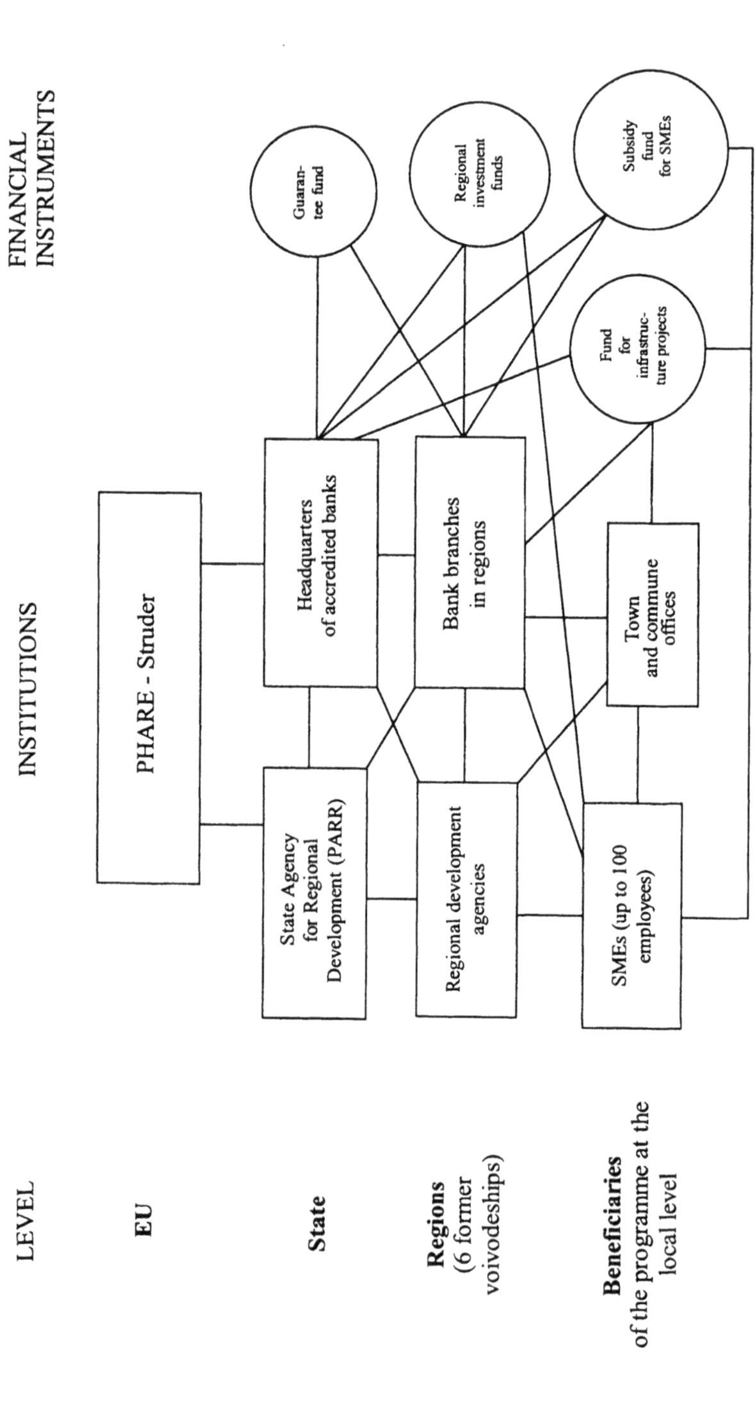

Figure 9.4 Network of links generated by the PHARE-Struder programme

Kozak (1996) emphasises that for all subjects involved in the Struder programme network its implementation has been an accelerated process of learning and mutual adaptation. Konopielko and Bell (1998) stress its particularly beneficial effect on the behaviour of regional banks, and its cultural effect as expressed in the slowly growing awareness that the European Union is taking a direct part in the restructuring of Polish regions most affected by the transformation crisis. What turned out to be the most significant barriers to the implementation of the programme were:

1. the underdevelopment of infrastructure (especially telecommunications), the banking system, and other institutions of the business environment;
2. the discrepancy between some EU regulations and the Polish legal system (e.g., in the field of financial operations of firms in foreign currencies), and
3. overestimating the powers of regional development agencies and regional branches of banks (especially in the initial period).

Thus, the thesis that 'institutional thickness' can be a significant hindrance to the networking of transition economies has been corroborated. Now the above-mentioned barriers are of diminishing importance.

Informal Transition Networks – The Case of the Distribution Network of the Clothing and Knitting Industries of the Łódź District Poland's historical experience has produced a specific feature of local social capital, namely a great importance of informal networks. Their high flexibility and adaptability to new conditions manifested themselves especially in the initial ('shock') stage of the transformation. As the rules of a fully formed market economy are growing in strength and EU standards are being adopted, the relative importance of informal relationships keeps declining; hence the case to be analysed now can be called an informal transition network.

This type of networking seems to be best reflected in changes that have taken place in the clothing industry. In 1990 the industry was especially hard hit by the fall in demand resulting from a drop in the population's incomes, competition from goods imported from Italy and the Far East, and the collapse of imports to the USSR, with no protective measures instituted by the government. In the former Łódź voivodeship, Poland's principal textile and clothing district, the output slumped by as much as 41 per cent in comparison with 1989 (cf. Matykowski, Stryjakiewicz 1992; Marczyńska-Witczak 1993; Markowski, Stawasz 1997). According to Korcelli (1995: p. 38), by the mid-1990s close to 30 thousand workers had been made redundant in 180 enterprises. This dramatic situation, however, triggered rapid adjustments on the part of small and medium-sized businesses in the private sector that had frequently been set up as a result of the break-up of large state enterprises, often employing their highly skilled staff, manufacturing premises, as well as machinery and other facilities; hence, one can speak of a specific, forced spin-off effect. As Chmiel observes (1997: p. 189), this sector has absorbed almost all the unemployed who had lost their jobs in the collapsing textile industry. A condition of such dynamic growth of the clothing industry was its close connection with the newly set up (or rather spontaneously emerging) distribution network displaying little formal institutionalisation – chiefly private wholesale warehouses and bazaars. The network (reconstructed in Figure 9.5) took full advantage of Poland's changed international situation (the opening of the borders, internal convertibility of the zloty, marked price differences between Poland and united Germany, market difficulties of the former USSR republics) and responded flexibly to market signals concerning the quality and range of goods.

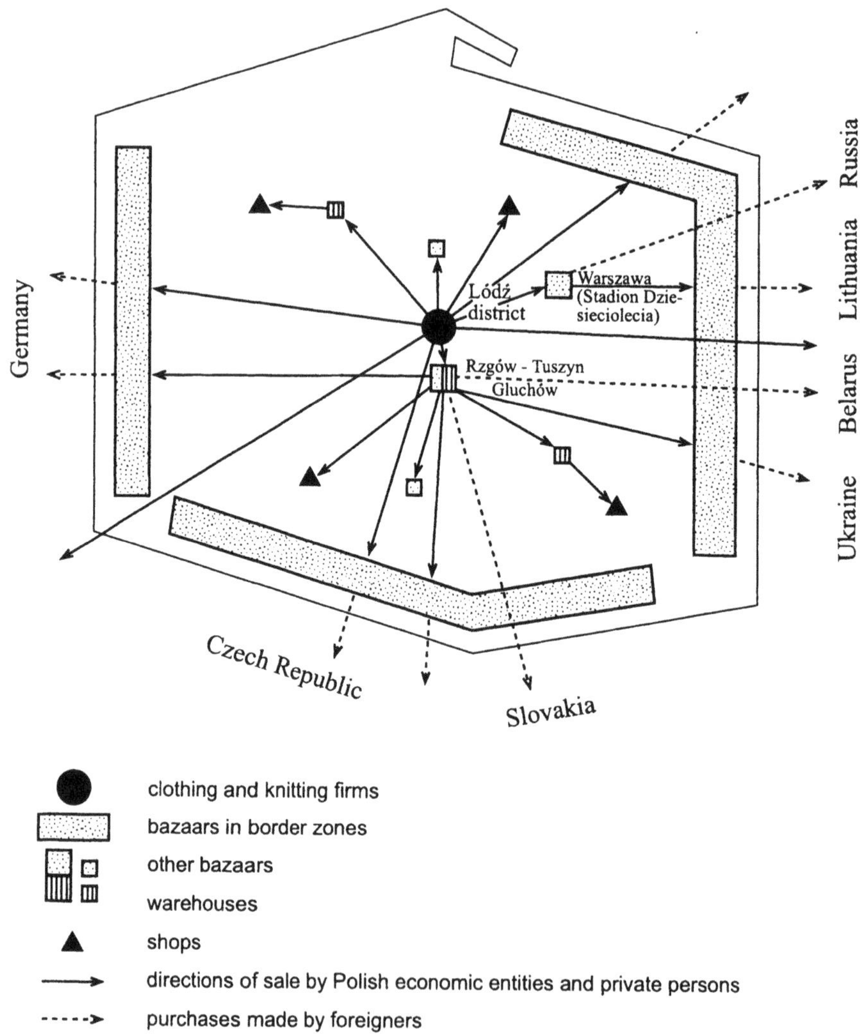

Figure 9.5 Distribution network of small and medium-sized clothing and knitting firms in the Łódź district – an example of a flexible network with a low level of formal-legal regulation

Equally noteworthy are its other characteristics:

1. it has developed independently of the networks of big enterprises, including those of multinational corporations;
2. it largely rests on loose coupling, i.e. short-term informal links which reduce transaction costs;
3. it is extremely sensitive to changes in regulations controlling trans-border traffic (e.g., visas, insurance charges) and currency exchange rates; and
4. its establishment and development can be an example of a 'non-standard' adjustment strategy of the private sector under the conditions of limited capital resources, restricted credit availability, and the European Union's trade protectionism.

This form of networking is clearly underestimated in regional studies, which as a rule treat highly advanced Western countries as a standard. Naturally, from the global perspective it is of little importance, but for some local economies the combined effect of such small-scale informal (or semi-formal) relationships has turned out to be really impressive, and this without any government help and subsidies. The biggest warehouse (or rather bazaar) complex of small clothing producers at Rzgów-Tuszyn-Głuchów near Łódź had 8 thousand stalls in the late 1990s, and through its supply and distribution links it ensured a living to about 100,000 people. A similar role was played by 'bazaar networks' in border regions (described in detail in Stryjakiewicz 1998 and Stryjakiewicz, Kaczmarek 2000). Although many Western experts have been predicting a headlong decline of this type of networking (e.g., after the mass entry of global commercial networks into Poland), they have survived, even though on a diminished scale, and they seem to be an inherent part of the local economy.

It is hard to anticipate the direction in which the evolution of the described network will go. Some clothing establishments are trying to use their earlier bazaar contacts and the experience and information collected in order to develop their own networks of direct buyers in the former USSR, thus recapturing the lost eastern markets. As a result, on the list of export earnings by commodity group (Solska 1997: p. 80), clothes ranked third (at US$1,069 million), overtaking the sales of ships, cars, and copper. This in a situation when large clothing enterprises show a decline in exports caused, among other things, by steadily diminishing orders for clothes made from the orderers' own materials and to their own designs. It seems, therefore, that just as in the initial period of the transformation, small and medium-sized firms are better able to break out of the crisis the large ones are still going through – this time in the field of internationalisation of the networks of direct distribution links. This development, however, will require much greater support on the part of state organs than what they have given smaller-scale business so far (for example in the field of export credit insurance) and on the part of local government. One cannot but agree with the thesis put forward by Raiser (1997) that informal institutions and social capital connected with so-called new entrepreneurship should not be the 'neglected dimension' of the economic transformation.

Grass-Roots Networks of 'The New Economy' – The Case of the Optimus Computer Firm Like the other countries of the Communist bloc during the command-economy period, Poland was an almost complete 'desert' from the point of view of innovation networks, including those based on new information and communications technologies (ICTs). In particular, the relationships between industry, R&D centres, and science were (and still are) very weak. Stryjakiewicz (2002) distinguishes two ways in which transition economies can catch up with the new, knowledge-based, global economy:

1. exogenous, connected with the activities of transnational corporations (TNCs), and
2. endogenous ('grass-roots'), utilising highly skilled local human capital.

So far TNCs have concentrated primarily on the development of their distribution networks, paying little attention to creating or joining local R&D networks. Among the few exceptions are ABB, Delphi, and Motorola. One of the consequences of this policy is the fact that the proportion of new and modernised products in the output of foreign firms located in Poland (6.7 per cent in 1998-1999) is lower than their proportion in the total output of Polish industry (7.3 per cent);[4] for details see Chmielewski et al. (2001).

A study by the Central Statistical Office (GUS) carried out on a sample of 8,273 enterprises in the years 1995-99 revealed that 1,318 (or 15.9 per cent) of them implemented technological innovations in their production. Out of this number 275 (or 20.8 per cent) came from abroad, and the chief channel of diffusion was the purchase of a licence (54.5 per cent), while the proportion of foreign direct investment in this transfer amounted to 16.7 per cent. This process reinforces regional disparities, as the enterprises that make use of foreign scientific and technological solutions are largely located in metropolitan regions (see Figure 9.6).

4 The situation is reversed in exports, where the proportions of goods classified in Poland as new and modernised are 38.9 per cent for foreign firms and 16.1 per cent for all Polish exports.

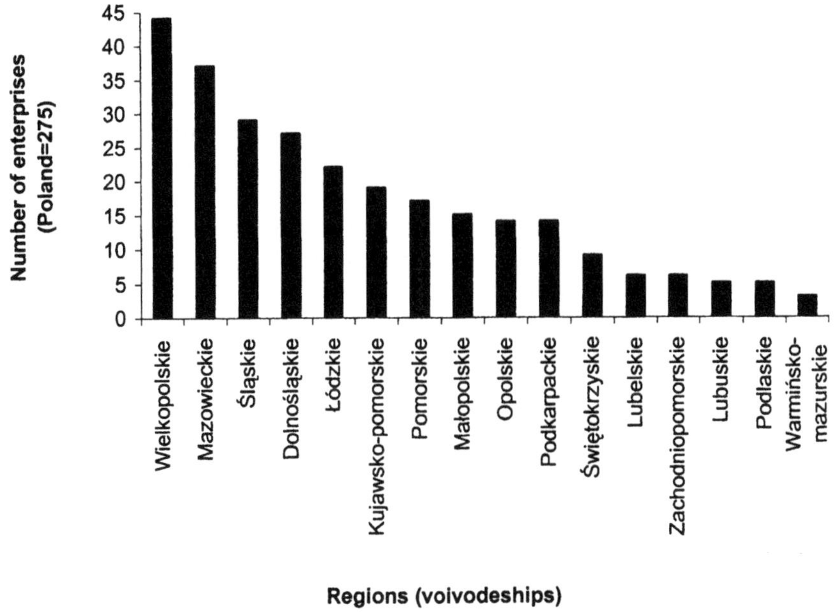

Figure 9.6 Enterprises using foreign technological innovations (1999)

In the context of the unsatisfactory role of TNCs and FDI in creating innovation networks and promoting the new economy, of special importance are local grass-roots ventures in this field, such as the case of the Optimus computer firm (cf. Stryjakiewicz, 2002). Its founder, Roman Kluska, a graduate of the Faculty of Cybernetics and Informatics of the Economic Academy in Cracow, started his business in 1988 by assembling computers in the attic of his parents' house. The economic transformation of Poland allowed him to establish a small computer company 'Optimus' in Nowy Sącz in which he employed young, dynamic people, some of them his old friends. By 1993 Optimus was already the biggest computer firm in Poland. The company was one of the first to be listed on the Warsaw Stock Exchange, thereby securing means for further development. Within a few years its staff was more than 700 strong and the firm started to diversify. Today, thanks to skilful mergers, co-operation with well-known strategic partners (such as Intel, Microsoft and Lockheed Martin), and substantial investment, Optimus is a holding company engaged in the production, assembling and distribution of computers, software, multimedia, and the Internet (regarding the latter, for example, it is the owner of one of the most popular Polish Internet portals, *Onet.pl*).

On 31 December 2000 the Optimus group comprised 38 companies, including 30 subsidiary and 7 associated ones. Its distribution network included more than 1,000 dealers. The capital-linked Optimus companies are located in almost all regions of Poland, and also in Slovakia (Optimus Košice), Lithuania (Ogmios Sistemos), Ukraine, and France (Printmark). The group's sales, in which service packets such as computer system integration and Internet technologies play an ever increasing role, are largely domestic market oriented (88 per cent in 2000). Effective market demand among both individual and institutional customers has grown rapidly (according to the Optimus 2000 issuing prospectus, at present about 57 per cent of firms in Poland have their own home pages, and 82 per cent have access to the Internet). The export destinations of Optimus products and services include Ukraine, Lithuania, Slovakia, the USA, Germany, and the Netherlands. Geographically, the regional context of the location of Optimus in Nowy Sącz is worth noting. Its location effect does not manifest itself, however, in the development of a cluster of local suppliers. It would be hard to find those anyway, because the firm's headquarters is situated in a typically agricultural and tourist region. Hence, sub-assembled components as well as software inputs are supplied by such well-known world corporations as Procomp, Microstar, Hyundai, Samsung Electronics, Fujitsu, Philips, Hewlett-Packard, and Microsoft (Table 9.1). As far as computer assembly is concerned, the only local location advantage is the availability of cheap labour in a region with a high unemployment rate, a factor that may have been a major consideration in the initial location decision.

On the other hand, this decision has produced another effect, namely, the attraction and development of highly qualified human capital. The town has been chosen as the seat of one of the best, most dynamically growing private international higher schools in Poland, with a strong informatics faculty, namely the Higher Business School – National Louis University. Many of its students do their internships at Optimus, and the best can find employment in the firm. Employees can further improve their knowledge through contacts with international corporations, which in turn benefits Optimus. For example, such combined sources of know-how have been influential in the development of the firm's new strategy of reducing the role of the assembly and distribution of computer hardware while focusing on software development, the implementation and integration of computer systems, and Internet services (e-commerce, e-finance). A separate company with an R&D centre does the designing of the firm's own licensed technological solutions, like computer-driven cash registers. It is too early to make a full assessment of the firm's regional impact and prospects, especially since its line of activity and the system of spatial links are still very fluid and keep diversifying. In 2001 the company, so far family-based, acquired a strategic investor in the shape of the Export Development Bank (BRE). The firm seems to be evolving towards a financial holding company specialising in electronic banking. Still, it is worth noting that, rather accidentally, a completely new entity has appeared in a traditional region and altered its economic performance. As a result, the former Nowy Sącz voivodeship has become a leader in terms of the proportion of new and modernised products in its industrial output, and it has been included in several knowledge-based national and global business networks. It is only to be deplored that the number of such cases in the transition economies has been insignificant.

Table 9.1 Main suppliers of hardware and services for OPTIMUS S.A. in 2000

Firm	Headquarters	Goods supplied
Fujitsu	Tokyo (Japan)	hard disks
Hewlett Packard	Palo Alto (USA)	printers
Hyundai	Seoul (South Korea)	monitors, memory modules
IFS Poland Ltd.	Warsaw (Poland)	class ERP/MRP software
Intel	Santa Clara (USA)	motherboards, processors, network cards, server subassemblies, memory modules, network accessories
LG Electronics	Beijing (China)	monitors, memory modules
Microsoft	Redmond (USA)	software
Microstar	City of Industry (USA)	motherboards
OPTIMUS IC	Nowy Sącz (Poland)	cash registers
Philips	Eindhoven (Netherlands)	monitors
Printmark Seiko-sha	Paris (France), Tokyo (Japan)	printers
Procomp	Washington, D.C. (USA)	motherboards
Samsung Electron-ics	Seoul (South Korea)	monitors, hard disks, CD-ROMs
SCO	New Jersey (USA)	Unix operating systems
Western Digital	Lake Forest (USA)	hard disks

Conclusions

The case studies presented above show that the experiences with networking in transition economies vary greatly indeed. What we are dealing with here is not only a mosaic of new forms of spatial organisation and regulation that enterprises are adopting under global pressures, but also with attempts to revert to the old mechanisms of governance (as exemplified in Poland by the re-nationalisation of the Szczecin shipyard or the growing number of concessions, which strengthens the centralisation of decision-making, is conducive to corruption, and perpetuates personal, non-market-based networks in which government or local officials play the dominant role). As Baczyński (2003) observes, the big business networks of the transformation period have often developed in contact with the state thanks to unclear privatisation procedures, licences, public orders, and intermediation. There are few businesses which have been set up as a result of innovation, inventions, risk, and the owner's passion.

What gives the Polish economy a chance to overcome the old is the country's EU membership and the implementation (and I do mean implementation, not a mere official adoption) of the Union's *acquis*; and on the other hand, taking up the challenge of rapid development of the new information and communications technologies (ICTs). Polish enterprises still employ these technologies to an unsatisfactory degree. Hence, despite the current crisis on the ICT market, the role of the 'new economy' networks (including e-commerce and e-banking) is bound to be growing. There should also be a strengthening of relations between industry, science, and the R&D sector, and hence of the importance of innovation, or knowledge, networks. Even now in many voivodeships (provinces) regional innovation strategies are being set up – with the assistance of selected European Union regions – that are primarily oriented towards small and medium-sized businesses.

The chief concerns about Poland's future place in global networks are the deficiencies of the 'core competence' of local enterprises/economic sectors/regions, as well as the level of popular acceptance of and support for the globalisation-oriented path of economic development. So far, for instance, the presence of TNCs has often been perceived mainly or exclusively in terms of job reductions, pushing local goods out of the market, and a menace to the local businesses. Although the societies of transition countries are taking a crash course in the market economy, real networking (including its social embeddedness) takes time.

The ultimate success of the transition depends, among other things, on whether, to what extent, and how fast it will be possible to weaken the state's omnipresence in the economy and the power of networks based on 'political clientism', while strengthening the role of independent, market-oriented businessmen and business organisations as the chief actors of the emerging new-economy networks.

References

Axelsson B., Easton G. (eds.), (1992) *Industrial networks: a new view of reality*. London-New York: Routledge.

Baczyński J. (2003) Kapitalizm (Capitalism). *Polityka* pp. 25, 81.

Bąk M., Kulawczuk P. (ed.), (1996) *Wpływ inwestycji zagranicznych na gospodarkę Polski* (The impact of foreign investment on the Polish economy). Warszawa: Instytut Badań nad Demokracją i Przedsiębiorstwem Prywatnym.

Boruc R. (1997) Agonia stolarza (The death throes of a carpenter). *Wprost* 7, pp. 40-41.

Buravoy M., and Krotov P., (1993). The economic basis of Russia's political crises. *New Left Review* 198, pp. 49-69.

Chmiel J. (1997) *Małe i średnie przedsiębiorstwa a rozwój regionów* (Small and medium-sized enterprises and the development of regions). Warszawa: ZBSE GUS i PAN.

Chmielewski R., Stryjakiewicz T., Twardowska J., Waloszczyk J., (2001) Innowacyjność przemysłu i jej zróżnicowanie w układzie wojewódzkim (Innovativeness of industry and its regional differences). In: Czyż T. (ed.), *Zróżnicowanie społeczno-gospodarcze w nowym układzie terytorialnym Polski*. Warszawa: Biuletyn KPZK PAN 197, pp. 59-101.

Conti S. (1993) The network perspective in industrial geography. Towards a model. *Geografiska Annaler* 75 B, 3, pp. 115-130.

Conti S., Malecki E.J., Oinas P. (eds), (1995) *The industrial enterprise and its environment: spatial perspectives*. Aldershot: Avebury.

Dei Ottai G. (1996) The remarkable resilience of the industrial districts of Tuscany. In: Cossentino F., Pyke F., Sengenberger W. (eds.), *Local and regional response to global pressure: the case of Italy and its industrial districts*. Geneva: ILO.

Domański B. (1999) Structure, regional distribution and selected effects of foreign direct investment in Polish manufacturing in the 1990s. Wien. *Wirtschaftsgeographische Studien* 24/25, pp. 71-88.

Domański B. (2001) *Kapitał zagraniczny w przemyśle Polski* (Foreign capital in Poland's industry). Kraków: Uniwersytet Jagielloński.

Durka B. (ed.), (1999) *Foreign investments in Poland*. Warszawa: Instytut Koniunktur i Cen Handlu Zagranicznego.

Dziemianowicz W., (1997) *Kapitał zagraniczny a rozwój regionalny i lokalny w Polsce* (Foreign capital and regional and local development in Poland). Studia Regionalne i Lokalne 20 (53). Warszawa: EIRRiL UW.

Grabher G. (ed.), (1993) *The embedded firm. On the socio-economics of industrial networks*. London-New York: Routledge.

Grabher G., and Stark D. (eds), (1996) *Restructuring networks in post-socialism: legacies, linkages and localities*. Oxford: Oxford University Press.

Grabher G., and Stark D., (1998) Organising diversity: evolutionary theory, network analysis and post-socialism. In: Pickles J., Smith A. (eds.), *Theorising transition. The political economy of post-communist transformations*. London-New York: Routledge, pp. 54-75.

Grotz R., and Braun B., (1993) Networks, milieux and industrial firm strategies: empirical evidence of an innovative SME environment. *Geografiska Annaler* 75 B, 3, pp. 149-162.

Hardy J. (1998) Cathedrals in the desert? Transnationals, corporate strategy and locality in Wrocław. *Regional Studies* 32/7, pp. 639-652.

Håkansson H. (ed.), (1987) *Industrial technological development. A network approach*. London: Croom Helm.
Håkansson H., and Johanson J., (1993) The network as a governance structure: interfirm cooperation beyond markets and hierarchies. In: Grabher G. (ed.), *The embedded firm. On the socio-economics of industrial networks*. London-New York: Routledge, pp. 35-51.
Implementation of the PHARE–Struder programme. Results and preliminary conclusions. (1997) Warszawa: PARR.
Konopielko Ł., and Bell J., (1998) Reinventing aid for SMEs in Eastern Europe: lessons from the implementation of the Struder programme. *Regional Studies* 32/3, pp. 290-294.
Korcelli P. (1995) *Regional patterns in Poland's transformation: the first five years*. Zeszyty IGiPZ PAN 34, Warszawa.
Kosonen R., and Salmi A. (eds), (1999) *Institutions and post-socialist transition*. Julkaisuja: Helsingin Kauppa - Korkeakoulun.
Kozak M. (1996) Regionally oriented financial instruments in Poland. In: Levitsky J. (ed.), *Small business in transition economics*. Warsaw: Intermediate Technology Publications.
Marczyńska-Witczak E., (1993) Przestrzenne i strukturalne aspekty łódzkiego rynku pracy (Spatial and structural aspects of the Łódź labour market). In: Parysek J.J. (ed.), *Krajowe, regionalne i lokalne rynki pracy w Polsce na początku lat 1990-tych*. Biuletyn KPZK PAN 161. Warszawa: PWN, pp. 63-80.
Markowski T., and Stawasz D., (1997) Łódź i region – przyczyny upadku i rozwoju (Łódź and its region - causes of downfall and growth). In: Gorzelak G. (ed.), *Przemiany polskiej przestrzeni*. Studia Regionalne i Lokalne 19 (52). Warszawa: EUROREG, pp. 182-204.
Matykowski R., and Stryjakiewicz T., (1992) The adaptation of industry and its labour force to the changing economic system in Poland. *Erdkunde* 46, pp. 149-157.
Matykowski R., and Stryjakiewicz T., (1993) Foreign enterprises in an 'economy in transition': the example of Poland. In: Schamp E.W., Linge G.J.R., Rogerson C.M. (eds), *Finance, institutions and industrial change: spatial perspectives*. Berlin-New York: Walter de Gruyter, pp. 162-182.
Mønsted M. (1995) Processes and structures of networks: reflections on methodology. *Entrepreneurship and Regional Development* 7, pp. 193-213.
Muent H. (2000) Localised co-operation between small firms in the socio-economic transition process in Poland. Two cases from the Poznań region. In: Parysek J.J., Stryjakiewicz T. (eds), *Polish economy in transition: spatial perspectives*. Poznań: Bogucki Wydawnictwo Naukowe, pp. 73-88.
Olesiński Z. (ed.), (1998) *Bezpośrednie inwestycje zagraniczne w Polsce* (Foreign direct investment in Poland). Warszawa: PWE.
Park S.O. (1996) Networks and embeddedness in the dynamic types of new industrial districts. *Progress in Human Geography* 20/4, pp. 476-493.
Pickles J. (1995) Restructuring state enterprises: industrial geography and Eastern European transition. *Geographische Zeitschrift* 2, pp. 114-131.
Pickles J., and Smith A. (eds), (1998) *Theorising transition. The political economy of post-communist transformations*. London-New York: Routledge.
Raiser M. (1997) *Informal institutions, social capital and economic transition: reflections on a neglected dimension*. London: EBRD, Working Paper 25.
Schamp E.W. (2000) *Vernetzte Produktion. Industriegeographie aus institutioneller Perspektive*. Darmstadt: Wissenschaftliche Buchgesellschaft.

Smith A., and Pickles J., (1998) Theorising transition and the political economy of transformation. In: Pickles J., and Smith A. (eds), *Theorising transition. The political economy of post-communist transformations*. London-New York: Routledge, pp. 1-22.

Solska J. (1997) Ranking eksporterów (Ranking of exporters). *Polityka* 25, pp. 78-80.

Stryjakiewicz T. (1998). The changing role of border zones in the transforming economies of East-Central Europe: the case of Poland. *GeoJournal* 44/3, pp. 203-213.

Stryjakiewicz T. (1999) *Adaptacja przestrzenna przemysłu w Polsce w warunkach transformacji* (The spatial adaptation of industry in Poland in the conditions of transformation). Poznań: Wydawnictwo Naukowe UAM.

Stryjakiewicz T. (2000) Implications of globalisation for regions and localities in an economy in transition: the case of Poland. In: Parysek J.J., Stryjakiewicz T. (eds), *Polish economyin transition: spatial perspectives*. Poznań: Bogucki Wydawnictwo Naukowe, pp. 7-28.

Stryjakiewicz T. (2001) Koncepcja usieciowienia w badaniach przestrzenno–ekonomicznych (The concept of networking in spatial-economic research). In: Rogacki H. (ed.), *Koncepcje teoretyczne i metody badań geografii społeczno-ekonomicznej i gospodarki przestrzennej*. Poznań: Bogucki Wydawnictwo Naukowe, pp. 37-47.

Stryjakiewicz T. (2002) Paths of industrial transformation in Poland and the role of knowledge -based industries. In: Hayter R., Le Heron R. (eds), *Knowledge, industry and environment. Institutions and innovation in territorial perspective*. Aldershot: Ashgate, pp. 289-311.

Stryjakiewicz T., and Kaczmarek T., (2000) Transborder co-operation and development in the conditions of great socio-economic disparities: the case of the Polish-German border region. In: Parysek J.J., and Stryjakiewicz T. (eds), *Polish economy in transition: spatial perspectives*. Poznań: Bogucki Wydawnictwo Naukowe, pp. 49-71.

Urbanek M. (2002) Jak wyssać pieniądze z państwowej spółki - przykład KGHM (How to pump money out of a state-controlled company - the example of the Polish Copper Mining -Metallurgical Combine). *Wprost* 10, pp. 24-27.

van Zon H. (1992) *Towards regional innovation system in central Europe*. In: FAST Dossier Continental Europe: science, technology and community cohesion. Brussels: Commission of the European Communities.

Zorska A. (1998) *Ku globalizacji? Przemiany w korporacjach transnarodowych i w gospodarce światowej* (Towards globalisation? Changes in transnational corporations and in the world economy). Warszawa: PWN.

PART IV
THE REORGANISATION OF 'LOCAL' PRODUCTION SYSTEMS IN A GLOBALISED WORLD ECONOMY

Chapter 10

Industrial Clusters in China: Embedded or Disembedded?

Jici Wang and Xin Tong

Introduction: The Importance of Territorially Embedded Industrial Clusters

Industrial clustering is by now a well known – if still inadequately understood – term for conceptualising the spatial agglomeration process of enterprises and related supplier and service industries. Over the last two decades the growing development thinking, based largely on experiences in Western countries, has re-evaluated the significance of clusters. 'Clustering matters in developing countries...However, public interventions are needed in promoting linkages to make the embryonic clusters grow' (Humphrey and Schmitz, 1995, p. i).

Although, as some recent critics have remarked, the term 'cluster' has become one of the 'fuzzy concepts' (Markusen, 1998) or 'policy panacea' (Martin, 2001, 2003) in current national, regional and local policy debate, we argue that the essential part of the so-called 'cluster theory' is the basic idea that links between firms, institutions and other economic agents that are located in geographical proximity to each other tend to generate advantages of scale and scope, e.g. development of general labour markets and specialised skills and enhanced linkages between suppliers and customers (Lloyd and Dicken, 1990). This leads to a further idea, namely, that geographical proximity and regional agglomeration may greatly facilitate the 'learning economy'. The new views related to this idea are those on industrialisation as a territorial process and on innovation as a social process (Asheim and Cooke, 1999). The thinking on agglomeration and the more fashionable term of clustering in industrial geography and related fields has revolved around the following concepts: industrial complex, industrial district, increasing returns to scale, competitive advantage and innovation system.

Many industrial complexes throughout the world, especially those related to heavy industry, were planned for the mixed purpose of regional balance and nationalism after World War II. However, such industrial complexes with huge vertically integrated companies in many countries ran into problems during the global economic crises in the 1970s. The successful story of the Third Italy in the late 1970s and early 1980s attracted the interest of mainstream regional researchers. Industrial district, a concept invented by Alfred Marshall, flourished again in academics to take advantage of the sustainable competitiveness which small and medium enterprises (SMEs) derive from agglomeration (Becattini, 1978).

Economic models were introduced to explain the role of increasing returns to scale in the spatial clustering process (Krugman, 1991). In business management, Porter created the well-known diamond model as the basis for a competitive cluster of industrial activity. In his view spatial proximity and the anchoring of firms in their 'home base' are important conditions for firm environments: 'competitive advantage is created and sustained through a highly localised process' (Porter, 1990 p. 19). Since Porter's influential work, the term cluster has attracted great scholarly and official interest from a variety of perspectives and found a number of different uses. In the system of innovation approach the cluster has been rediscovered as a characterisation of innovation as an interactive process and a reconceptualisation of the firm as a learning organisation embedded within a broader institutional context (Nelson, 1993; Braczyk, Cooke and Heidenreich, 1998; Mytelka and Farinelli, 2000).

The growing interest in clusters can be traced back to a number of changes in the competitive environment of firms, e.g. the growth of knowledge intensity in production and the emergence of innovation-based competition and its globalisation. These changes have placed a greater burden on SMEs to engage in a continuous process of innovation (Mytelka and Farinelli, 2000).

Globalisation is best understood as set of processes, rather than the outcome (Hsu and Saxenian, 2001). Regardless of the lack of a commonly accepted definition of globalisation, no one can doubt that international economic interaction has increased in recent decades. Investment and innovations in transport and communication systems and the success of governmental efforts to remove of the most former barriers to trade have jointly made national frontiers more porous (Maskell, 1999). As the process of globalisation gradually converts most properties and many created localised capabilities into ubiquities, an increasing number of firms have to meet the challenges by no longer chiefly aspiring to obtain competitiveness through cost reduction, but by evolving inter-organisational routines for knowledge creation (Maskell, 1999). This poses a formidable challenge to the firms in developing countries in their efforts to develop new technological capabilities, to innovate and to ensure sustained industrial growth. This underlines the strategic importance of clusters for the promotion of innovation and economic growth, which are based on a use of both formal, codified and informal, tacit knowledge. It also emphasises the importance of territorially embedded regional innovation systems. The rationale for embedded systems of production and innovation is to provide bottom-up, network-based support so that the firms and institutions are interactive. Primarily the essential nature of local linkages has been stressed. However, global linkages, which bring both opportunities and challenges, has been overlooked to some extent.

A dilemma for developing countries is that the global linkages generating local agglomerations do not always fit the local social economic context. At present, MNCs provide vital linkages to the global economy. Experience shows that global firms hold the power in bargaining with host countries and areas (Malecki,

Oinas and Park, 1999). Global linkages provide access to knowledge not available in the locality, and they may facilitate incremental innovations in technologies already in use. However, with increasing globalisation it is rather doubtful whether incremental innovations will be sufficient to secure the necessary competitiveness of clusters. What is needed to stay competitive in a global economy is to develop core competencies through substantial efforts and embedded localisation. It is not yet clear which kinds of innovation processes or activities are dependent on proximity and which can be carried out over a long distance.

The purpose of this chapter is two-fold: to contribute to the understanding of local clustering in China and to discuss the factors governing the process of local clustering in China. The intention is to answer the question: how are clusters developing in China, and does the case of China's clusters provide some new insights into the development of developing countries in general? For the various cases described in this chapter, the concept of 'cluster' adopted hereafter refers to a local concentration of a manufacturing sector as the common landscape character, a local concentration of interrelated industries through supplier linkages as the underpinning local production system, and a local concentration 'embedded' in social capital as the socio-economic context. These meanings are interrelated in the different cases. The chapter deals with the trajectories of the local clustering process in a transitional economy based on our extensive field research on several newly emerging clusters in China. It will be of interest to find out how different regions responded differently between the increasing enthusiasm for bottom-up development and the tentative reforms in China since they started in the late 1970s.

Industrial Clusters in China – Several Cases

In China the notions of competitiveness and of industrial clusters have already made their appearance. As we can see, in some Chinese regions clusters are forming the institutional and social links necessary for a competitive response to globalisation. The territorially embedded regional economy is expected to develop in these clusters where the legacy of the top-down bureaucratic system is weak and the local entrepreneurship is strong.

The chapter intends to find the implications of the insights on clusters for those latecomers in the global economy who have made efforts to modernise their local traditional society by importing and adapting technology through extending global linkages with the industrialised centres of the world. Yet adapting those theories developed by Western scholars is problematic. There are three reasons to believe that the industrial clustering in China has its uniqueness.

The first is that both in old industrial districts dominated by state-owned firms and in the new development zones controlled by the state government, the trend towards overly heavy state intervention has always been a barrier to the growth of clusters. Bureaucratic networks sometimes have only one-way, top-down links. Since the government still plays a central role in the economy, the government interventions go beyond the state-owned sector and affect the newly developing private sector in high tech or rural industrial districts in China, as shown in the following cases.

Secondly, investment from the outside world strongly affects the structure of the local production network. The global linkages between local Chinese firms and firms in advanced countries are stronger than the local linkages among local Chinese firms in the clusters.

Thirdly, most of the clusters in China have emerged since the economic reforms and have only a short history of less then 20 years. China has a long way to go to build mature market mechanisms to facilitate competitive clusters with rich social capital in the interaction between firms.

This chapter will develop a typology that differentiates China's clusters in terms of their potential for dynamic change. Two types of clusters in contemporary China can be distinguished on the basis of their territorial embeddedness: embedded clusters and disembedded clusters.

The origin of the former type of clusters can be traced back to 1978, when economic reform began in China and private economic activities were officially sanctioned in both rural and urban industry. The sectors that the embedded clusters are involved in seem to have two different trajectories – high-tech on the one hand and traditional on the other. That is to say, the clusters based mainly on non-state-owned technology firms and township firms can be presumed to have the same characteristics: territorially embedded and entrepreneurship oriented. in reality entrepreneurs are not lonely Shumpeter individuals, but are embedded in the social context (Granovetter, 1985). For the cases in this chapter, we distinguish the entrepreneur's locally embedded linkages, including local production linkages and local social ties, from those top-down hierarchical relations typical in the formal central planning system or from pure trade relations between local firms with their overseas customers. Territorially embedded non-trade interdependence is essential in developing regions (Storper, 1998).

The disembedded clusters in China can be divided into two groups as well. One is the clusters caused by inward investment from overseas after 1978. The other is the complexes established by investment from the central government three or four decades ago, where central planning forces are still very much in effect. They face a painful institutional transition.

In the following sections of this chapter, four cases will illustrate the different types of industrial cluster in China. A general description and analysis of the global-local linkages and embedded-disembedded characteristics of each case will be included. The analytical framework has four points. Firstly, it shows the evolution of internal linkages, including vertical relations such as relations with subcontractors and specialised suppliers and horizontal relations such as co-operation and competition with fellow firms. Secondly, it details the external linkages, such as relations with foreign investors or buyers. Thirdly, it reviews at length the local social-cultural background, which is important for illustrating the embeddedness and disembeddedness of economic activities. Lastly, it explicitly examines the effect of government that plays a key role in setting up collective service centres and promoting innovation through collaboration among clusters.

Clusters of Endogenous Firms of Traditional Industries: A Case in Zhejiang Province

Since 1978, when China's economic reforms began, Chinese rural entrepreneurs in several provinces such as Zhejiang and Jiangsu have succeeded in penetrating into light industry. While the top-down public strategy in China is to stake the future on a few increasingly large firms and on planned technology parks operated under special treatment for FDI, the people from below are grasping the opportunities left for rural firms and market niches in traditional industries. These local businesses are integrated with the state economic sector and are directly involved in China's transition to a market economy, finding and filling various employment niches left by the structural defects of the command economy.

Set up by rural households and starting with simple products like food, clothes, or ballpoint pens, small manufacturing firms clustered in neighbourhoods, usually around a market place for their products. After more then ten years, some of the rural grassroots firms began to create brand-name products for home and abroad. This process of spontaneous bottom-up industrialisation in rural areas was usually referred to as a job generator for surplus rural labour and as an engine to raise the income of local residents. The inherent competitive advantages of such clusters were not recognised until very recent years, because they are inconsistent with the experience of rapidly growing non-locally embedded high-tech branch plants.

Clustering in traditional industries from below can be found in many Chinese provinces, such as Zhejiang, Fujian, Guangdong, Jiangsu, Jiangxi, Hebei and Henan. The examples include the cluster of sport shoes manufacturers in Jinjiang in Fujian, the furniture manufacturing cluster in Lecong in Guangdong, the firework making cluster in Wanzai in Jiangxi, the cashmere making in Qinghe in Hebei, etc. At the provincial level, a distinguished example of industrial clusters of traditional industries developing from below in rural areas is the case of Zhejiang, a province in East China. Growing from township firms, the success of non-state-owned firms in many clusters in Zhejiang province is at least in part attributable to the numerous localised industrial networks. More than three hundred industrial clusters specialised in various light industries (textiles, garments, shoes, socks/stockings, badges, neckties, fasteners, slide fasteners, fitments, furnishings, furniture, pens) have proliferated since 1985 and further fuelled the growth of the province's economy. These clusters have been anchored to a network of small and medium-sized towns somewhere between the large cities and the deep countryside. With the setting up of industrial parks based on such clusters by the local governments, urbanisation progressed quickly.

The Trajectory of a Local Cluster at a Crossroad Let's take Datang, a county in the east of Zhejiang province, as an example. Socks/stockings firms dominate the local communities. More than 150,000 people from 10,000 households in 120 villages are involved in sock-stocking-related industries, 80 per cent of which are small processing units (artisan firms). About ninety per cent of the GDP of Datang is created by the socks/stockings industry. Among the local firms, over 1000 are engaged in material processing and about 400 in material trade; nearly 8000 are socks/stockings manufacturers; there are over 300 suturing shops, 5 print works, about 100 shaping shops, 300 packaging firms, 200 machinery shops, 600 socks/stockings wholesalers and retailers and 100 forwarders shops.

The sock and stocking industry cluster in the Datang area actually started by accident. At the end of the 1970s a passenger on a train, a staff member of a township firm in Datang on his way to Shanghai, happened to meet a director of a Shanghai sock and stocking factory. As they chatted, he recognised that making socks/stockings could be a big business, and that it does not require large investments and state-of-the-art equipment and skills. In Shanghai he took the opportunity to visit the factories there and brought back some outdated knitting machines from Shanghai. Then some people from the village were sent to learn skills in Shanghai. After their return, the business of socks/stockings started in the Datang area. They peddled their products in the area between two main roads (Hangzhou-Jinhua and Shaoxing-Datang), which caused the market to flourish and promoted the further growth of the industry. The concentration of industry, once established, tends to be self-sustaining (Krugman, 1991). During the last decades this cluster has made amazing progress, being transformed from an agricultural area to China's largest concentration of socks/stockings industrial activities.

One case in point is that many clusters, just like Datang, have formed a 'collective brand' in their specialised product, but have not had any particularly well-known brand of an individual company. Moreover, high-tech or large firms are scarce. Related activities required to produce final products are conducted jointly by local SMEs, and many companies outside this region, including MNCs with world famous brands, are attracted to subcontract their manufacturing processes in these areas.

From the experience of the clusters in developing countries, Humphrey & Schmitz found that market agents and institutions such as trade fairs have been of great importance in providing SMEs with access to fashion and product design information and general market intelligence. This is not to suggest that holding trade fairs is in itself a panacea for achieving export growth for SMEs, but that SMEs have to interface closely with their markets, at home and abroad, in order to remain competitive and to grow (Humphrey & Schmitz, 1995).

With their catalytic effect, trade fairs in Zhejiang Province are also the leading source of export marketing support and technical learning. Several large exhibitions and wholesale centres for garments, cloth, machines, and small commodities are established in Zhejiang. For example, a local market town for socks/stockings and related materials has been built in Datang and acts as hub for related commodities and for information on the local products. Besides sales in the local market places, the finished products are also distributed all over the country and exported from there.

The clusters of traditional industries in Zhejiang are an example of 'development from below' and through self-organising processes, which is quite different from the centrally-planned industrial complexes dominated by state-owned enterprises (SOEs) in other places of China. Embeddedness has contributed to the emergence and development of industrial clusters in many rural places in Zhejiang. The actors undertake their activities with the expectation that they will share a framework of action with others engaged in the same activity. They play according to the same rules of the game, not necessarily because they like to do this (Storper, 1997). Dense interpersonal networks facilitating knowledge exchange among actors exist in the clusters. The collective efficiency and the synergy of local production networks are achieved by co-operation between local firms, which share the same understandings and meanings. This is the key difference between spontaneous industrialisation in Chinese rural areas and that supported by central commands.

The Embedded Local Government The local government played an important role in promoting local industrialisation. Since most of the staff in local governments run their own businesses or lend a hand in their family-based workshops, the role of the local government is a hybrid of developmental and corporate aspects. As the local financial strength increased, some government-sponsored institutes were established to provide local firms with information on technology and markets. The local government also provided public services for local enterprises, such as festivals of local products, shows at international exhibitions and homepages on the Internet to promote local brands in national or even global markets.

The export-oriented development in many of the newly industrialised regions tends to break down the local social networks, lead to dependence on foreign technologies and make the regional economy vulnerable to the dynamics of global the market. However, the industrialisation in rural places in Zhejiang is quite different from this formula. The economic performance in Zhejiang was based on local entrepreneurship with no subsidies or support from the central government. The local governments, especially the administrators at the level of the county or village, share the burdens and risks of reform with the local entrepreneurs.

After China's entry into the WTO, the industrial clusters, most or which deal with light industries such as textiles, clothing and daily necessities, face more difficulties. On the one hand, the firms in the clusters are able to export more products into the international market. However, the clusters are susceptible to external forces (e.g. environmental regulations). In such a case, the actions of the bigger organisations and the pattern of power relations within the clusters will change. While policy tends to be directed towards enhancing the collective efficiency of local SMEs in the clusters, the local inter-firm linkages should not exclude large firms, and the non-local learning activities are also not excluded.

A Cluster of Endogenous High-Tech SMEs: The Case of Zhongguancun, Beijing

The technology-based firms and the township firms are said to be the two Chinese wings for the painful post-cultural revolution takeoff. Thanks to the more liberal attitudes towards entrepreneurship in the late 1970s, the township firms in the rural areas boomed and became one wing for the takeoff. Since the 1980s urban and industrial reform have been in full swing, along with the start of small but growing technology-based firms – the second wing for the takeoff. A number of electronics firms that were initially established by intellectuals and agglomerated on Zhongguancun Street in the Haidian district of Beijing represented a breakthrough in the Chinese economy (Wang and Wang, 1998). The growth and clustering of technology-based firms can be regarded as a process of spin-off from state-run research institutes and universities. Spin-off activities have been most intensive in the Chinese Academy of Sciences (CAS) and the major universities of China, such as Peking University and Tsinghua University.

At the beginning of the reforms of the 1980s the Chinese government generated new dynamism by commercialising research results as new incentives to technological development and diffusion. The basis for this entrepreneurial approach is a belief in market forces as the prime regulators of advanced technological change. According to the 'four-self principles' (Siziyuanze) of non-interference created by Chinese authorities, the fledgling new-tech firms enjoy the right of choosing partners by themselves, financing for themselves, operating by themselves, and being responsible for all losses caused by the venture (Ziyou jiehe, Jichou zijin, Zizhu jingying, Zifu yingkui). They are listed outside of the national plan (Jihua wai) and allowed to look for profitable projects by themselves while receiving no increasing financial support from the government.

The Development of a Local High-Tech District between the State and MNCs The original toponym of Zhongguancun refers to a small suburb in which the plan of the streets and building sites was developed for scientific research and tertiary education in 1950. When economic reforms began in 1980, an innovative atmosphere arose and a bottom-up trial started. The area became an unintended high-tech cluster.

The formation of a cluster of high-tech SMEs in Zhongguancun was in a sense a technology-driven process. In 1980, a few professionals acted as risk-takers and devoted themselves to an early experiment for establishing non-state-owned firms based on their technology skills. After witnessing the success of Silicon Valley, an intellectual set up the first start-up emanating from the Institute of Physics of CAS. Since 1983, when the central government stepped in because of a positive assessment of his business, more and more technology-based firms have been set up in Zhongguancun. As an immediate response, the name 'Electronics Street' appeared in the media (Wang and Wang, 1998). Afterwards, many firms spun off from the CAS, Tsinghua University and Peking University as well as other state-owned research institutes.

On the other hand, a market-driven process also pushed the Zhongguancun case. Most of the firms were involved in commercial activities. They started by cloning computers and computer components, and doing services related to the products they sold. They became clone makers in computer-related products, such as printers, UPS, materials for computer offices, word processing for Chinese characters, scanners, etc. The population of newly created firms increased from 90 in 1985 to 527 in May 1988. This trend illustrates a distinctive Chinese way of developing new-tech industry from scratch and with a peculiar spatial phenomenon - the cluster of new-tech SMEs in 'Electronics Street' and its surrounding area.

The business ties of the early entrepreneurs in Zhongguancun are mostly locally embedded. The expertise of the state-owned research institutes and of the universities was used to serve the local market, and most of the start-up firms had extensive connections with their mother institutes in terms of finance and staff. However, global ties are critical in terms of technology innovation resources. This is because local firms often act as sale agents for MNCs and provide system-integrated services to local users. Such global-local ties played an important role in the technological change and institutional transformation in the high-tech cluster of Zhongguancun. Since the 1990s, the firms in Zhongguancun have linked up with multinational firms for alternative support. Sino-foreign joint ventures have been approved and sole-foreign-invested firms had been registered. The famous MNCs related to IT industries have entered or plan to enter.

The Role of the Government in the Cluster Upgrading Process – A Criticism Since 1985, when the 'Determination on the reforms of S&T system' issued by the Central Committee of the Communist Party encouraged 'selecting several intelligent-intensive area in the country and adopting the special policy there, forming new-tech development zones with different characteristics', China has established 53 national New- and High-tech Industrial Development Zones (NHTDZs). Zhongguancun in Beijing is the original model for this national policy and typical of such high-tech clusters in China.

In 1988, Beijing Municipal Government established the Beijing Experimental Zone for the development of New Technology Industry (BEZ) around 'Electronics Street'. Over the past decade, Zhongguancun has contributed a great deal to China's IT industry while enjoying tremendous growth itself. Compared with 527 enterprises in 1988, there were 5657 enterprises in the BEZ by the end of 1997. The software industry in the BEZ has a 40 per cent share of the national domestic market, and technical services a 47 per cent share (Lin, 1998). The most important contribution of Zhongguancun to China's high-tech industry is that it bred a generation of entrepreneurs who are technically competent, understand the market situation and experienced in such an emerging market environment. These entrepreneurs have played an important role in China's latest IT revolution. Since 1992, the geographical areas of NHTDZs in China have been continually enlarged. They are characterised as 'one zone consisting of several parks (sub-zones)' (yiqu duoyuan), such as 'one zone consisting of three parks in Beijing', 'one zone consisting of six parks in Shanghai', or 'one zone consisting of four parks in Tianjin'.

The year of 1999 can be labelled as 'The zero year of Zhongguancun'. In the era since 1999 the embedded cluster has completely changed. China's central authority endorsed an application to build a science and technology park around Zhongguancun and designated Zhongguancun as the high-tech centre of the country. The programme 'Zhongguancun Science and Technology Park' is characterised as 'one park consisting of five sub-parks' (yiqu duoyuan), covering an area of 280 square kilometres in the central park (original BEZ) and four satellite industrial parks in the suburbs of Beijing. The original Zhongguancun area received the largest state boost. Zhongguancun 'Electronics Street' disappeared and was replaced by the broadened Zhongguancun Road. Besides the road an ambitious and luxurious construction project covering one million square metres is to be completed in the next ten years. It is claimed to be China's Silicon Valley.

Is the 'science park' solution a panacea that will provide the answers to the problems and dilemmas confronting Beijing? The problem is that the Zhongguancun high-tech cluster has little basis for self-sustaining growth. The price of land in the central park is so high that it is forcing indigenous SMEs to leave. The authorities put more attention into inducing foreign investors instead of encouraging co-operation and communications within the local technological community. There are fewer and fewer new homegrown firms and fewer links between newcomers and indigenous industries. The original embedded start-ups are losing their competitiveness when competing and co-operating with MNCs in the domestic market.

This phenomenon is mainly a legacy of the bureaucracy of central planning, which can be illustrated with the Chinese saying, 'putting on new shoes, taking the old road' (Chuanxinxie, Zoulaolu). In the years before 1999, the government created a new institutional arrangement for BEZ to cut red tape and promote high-tech industry and help local private technology-based firms. However, 'Zhongguancun Science and Technology Park' is dominated directly by the central government now. The higher level of domination in the bureaucracy does not seem so helpful for solving the problems of local firms at present.

First, co-operation among various actors, such as firms, universities and research institutes, was obstructed by the problems of demarcating property rights between the institutes and their spin-offs; consequently the spin-offs can hardly reach an agreement as to how to dispose of the state-owned property in the business. This is a legacy of the former central planning system that was left to the comparatively large companies established over 10 years ago. It is difficult to clearly distinguish the contribution of the entrepreneurship of the managers from the original sponsorship of the mother institutes, since there were no formal agreements when the tentative reforms began.

Second, specialisation is less well developed among local technology-based firms. One reason was a lack of social capital, the stocks of social trust and norms as well as networks that people can draw upon to solve common problems. The transaction costs between customers and suppliers increased, and large companies would rather internalise most of the business activities than to co-operate with other firms. The SMEs find it almost impossible to focus on their speciality in such a milieu.

Third, the competition between MNCs and the local SMEs for qualified technicians and management talent increased. Many local entrepreneurs complained about a lack of high-level training for technicians and managers. The cost of such training is formidably high for local SMEs, and the high rate of turnover in employees discourages the companies from investing in such training.

As Saxenian pointed out, there are social barriers to overcome if other regions are to dramatically spur their economic growth either through high tech itself or by taking advantage of the Internet's new opportunities. 'The key here is not simply a decentralised industrial structure, but also a set of social structures that grew up alongside it – social networks, gossip… While good schools are essential, people are also learning through this community, so you can generate more net new ideas through this networked kind of system' than with the same quantity of un-networked human talent. 'Creating a dynamic, high tech region is not a matter of combining ingredients... It is (the) relationships between the individuals, firms and institutions that matters, not their simple presence' (Saxenian, 1998, p. 2).

In sum, the development process of the Zhongguancun cluster is a mixture (Wang and Zhou, 2001). Since it started from the bottom, the development could be seen as endogenous. As the City and the Central government were involved in planning a luxurious S&T Park and aimed at attracting MNCs, the development of this cluster diversified from its original track onto a more exogenous route. The efforts of the government would be justified only if the local firms could benefit in an open competitive/co-operative milieu with MNCs. However, the overheated estate development following the ambitious plan of the government to build the 'Silicon Valley of China' has already driven many local SMEs to leave this area to find cheaper peripheral locations. The embedded local network has to face the disembedding forces resulting from government intervention.

Clusters of Taiwanese and Foreign Invested Firms: A Case in Dongguan City

Since China opened to the outside world, especially since it joined the World Trade Organisation (WTO), more and more investors have entered from outside Mainland China. Some of the outsiders have become the anchors of clusters around which homegrown firms developed; others brought their own satellite firms from all over the world. As a result, a kind of cluster, which is centred on Taiwanese or foreign firms, came into being.

Taiwanese-based clusters first appeared in the Pearl River Delta (PRD), a fast developing region in South China. Not having had much industrial accumulation before the opening and the reforms, this region is mainly foreign investment driven. Due to its proximity to Hong Kong and the preferential development policy for coastal regions, the PRD, especially its Shenzhen Special Economic Zone, has become the brightest spot of FDI in China since the early 1980s. Like many other developing areas, the rapid industrialisation of the PRD was based on exploiting lower costs of land and labour. As more and more industrial firms clustered in this area, the export processing activities expanded from assembly to more advanced stages in the production chain. Local supply networks also developed. Clusters of shoe makers, electronics producers, furniture makers, garment manufacturers, fresh flower processors, and toy plants, etc., are scattered around the newly urbanised land, with workers coming from other regions of Mainland China and products being exported through Hong Kong to the world. The landscape of quick industrialisation and urbanisation is quite impressive.

However, as China joins the WTO and the hinterland opens further, these industrial clusters in PRD will face a changing market environment. The advantage of tax incentives and low costs of labour and land in this area, which contributed to the fast growth in the past decade, will vanish in the near future. The sustainable growth of this region depends on the innovation capacity of local industries.

The Local Clustering Based on Export Processing Activities The development of a PC-related manufacturing cluster in the Shenzhen-Dongguan-Huizhou area is a representative example in the PRD region (Tong and Wang 1999). Dongguan was a rural county in South China neighbouring Hong Kong and Shenzhen. Thanks to the relocation of PC companies from Taiwan to low cost regions in the Mainland and the booming PC market in Mainland China, over 2000 IT firms have agglomerated here. Numerous PC parts and components needed for assembly, except CPU, memory chips, or other kind of ICs, are manufactured and can be bought locally within a distance of 100 kilometres. Both famous PC-related companies, including IBM, Compaq, Dell, Acer, and the domestic companies Legend and Founder have established plants or purchased parts in this area.

In the late 1980s and early 1990s, based on low cost labour and land, the initial wave of PC-related investors established (or rented) assembling plants with strong links to the overseas market. Linkages with local plants were weak, however. After the mid-1990s, the PC-related industry in Taiwan kept moving offshore, and Dongguan became the Taiwanese PC makers' gathering place, one of the densest concentrations of Taiwanese PC makers on the Mainland. During this period, the local PC production network became stronger. The interface between global and local production networks became more complicated.

Most outside PC-related investors in Dongguan have built plants for export in co-operation with local Chinese partners. First, the government offered inducements to MNCs to export. Second, most of Dongguan's input materials had to be imported. Most companies found it was easier to import inputs from overseas instead of searching for local suppliers. Third, a lack of confidence in China's political stability limited the development of local linkages.

As the clustering process progressed, the local backward and forward linkages were also evolving. Local firms were eager to keep up with the fast changes in technology in this industry and sharpen their competitive advantages in manufacturing processes through local interactions with MNCs and close linkages with the global market. Subcontracting activities are extensive in the local PC production. Many local suppliers felt the pressure to improve the quality and speed of reaction while co-operating with big buyers. In this regard, the quality management departments of buyers play an important role in the technological transfer between suppliers and customers. This interaction could probably lead to further networking of the local production.

The growing PC domestic market in Mainland China is another stimulus to the development of PC-related production in this area. The huge potential of the domestic market will also affect the network pattern of the local PC production system in the near future. At present, the PC industry in this area is still mainly export-oriented, although the local markets have had a great influence on the IT market of the whole country. In this regard, the PRD has advantages in being able to use its manufacturing agglomeration based on export-oriented processing activities to meet interregional competition on the Chinese PC market. There are relatively advanced manufacturing technology and management methods in this region. Subcontracting networks, just-in-time production and inventory management are also well developed among all kinds of firms, both foreign controlled and domestic PC corporations. The intensive intermediate production chains have also attracted PC assemblers aiming at the domestic market.

The Interactions Between the Local Government and PC Makers The interactions between local producers and the government also played an important role in the development of the local PC manufacturing cluster. Local governments have strongly supported overseas investors. Among overseas investments induced by the local government, the PC industry, being a high-tech sector, received priority. This helped overseas investors, especially small and medium-sized ones, to reduce the risks of business. The unstable import and export regulations in Mainland China have created uncertainties in addition to those associated with technology and marketing. The local governments often helped by providing access to information on adjustments to regulations. In such adjustments the personal relations between entrepreneurs and local officials play an important role.

Because the PC-related products in the PRD are mainly keyboards, mice, cases, and monitors, which are mature commodity lines usually bringing in little profit, the local government also made some efforts to promote the technological upgrading of the local industry. An Advanced Manufacturing Centre was established in 1999 to help local SMEs adopt information technology in their operations. Public funds were also raised as incentives to local entrepreneurs investing in R&D activities.

Problems in Dongguan's Clusters The institutional bottleneck has hampered the spontaneous interaction between local firms. Because export was the premise for preferential tax treatment in Dongguan, many makers just export their products to Hong Kong and sell them back to customers on the mainland again. This wastes too much time. The customs officials faced the dilemma of trying to prevent smuggling while simplifying the customs procedures for products, while the producers found it difficult to arrange production so as to respond rapidly to changes in the market without violating the regulations (Tong and Wang, 2000).

As the entrance barriers to the domestic market decrease, many PC-related firms, especially Taiwanese PC firms, are transferring investment from Dongguan city to the cities in the Yangzi River Delta, such as Suzhou, Shanghai and Ningbo. They do so because of the proximity to the majority of the domestic consumers and to reduce the costs of local recruiting. Dongguan was a rural area where specialised education was lacking. In addition, some people with a high-level educational background do not want to work in such a 'cultural desert'. Hence the cost of local recruiting is high. The new generation of PC-related investors, the more law-abiding Taiwanese firms, are interested in sites where the local government acts according to law rather than for opportunities for smuggling. Consequently they are leaving Dongguan for Suzhou for the sake of governance by a more efficient local government. In this regard, the most important task for upgrading Dongguan's PC-related clusters is institutional reform. However, the principal agenda of the city government is to invest heavily in constructing a big science and technology park to attract foreign investment. As the case in Beijing showed, the 'science park' solution is not a panacea that can provide the answers to the problems confronting Dongguan.

A Cluster Under the Central Planning System: A Case of Shipbuilding in Shanghai

For the last style of clusters in China, the clusters of state-owned enterprises established under the central planning system, we will take shipbuilding in Shanghai as an example. According to the definition of an industrial cluster, the former industrial complexes planned under the central planning system in China can hardly be referred to as real clusters. The original planning of these complexes was based on planned input-output relations between different enterprises located in different regions. The spatial allocation of industrial infrastructure and facilities by the central government also reflected considerations of regional balance and national defense. Most of this kind of industrial agglomerations face increasing difficulties under competitive market conditions. The reform of state-owned enterprises is one of the major projects in China.

Clustering Out of the Industrial Complex Legacy Chinese shipbuilding and overseas trading thrived during the Yuan and Song dynasties. The policies of the Ming and Qing dynasties forbidding sea transport, however, caused the development of private shipbuilding to stagnate (Li, 2001). During the late part of the Qing dynasty, the government invested heavily in shipbuilding for purposes of defence, and the overall pattern of the Chinese state-owned shipbuilding industry since then dates to that period.

In the early 1980s, the state-owned Chinese shipbuilding industry began to transform to compete in the global civil-use shipbuilding market. Since shipbuilding is by and large a labour-intensive industry, China has comparative advantages in terms of labour costs on the world market. Therefore, China's market share in shipbuilding increased rapidly in the last 20 years. China is now the third largest shipbuilding country in the world, following Japan and Korea. It accounts for 12 per cent of the world market and is second in the world in speed of growth, only behind South Korea (Chen, 2002).

In the reorganisation of enterprises in 1999, the state-owned shipbuilding enterprises were divided spatially into two parts, China State Shipbuilding Corporation (CSSC), or the South group for short, centred in Shanghai; and China Shipbuilding Industry Corporation (CSIC), or the North group for short, centred in Dalian. The production of these two groups accounted for 77 per cent of the national shipbuilding industry in China in 2001 (Cheng, 2002).

Shipbuilding in Shanghai has a history of over 100 years. From its very beginning in the late years of the Qing dynasty, it was established and controlled by the central government, and the production was mainly defence oriented. Before China's economic reforms and opening to the outside world in the late 1970s, the financial budget from the central government was already insufficient to sustain the military-based shipbuilding industry. The enterprises began to build ships for civil-use in the global market to make up for the cost of running all the facilities. The government invested much less in this industry during the same period, however, compared with many specially selected and supported 'pillar industries', such as the automobile and electronics industries. Lacking the brilliance of the high-tech industry, shipbuilding is almost absent in the national industrial policy, although it employs thousands of workers and breeds a large number of suppliers around it.

As an assembly activity with tremendous linkages with upstream sectors, including the sectors of metal and steel, machine and engineering equipment, and even construction and furniture, shipbuilding played an important role in the industrialisation of many countries, such as the UK, the Netherlands, Germany, the US, Japan and Korea. However, shipbuilding has decreased dramatically in the developed countries and shifted to the developing countries. Nevertheless, the developed countries still have the technical advantages to fill special niches in the global shipbuilding market. In contrast, comparatively speaking, shipbuilding in China lacks the support of local subsidiary industries for the competition in the global market.

In Shanghai, several major state-owned shipbuilding enterprises are clustered along the Huangpu River, such as Hudong, Jiangnan, Shanghai and Zhonghua, most of which have a history over half a century. Take the example of Hudong in the Pudong Development Zone. It has over 10,000 workers and occupies about 87 km2. According to its building capacity in 1994-1995, Hudong was among the top 40 shipbuilders in the world (Shao, 2000). The network of suppliers expanded around the Yangzi River Delta and extended even to the hinterland of the country. However, fewer and fewer suppliers exist nearby in Shanghai due to the higher and higher land prices in the metropolis.

Beyond the well-equipped state-owned enterprises, there are a lot of private small and medium-sized shipbuilders in China. The technical and managerial gap between these private SMEs and the state-owned enterprises is quite large. Their target customers also differ. The large state-owned enterprises focus on international buyers and the national defence budget. The transactions between the buyers and producers are formal and restricted, generally in accordance with the international conventions on maritime affairs. However, the private SMEs mainly focus on local private users, and lack professional skills and knowledge of international conventions in shipbuilding contracts. It is almost impossible to establish business linkages between these private and state-owned enterprises. This causes the industry to have a dual structure.

The Government-Driven Restructuring of State-Owned Enterprises The restructuring of the state-owned shipbuilders has been undertaken in quite a difficult manner. The large enterprises have to be disintegrated into many units, so that each of them can grasp market opportunities autonomously. Reshaping the governance structure will lead more firms to spin off from the mother enterprises and specialise in different production activities. But the process of transformation is difficult and painful, since a large number of workers will probably lose their job during this process. For example, there are over 20,000 workers in the old Hudong Shipbuilding Company, while Waigaoqiao, a larger newly established company with similar production capacity also in Shanghai, will have only about 3000 workers. The increase in labour productivity without sufficient mobility of labour forces could lead to a net loss of jobs in the city, which will cause other social problems in the country. The slow and painful reform of the state-owned shipbuilding firms makes it difficult to transform the dual structure into an integrated one.

The presence of MNCs is a critical challenge to the local shipbuilding enterprises. For example, a Japanese shipbuilder established a joint-venture plant in Nantong, a city to the east of Shanghai. By establishing plants in China, not only can it exploit the low cost labour, but also foster its supply networks of local SMEs. China made efforts to establish a full national industrial system under the central planning system from the 1950s to the 1980s. In contrast, bottom-up industrialisation has thrived within non-state owned sectors in the last two decades. Bridging the gap between the two systems is important for promoting connections between the various actors in local economic development. As the case of shipbuilding in Shanghai shows, not only should local clusters in developing countries strengthen their global linkages with overseas sources of innovation and market demands, but the local linkages among producers under different authorities should also be promoted.

Conclusion: Solution to the Dual Structure

Clusters in China, which were created under different conditions from those in the West, have various characteristics. The cases in this chapter range from high-tech, notably IT sectors, to traditional sectors, including rural SMEs and state-owned-enterprises. The relationship between embeddedness and disembeddedness varies in these cases. The Zhejiang case addresses the question of upgrading and restructuring in local embedded rural industrialisation; the case in Zhongguancun shows how an 'embedded' region is threatened by new disembedding tendencies; the case in Dongguan draws attention to the danger that disembeddedness may result in out-migration of firms, and the case of the shipbuilding complex in Shanghai discusses the problem of how a state owned production complex could be transferred to an embedded local cluster. We will analyse these cases from the perspective of their global-local ties according to our embedded-disembedded framework presented above.

The trajectories of the four cases above are typical of the alternative pathways towards industrialisation in different regions. They are all within the transitional context in China, each with a unique local background. They reflect the interactions and tensions between the top-down thrust of national goals of modernisation, the inflows and involvement of international capital and technology, and the bottom-up initiatives from the grassroots. As China is opening to the world, the local firms access and pursue high-tech development with government supports, while the thriving labour-intensive manufacturing activities represent the true comparative advantages of China. The dual structure is prevalent in all the cases we discussed.

From the cases in this chapter, we can see the extreme difficulty in promoting local linkages among MNCs and local actors in the industrial cluster. The dual structure, in which export-oriented production is separated from domestic-oriented production, and in which the state-owned firms are separated from the private firms, serves as an obstacle to networking of the local industrial system. However, in all the local communities we can also see enthusiasm for bottom-up development. The intelligence of entrepreneurs coming out of the state-owned research institutes, the rural villages, the overseas Chinese, manifest themselves in different ways in different settings. As China's tentative reform goes on, industrial clusters in different regions transform themselves constantly in the process of industrialisation and upgrading, which can hardly be simplified into any models of development.

The development of industrial clusters in China could still have great implications for other developing countries. According to UNCTAD's World Investment Report 2001, China is among the 'brightest spots of FDI in the world' in recent years. The local clusters are becoming more and more important for attracting foreign investors. Since the local communities and the foreign investors do not always share the same goals in local development, it is essential for local governments to help establish a win-win relation in the construction of global-local ties. Although the context in developing countries is very diverse and the trajectories of development in different regions can hardly be similar to each other, some strategic tools have been proven to be useful for attaining goals in developing countries.

Firstly, public policies on linkage promotion can be helpful for the private SMEs in local clusters. The local SMEs face much higher transaction costs when searching for business partners outside of their daily life community. Running business in a restricted network can decrease the risk of investment, but also restricts the growth of individual firms. Public policy should focus on promoting communication among different actors. However, from the experience of China, such promotion should be done in a gradual way based on the understanding and willingness of individual entrepreneurs.

Secondly, customer-oriented supports can help local firms to access markets over a long distance and to meet the demands of customers. In China, most local firms initiate their business by focussing on the huge home market. This is quite different from most of the East Asian development models, which depend heavily on long distance markets and overseas buyers. However, the comparatively mature markets in developed countries still play a significant role in improving the performance of local companies, especially when they compete and co-operate with each other directly.

Thirdly, the strength of the cluster can be embedded into the local network, instead of putting all stakes on a few disembedded large companies. Collective efficiency is most important for industrial clusters in developing areas, because most of the individual firms within local clusters are small and lack strength. It is quite difficult to establish formal co-operation among local actors when large numbers of local firms compete for partnerships or contracts with the stronger MNCs. However, small firms still depend on their local social network to access the economic resources they need in an implicit way. Such an intangible network is subtle and fragile in face of intervention from outside, but can strongly affect the behaviour of local actors.

References

Asheim B. and Cooke P., (1999) Local learning and interactive innovation networks in a global economy. In Malecki E and Oinas P (eds.): *Making Connections: Technological learning and regional economic change.* Aldershot: Ashgate, pp. 145-178.

Becattini, G. (1978) The development of light industry in Tuscany: an interpretation, Economic Notes, Vol. 2-3, pp. 107-23.

Bergman, E. and Feser, E. (1997) *Industrial and Regional Clusters: Concepts and Comparative Applications*, Internet book, http://www.rri.wvu.edu/WebBook/Bergman-Feser/contents.htm.

Braczyk, H. J., Cooke, P., Heidenreich, M. (eds.) (1998) *Regional Innovation Systems*, UCL Press.

Chen, G. (2002) Shijie Zaochuanye Fenxi (Analysis of World Ship-building). *Hanghai Jishu* (Marine Technology), 4, pp. 76-79.

Cheng, J. (2002) Fazhan zhong de Zhongguo Zaochuanye (Chinese Shipbuilding in Developing). *Shuiyun Guanli* (Marine Management), 6, pp. 34-35.

Enright, M. (1996) Regional clusters and economic development: A research agenda. In *Business Networks: Prospects for Regional Development,* edited by U. H. Staber et al., Berlin, Walter de Gruyter.

Gordon, I. R. and McCann, P. (2000) Industrial clusters: Complexes, agglomeration and/or social networks? *Urban Studies*, 37(3), pp. 513-532.

Granovetter M. (1985) Economic action and social structure: the problem of embeddedness. *American Journal of Sociology*, 91, pp. 481-510.

Hsu J. and Saxenian, A. (2001) The Dynamic Institutionalism of Taiwan's Late-industrial District in the Globalization Process, Paper for the 97th Annual Meeting of the Association of American Geographers, New York, February 27 – March 3, 2001, Available from the author.

Humphrey, J. and Schmitz, H. (1995) Principle of promoting clusters and network of SMEs. UNIDO report.

Krugman, P. (1991) *Geography and Trade*. Cambridge: MIT press.

Li S. (2001) Lun MingQing Shiqi de Zaochuan yu Waimao (Shipbuilding and Trade in Ming and Qing Dynasty), *Dongfang Luntan* (Oriental Forum), 1, pp. 62-66.

Lin W. Y. (1998) Beijing: constructing worldwide business partnership towards global economy, Speech at XV IASP World Conference on Science & Technology Parks, Perth, Australia, 21 Oct., available from Internet, http://www.bed.gov.cn.

Lloyd, P. and Dicken, P. (1990) *Location in Space: A Theoretical Approach to Economic Geography*, 3rd edition, New York: Harper and Row.

Malecki, E., Oinas, P., and Park, S.O. (1999) On technology and development. In Malecki E. and Oinas P. (eds.): *Making Connections: Technological Learning and regional economic change*. Aldershot: Ashgate.

Markusen, A. (1996) Sticky places in slippery space: a typology of industrial districts, *Economic Geography*, 72, pp. 293-313.

Markusen, A. (1998) Fuzzy concepts, scanty evidence, policy distance: The Case for rigour and policy relevance in critical regional studies, *Regional Studies*, 33(9): pp. 869-884.

Marshall, A. (1920) *Principles of Economics*. 8th edition, London: Macmillan.

Martin, R. (2001) Geography and public policy: the case of the missing agenda. *Progress in Human Geography*, 25(2), pp. 189-210.

Martin R. and Sunley P. (2003) Deconstructing clusters: chaotic concept or policy panacea? *Journal of Economic Geography*, vol. 3 (1), pp. 5-35.

Maskell, P. (1999) Globalization and industrial competitiveness: the Process and consequences of ubiquitification. In Malecki E and Oinas P. (eds.): *Making Connections: Technological Learning and regional economic change*. Aldershot: Ashgate, pp. 35-60.

Mytelka, L. and Farinelli, F. (2000) Local clusters, innovation and sustained competitiveness, Discussion paper series #2000-5, The United Nations University, Institute for New Technologies.

Nelson, R. (Ed.) (1993) *National Innovation Systems: A Comparative Analysis*. New York and Oxford: Oxford University Press.

Porter, M. (1990) *The Competitive Advantage of Nations*. London: Macmillan.

Saxenian, A. (1998) *Silicon Valley's New Immigrant Entrepreneurs*. California: Public Policy Institute of California.

Shao, T. (2000) Zhongguo Zaochuan Xitong de Tedaxing Qiye – Hudong Zaochuanchang (Titanic Ship-builder in China – Hudong). *Caiyouji* (*Diesel Engine*), 2, pp. 49-50.

Storper, M. (1997) *The Regional World: Territorial Development in a Global Economy*. New York: Guilford Press.

Storper, M. (1998) Industrial policy for latecomers, products, conventions, and learning, in Storper M. and Stavros B. et al. ed. *Latecomers in the Global Economy*, London: Routledge. pp. 13-39.

Tong, X. and Wang, J. (2000) Global-local Networking of PC manufacturing in Dongguan, China. Paper presented on IGU conference 2000 in Dongguan, China.

UNCTAD. *World Investment Report* 2001. Geneva.

Wang J. (1998) In Search of Innovativeness: The Case of Zhong'guancun, In Malecki E. and P. Oinas (eds.) *Making Connections: Technological Learning and regional economic change*. Aldershot: Ashgate pp. 205-220.

Wang, J. (2001) *Chuangxin de kongjian: Qiye jiqun yu quyu fazhan (Innovative Spaces: enterprises clusters and regional development)*. Beijing: Peking University Press.

Wang, J. C. and Wang, J. X. (1998), An Analysis of New-tech Agglomeration in Beijing: A New Industrial District in the Making? *Environment and Planning* A, 30, pp. 681-701.

Wang, J.C. and Zhou Y., (2001), Beijing Zhongguancun dianzi xinxi chanye de qiye jiqun fenxi (Analysis of Beijing Zhongguancun information technology firms regional cluster), Ch. 6 in *Chuangxin de kongjian: Qiye jiqun yu xuyu fazhan (Innovative spaces: Enterprises clusters and regional development).* Beijing: Peking University Press. pp. 220-233.

Wang, J.C., Zhu, H., and Tong, X. (2001), Districtization in Zhejiang Province of China: with reference to Datang sock/stocking industrial district. Paper presented at the residential conference of the IGU Commission on the Organization of Industrial Space, Turin, Italy.

Chapter 11

Cross-National Ethnic Networks in Financial Services: A Case Study of Local Banks in Singapore

Shuang Yann Wong

Introduction

The 1997 Asia Economic Crisis destabilised a number of countries in Southeast Asia. For Singapore, the crisis offered an opportunity to expedite the process of changes that it has long planned. Today, almost one-third of the nation's GDP and employment comes from the services sector, where finance and banking in particular stands out above others. The fact that Singapore's banking and finance sector was less affected by the 1997 crisis compared to its regional counterparts is an evidence of its strength.

The development of the city-state's economy has been written about by many authors (e.g. Lim et al., 1988; Low, 1999; Low & Johnston, 2001; Perry et al., 1997). The industrialisation of the republic, and in particular the manufacturing sector, has been widely studied in relation to the dominant roles of the State, foreign direct investment (FDI) and multinational companies (MNC) (e.g. Chia & Robson, 1998; Haggard and Low, 2002; Toh & Tan, 1998; Rodan, 1989). Services sectors such as finance and banking are discussed actively as a consequence of the 1997 financial crisis in the region, rising competition from regional countries and liberalisation of the sector (e.g. Chia, 1999, 2000). Increasingly the research focus has shifted towards analysing strategies that can sustain the competitiveness of the city-state.

Diversifying and deepening the services sector, especially the high value added services such as finance and banking, is one of the major keys to leading a knowledge-based economy (KBE). Various measures are put in place to inject new activities into the finance and banking sector. Local banks are asked to take drastic moves such as the merging or hiring of foreign talents and expertise to compete in the new global economy. The relevance of such attempts is mainly examined from a macro- or micro-economic perspective. In the UK, the issues of social and economic exclusion that have arisen from the restructuring of banks have been actively pursued. In Singapore, study of the changing roles of local banks in nation building, their contributions to the local economy and their potentials to forge the local-global nexus seems to have been neglected. The ramifications for small consumers and businesses in relation to bank enlargement are hardly explored. Some of the more fundamental questions remain to be examined more fully. Is the local banking sector an outgrowth of the success of the Republic's manufacturing sector? How do these banks come to accumulate today's strength in asset standing, profits performance and credibility? What kind of economic, social and cultural networks do these banks have, allowing them to reach their current strength? Can such networks withstand the contemporary forces of competition as the financial sector becomes increasingly liberalised? What strategies will these banks have to employ to sustain their dominance in the city-state? How would the existing networks function from now on under such circumstances? Before answering these questions, it is essential to have an understanding of the background of these banks, their operating networks and their behavioural patterns. Such knowledge will enable one to discern the future directions of these banks as they try to sustain their territorial space at home, while at the same time struggling to contest for a broader market space abroad.

This chapter is divided into three parts. First an attempt is made to construct a conceptual framework in order to explain the evolutionary processes and social-cultural backdrop that help foster ethnic relations and the subsequent interstitial functional economic networks that culminate in the formation and strengthening of local banks vis-à-vis foreign banks. Second, it traces the development process and the social, cultural, political and economic forces that have led to the formation of local Chinese banks. It argues that the emergence of these banks is very much related to the compelling needs of the local and regional sectors for the exchange of goods and services, opportunities provided by the colonial government and the ability to canvass support from local communities and government. It is also argued that the territorial space and social networks of these banks is very much an outcome moulded by the key actors involved in forging the transactional links between the various trades and local banks. Lastly, the future prospects of these local banks are discussed.

An Explanatory Framework for the Evolution of Local Banks

In explaining the economic development of Singapore, many have referred to the dominant role of the State. This is particularly so for sectors that have been strongly introduced and supported by the State in post-independence Singapore such as exported-oriented, MNC-led and FDI-based manufacturing, modern transport and communications. Since the early twentieth century, more than a dozen local banks have been established in Singapore. Some ceased to function while others managed to survive the uncertainties of commodity trades, the turmoil of the two world wars and the change of government in the post-world war years. Few succeeded in becoming local giants such as the Overseas Chinese Banking Corporation (OCBC) and the United Overseas Bank (UOB), which are owned by local Chinese inhabitants.

The social, cultural and political economy perspective (e.g. Hamilton, 1991; Thrift & Olds, 1996) may help in explaining the overall pattern and organisational tendency but is inadequate for tracking the processes and actual forces that have led to the emergence of local banks today. A city-state like Singapore, which began as an immigrant foothold and was ruled by two different political systems (British and Malaysia), will not have the same kind of experience. The inter-ethnic, inter-agency relations established in the pre-independence days are built into the inter-corporate relations of local banks today. The existence or viability of Singaporean local banks is not just subject to the networks that they belong to but also to the levels of spatiality, reflexibility, territoriality and accountancy (Dodd, 1994) that they command. Increasingly, it is the high level of accountability that has gained the city-state its credibility as a regional financial hub and its resilience to withstand the 1997 regional financial crisis. Achieving the desired levels will depend on the dexterity in interacting with the various actors and orchestrating resources as well as embeddedness within the multi-layered and multi-faceted sets of operational networks.

There is evidence to suggest that the growth of local banks is very much related to the prevailing market, social and cultural factors, whereas their organisational structure and changes are shaped by authority patterns and legitimating strategies. Two kinds of mechanism are discernible, the *laissez-faire* system before independence and the guided or regulatory system in the post-independence period. I would not regard the latter as a hybrid of the first or the West, as each has its own specific traits that were born out of and adjusted to the needs and interests of the respective period. Neither is the latter decoupled from the previous system. The local banks of today have evolved from the social, cultural networks and political backdrop of the locality, reinforced by leading actors who are integral parts of the networks. These banks thrive on the economic functions they provide at a particular point in time and space.

The functions of the banks are very much dependent on the interactions between the actors, the prevailing activities and resources (Figure 11.1). Their functions can be guided or regulated by a legitimised authority. An interventionist, independent state may provide more resources to local banks but their activities may be checked if priorities are given to designated sectors of the economy. Local banks that rely mainly on domestic social networks and work close to the needs of the State, which in turn protect their interests, may thrive within the confines of the domestic economy. However, they may not be able to compete globally unless the social networks that they have been tapping into can transcend national and regional territories.

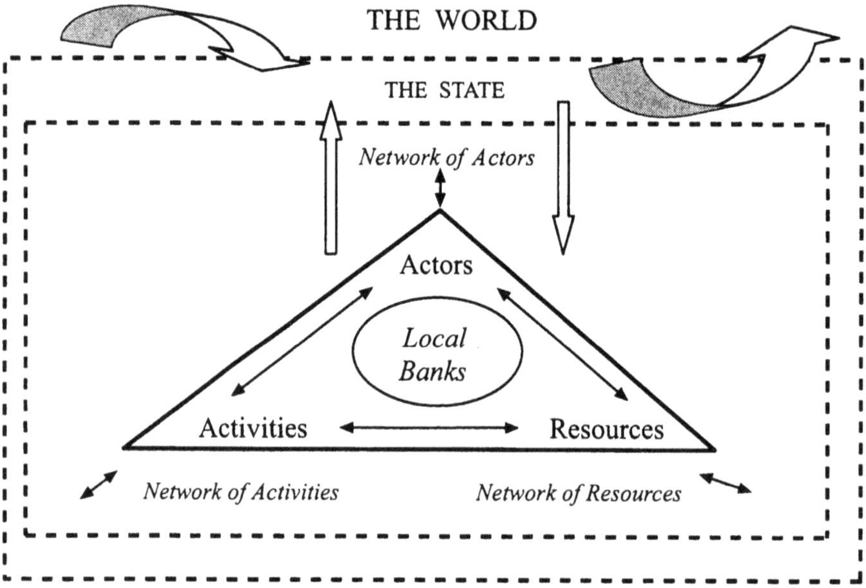

Figure 11.1 Network of banks

The two systems operate on different sets of relations depending on the kind of relations that the actors or agents have with each other (Figure 11.2). The proximity and connections of actors within the network can be discerned from six dimensions, namely: spatial, social, organisational, capital, transactional and political. Relations and actions are space-bound. People often congregate in the same place at the same time to act in concert. Spontaneous actions diffuse spatially. Individuals become aware of the information because they are near it. However, space is often socially organised. People are usually segregated by language, class, ethnicity, political orientation, education and religion. As a result, different kinds of people are found in different kinds of public spaces. Rules of social etiquette regarding class, ethnicity and gender, as well as language differences, may create communication barriers that are the practical equivalent of great distances. A wide variety of routine social structures can create network ties. 'Little India', 'Chinatown' and 'Kampong Glam' are examples of such creations created in the past in Singapore. The low level of education and poor communications in the underdeveloped economy further enhance the values and practice of such relations. Hence the capital and transactional relations of local banks in the early days often reflect cultural bias and preferences in the allocation of resources due to such differences.

When the economy advances and the educational level is raised, other social relationships emerge that go beyond proximity relations built on similar race, language or religion. People can be members of alumni associations, recreational clubs, business associations and other organisations that promote either cultural or political activities. Within spaces, actions may thus also be coordinated through political and/or organisational ties. Different organisations with similar goals would tend to have positive ties although some element of competition may exist between them. Such organisational ties are meant to minimise competition, search for market niches and exchange information on a reciprocal basis. The more organised they are, the better is their communication, resulting in greater exchange of information and closer relations.

Organisational and other business ties help open up opportunities and the requisite resources to advance the interests of banks to a certain extent. Political links are also essential in order to fully grasp the institutional structure and strategies of control which will determine the directions of the banks. Conflicts with the government or power institutions are often avoided by accommodation. To gain greater access to information on policy regulations, members of government are often invited to sit on the board of directors. Similarly, to better understand the banks' performance, the government may invite some to sit on the economic council as regular consultative or advisory members. These relations are transparent and formal.

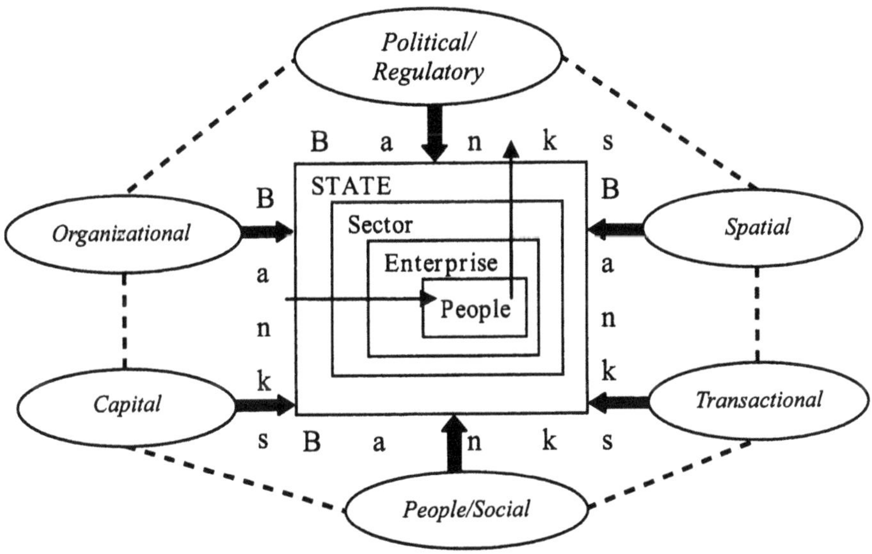

Figure 11.2 Relations in a bank's network structure

The relations tend to be horizontal when the influence of the State is not so strong, as exemplified by the *laissez-faire* system during the colonial period. In the early days, local banks were the major actors in supplying the financial services needed to local businesses. They capture the local market by relying on the levels of sociality and accountability that they command in the community through their active participation in socio-cultural activities as well as their ability to mobilise large amounts of resources in addition to their own cumulated wealth in private enterprises. Hence, local banks or Chinese banks operate on a people-to-people and enterprise-to-enterprise basis and the nature of operations tends to be personal, organisational and transactional. The market territories of the banks are subject to the manoeuvring social space of the actors, but in close association with the kind of businesses or enterprises that they are dealing with. Since local businesses of the colonial days were mostly centred on the Southeast Asia region and China, the network structure also orbited around the relations with the two market territories. The social and economic spaces are very much intertwined and woven through layers of personal and organisational relations that often transcend national borders. There is, however, no single central actor in this early network structure.

When the influence of the State became strong after independence, the embedded relations at the business level were forced into another structure that turned out to be vertically integrated. The State is now held accountable for the performance of the economy. Strict governance in the form of tight regulations is transmitted to the banks. A star network has taken shape as the State has become the central actor in directing and influencing the banks. Actors within the network respond to information about others who have acted recently. Interestingly, a similar structure may be observed in the respective bank network as the Chairmen of some local banks occupy top positions in many influential organisations, private or public.

The *laissez-faire* mechanism in the pre-independence era was related to the inextricable links between the social, cultural and business networks of the respective ethnic groups. During this period a dualistic economy prevailed, a Euro-centric sector that was supported by the colonial rule and an Asia-centric sector that was largely led by immigrants from China and India. While the colonial government provided the basic physical infrastructure to let the *entrepot* trade thrive, the supplies of other basic needs were very much left to the immigrants themselves. The credit facilities offered by British or other foreign banks are inaccessible in terms of costs and due to the lack of collaterals. Some may have access to foreign banks through the compradors.

However, networks of individuals and organisations help to cater to local needs through the pooling of resources or the building up of resources on a collective base. The network of individuals comprises individuals who have displayed enormous achievements in their respective businesses in order to be able to contribute and support community activities either in finance, education, business or clan associations or clubs. They can also be established professionals such as doctors and lawyers. They become directors of local banks and deposit their money as a demonstration of trust and moral support. To cater to the needs of his own dialect group, an established Chinese merchant would pool his own resources together with that of other business leaders to set up a bank. Since each dialect group tended to specialise in a different line of business, these banks serve to complement each others' needs rather than compete among themselves.

In comparison to European trades and commerce, the operations of Chinese immigrants were of a much smaller scale and locally or regionally oriented.

These were mainly in retailing services and the re-distribution of tropical products marketed by smallholdings and miscellaneous consumer items such as staple foods and beverages and basic clothing imported mainly from China. Hence local banks are of smaller size and actually run on a less formal basis and tight liquidity (Figure 11.3).

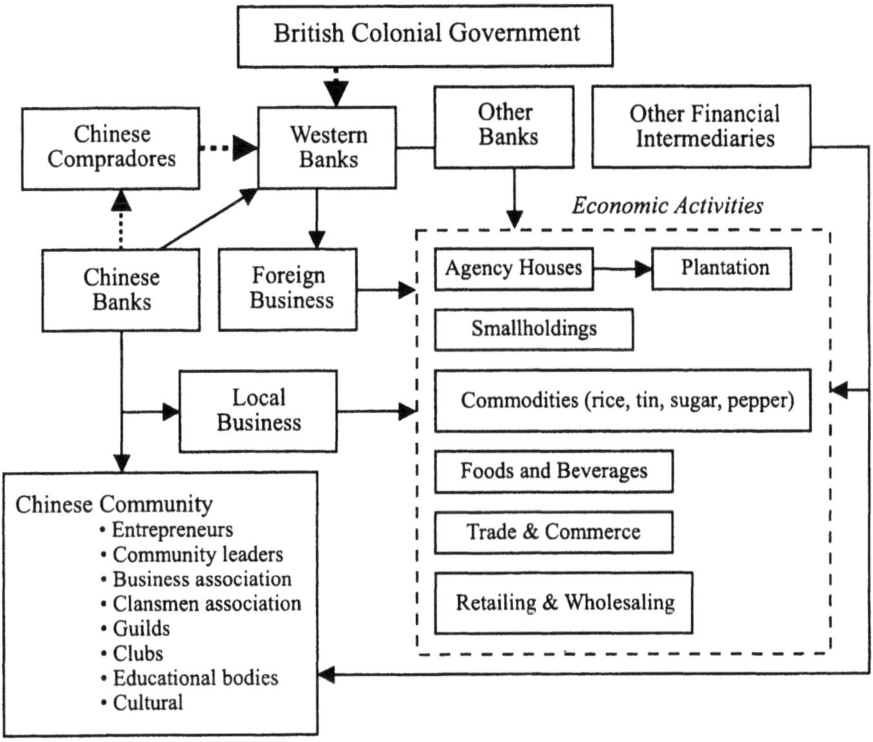

Figure 11.3 Networks of Chinese banks during colonial rule (laissez-faire system)

The guided system generates forces that compel banks to modernise and accommodate the needs of the State. The stark dichotomy in the supply of services between foreign and local banks is blurred as local enterprises also participate in export-oriented activities and borrow from foreign banks. The market becomes segmented as the division of labour increases. The embedded social networks of local banks persist but more formal networks appear and more relations are established with the authority as they participate more widely in the national economy. More resources and activities are added to the litany of banks' functions and services as they help finance not just the retail and wholesale trade and commerce sectors but also manufacturing, transport and communication, property and real estate development (Figure 11.4). The networks of individuals shrink as competition toughens while the networks of organisations expand. A hierarchical framework is forged whereby local banks will act under the directions of the State. Government banks will set an example for them to follow. Banks may not be able to stretch to their maximum due to regulatory checks on integrity and credibility. Old networks that do not fit in with the modernised banking practices and management routines may become redundant.

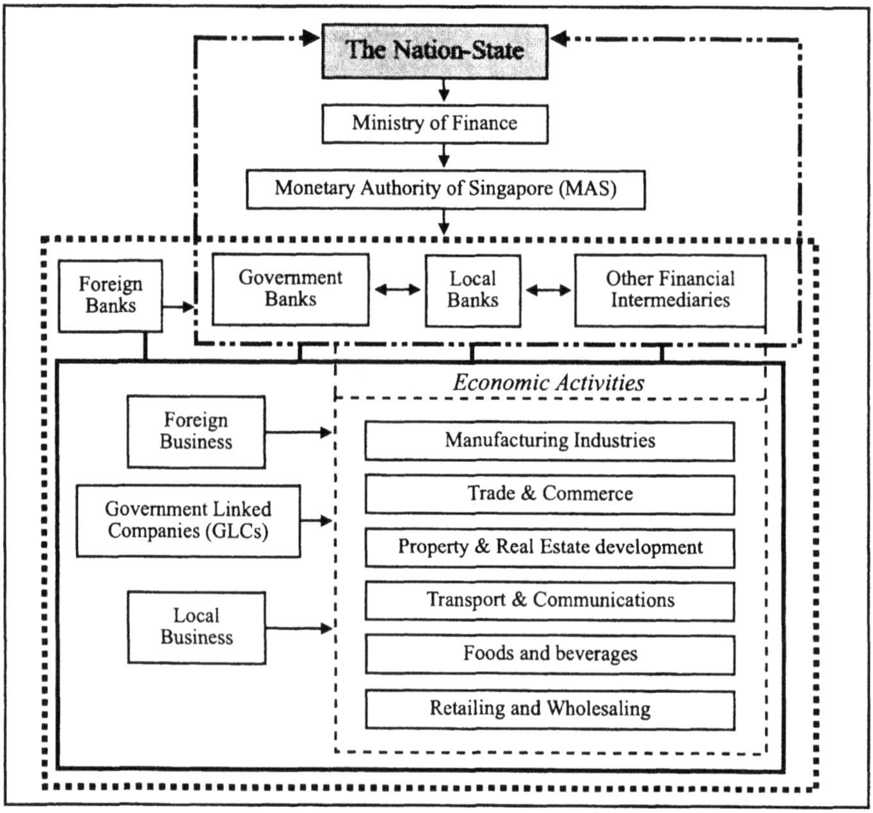

Figure 11.4 Networks of local banks after independence (developmental state system)

Evolution of Local Banks: Processes, Development and Change

The State-led development strategy has expedited the process of development of the small city-state. Within a short span of about thirty years, it was labelled a developed country. As with any industrialised economy, about a quarter of its GDP is derived from manufacturing. Deprived of a primary resources sector, the rest of its economy is driven by the tertiary sector or services sector. Transport and communications, tourism and construction are its other major sources of income as the importance of *entrepot* trade declined. From the 1970s, finance and banking was promoted to become one of the key sectors, raising its relative importance in the nation's GDP, even superseding the manufacturing sector in subsequent years (Figure 11.5).

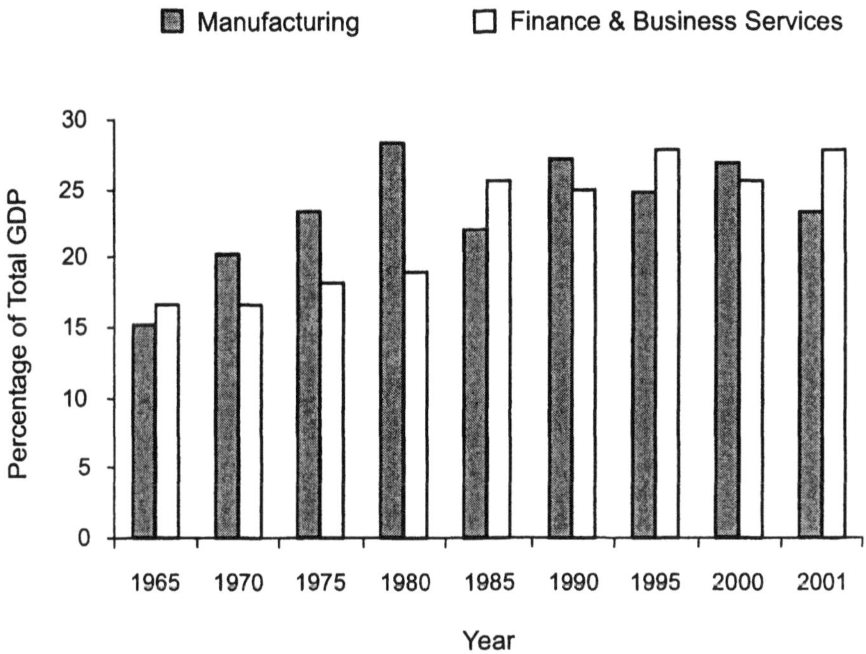

Figure 11.5 GDP contribution by industry, Singapore, 1965-2001

Source: Yearbook of Statistics, Department of Statistics, Singapore.

There were about 136 commercial banks in Singapore in the year 2000 (Table 11.1). The number of local banks (all are full license) stagnated at around 12 until 1995, which is about half the number of its foreign counterparts (Figure 11.6 and 11.7). These local banks are however dwarfed by foreign banks which total more than 200, including offshore banks.

Table 11.1 Commercial banks in Singapore by number and type, 1970-1998

Year	Local Full Li-cence	Foreign Total	Foreign Full Licence	Restricted	Off-shore Banks	Total No. of Offices	Foreign Rep. Offices
1970	11	26	26	0	0	37	8
1975	13	57	24	12	21	70	38
1980	13	78	23	13	42	91	44
1985	13	116	23	14	79	129	50
1990	13	128	22	14	92	141	45
1995	12	128	22	14	92	140	57
2000	8	128	22	14	92		

Source: Monetary Authority of Singapore, Annual Reports, various years.

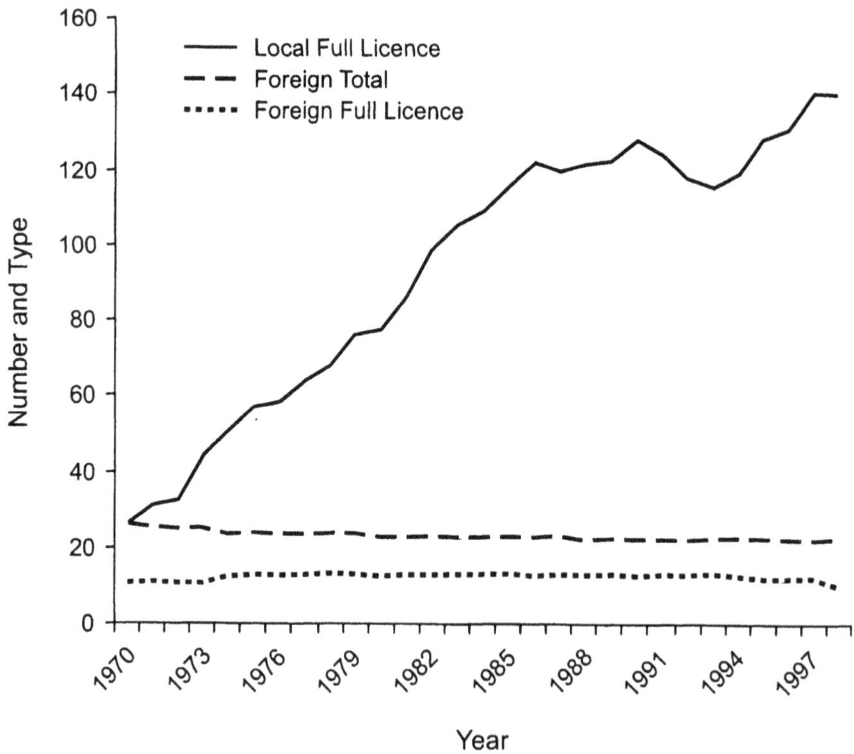

Figure 11.6 Commercial banks in Singapore by number and type, 1970-1998

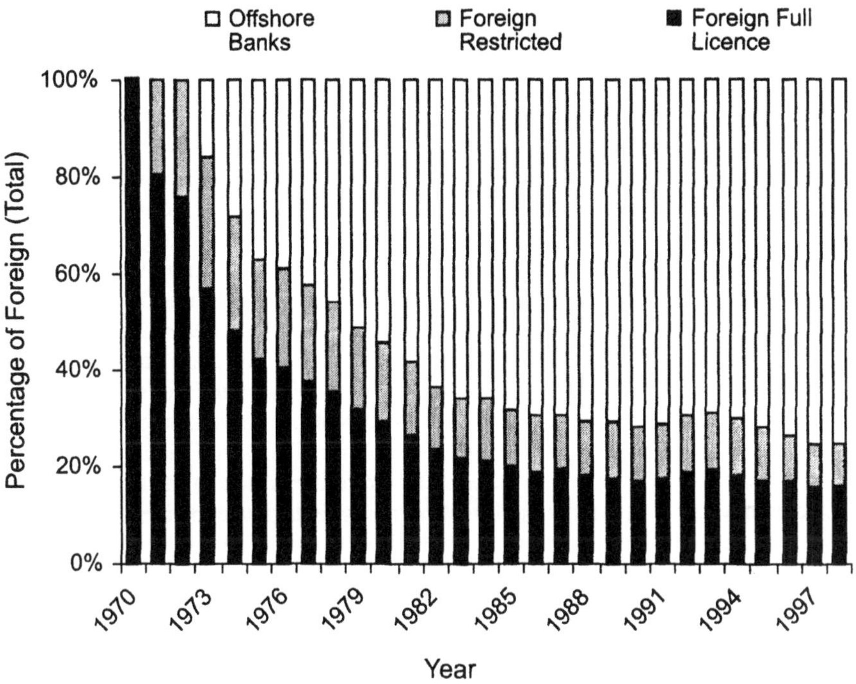

Figure 11.7 Foreign commercial banks in Singapore by type and percentage, 1970-1998

Despite the dominant presence of foreign banks, local banks showed remarkable performance over the past four decades and offer services comparable to those of foreign banks. The State's protection and support contribute greatly to the strength of these banks. Among the local banks, the government development bank or the Development Bank of Singapore (DBS) is the most aggressive and innovative. DBS stands out from the rest in terms of resources, management, and investment practices. It is one of the few banks that are ranked with others in the top 500 banks in the Asia Pacific region in terms of assets and sales performance. The government's share of the bank was reduced from 58 per cent to 38 per cent in 2001 as a gesture of privatisation. To set an example to others in promoting regionalisation and infusion of foreign talents, they have expanded their services furiously to non-traditional hinterland areas and now employ foreign professionals to head the top management positions. Though differentiated by business areas, management style and ownership patterns, DBS and local Chinese banks share the common problems of not being able to make full use of their excess resources. The endogenous communal-based Chinese banks and institutionalised bureaucratic DBS make them conservative and bereft of the ingenuity to undertake more risk-taking activities. Despite this, DBS has aimed to increase the contribution of its overseas profits from 10 per cent to 50 per cent in three to five years' time (Far Eastern Economic Review, 29 March 2001).

Unlike DBS, the Chinese banks are *de facto* local banks. As in the West, local banks grow out of the needs of a booming domestic economy. These banks were formed by Chinese immigrants, who formed the majority of the population since the early part of the twentieth century. Banks were also established by Indians, but these were short lived. The critical mass of the population size and the flourishing trading and primary activities that they engaged in might have paved the ground for the formation of such banks. Today many of these Chinese banks are either closed or have merged with stronger ones. Their activities have also gone beyond the trading and production of rubber and tin as they venture deeper into other financial and non-financial sectors. How exactly have these banks emerged? Can it be explained by relating to communal links? What made these banks different from government banks' like DBS? Why have other communal banks failed to grow or exist? This chapter will focus more on these banks and highlight the processes, backdrops and forces that make them thrive in relation to the social, cultural and economic networks and the political economy factors as explained earlier.

To rise to the challenge of liberalisation and globalisation, local banks have been asked to combine resources. MAS chairman, Deputy Prime Minister Lee Hsien Loong exerted pressure by stating that there was room only for two major local banks (Sunday Times, 28 March 1999). To lead the merger movement, DBS and PosBank were merged in 1998, elevating DBS' world ranking from 90th to 65th (Straits Times, 25th July 1998). In 2000, there were 8 local banks (Table 11.2). Following a series of mergers and acquisitions, only three local banks are identifiable: DBS, OCBC and UOB. It is a good question whether there will eventually be only two banks in the Republic.

Table 11.2 Major local banks in Singapore in the year 2000

Ranks	Institution	Assets ($mn)	Profits ($mn)	Profits % of		500 Rank
				Assets	Equity	
1	Development Bank of Singapore (DBS)	64,982	802.1	1.2	13.2	31
2	United Overseas Bank (UOB)	38,304	527.2	1.4	13.5	56
3	Overseas Chinese Banking Corporation (OCBC)	34,484	485.1	1.4	10.3	63
4	Overseas Union Bank (OUB)	26,915	314.8	1.2	10.6	79
5	Keppel Tat Lee Bank	15,007	174.1	1.2	10.8	131
6	Industrial and Commercial Bank	2,229	18.2	0.8	4.1	347
7	Far Eastern Bank	411	5.0	1.2	7.2	447
8	Bank of Singapore	220	7.7	3.5	4.3	483
	Total	182.6bn	2.3 bn			

Source: Far Eastern Economic Review, 2000. Formative Years before Independence.

Singapore was a British colony and part of Malaya before achieving independence in 1965. Its geo-strategic location in Southeast Asia, sheltered harbour and proximity to main centres of production gave it the natural advantages to be developed into a thriving entrepot. This allowed it to facilitate the trade of raw commodities and manufactured products between the East and West. However, it is not only located favourably geographically but the production, exchange and transhipment services of the commodities made possible by immigrants, local merchants and foreign interests catapulted the city-state to its later success. The volume of trade and shipping demand gave it the economies of scale to expand and excel. The interconnected and closely-knit network of business interests with the hinterland region further reinforces its status as the business hub of the region. When rubber became the major commodity export of colonies under the British and Dutch regime, Singapore became indispensable as it drew on an extensive range of services – management, processing, marketing, financing and investment.

The booming entrepot trade demanded the ready supply of credits to finance and facilitate the exchange of goods and services. The first bank was thus established in Singapore in 1840, the Union Bank of Calcutta. The British colonial government also set up the Post Office Savings Bank (POSB) in 1877. Later, large British banks operating from India also set up their branches locally to provide financing to trading merchants. Singapore's three main European banks (Mercantile Bank (1856), Chartered Bank (1859) and Hong Kong and Shanghai Bank (HSBC) (1877)) had been established prior to the expansion of commodity production, so there was no difficulty in financing such exports. The First National City

Bank of New York, an American bank, also joined in 1902. The booming re-export trade also attracted foreign capital and promoted the development of insurance, shipping and communications, which were integral to the banking industry. The British colonial government set in place in both Malaya and Singapore an efficient banking framework, imposing its laws and regulations.

To function as the centre of commodity exchange of the region and for the British Empire, the island was henceforth introduced into the London money market. Short-term credit for trade and commerce came primarily through the branches of a few European banks headquartered in London. Under the British regime, institution of the sterling exchange standard made the currency of a colony like Singapore freely convertible at a fixed rate of exchange with sterling.[1] By linking with the sterling system, debts were payable and capital could be sourced from the United Kingdom.

British banks adhered strictly to a policy of short-term commercial lending while long-term finance was partially filled by the managing agency system. The agency system was mainly used for European estate agriculture that required high investment costs and a long period of gestation before maturity. From the turn of the twentieth century, Singapore merchant houses transformed themselves into agency houses and were central to the establishment of a large European rubber estate sector and assumed a major commercial role in it.

The Chinese on the other hand tended to engage in peasant or smallholder rubber export production that required smaller capital investment compared to the European estate system. The smallholder outputs were often acquired by local Chinese dealers who then sold it to the agency houses. Chinese traders and rubber millers in Singapore built up a complementary relationship with dealers in hinterland Malaya and Indonesia. The arrangement was an interlinked transaction as dealers who marketed primary commodities through Singapore also obtained consumer goods and credit as part of the same exchange. The credit from Singapore Chinese, usually in the form of goods, worked its way through a network of traders in British Malaya and the Dutch East Indies to finance growers. Singapore's role as a centre for such interlinked transactions reflected the fragmentation of hinterland export trade and the absence of well-developed hinterland markets for consumer goods and credit.

By 1900 Singapore became the world's main centre for milling smallholder rubber and tin smelting. The expansion of hinterland production made possible the introduction of specialised marketing facilities on a scale sufficiently large to take advantage of the considerable economies of scale associated not just with port and storage facilities but also trading and financial services. Hinterland producers looked to the Singapore port as the focus of a characteristically complex network of dealers, commission agents, brokers and exporters, who purchased and assembled the commodity for shipment abroad. Singapore thus linked the

---

1 For Singapore, this standard was adopted in 1906 when the Straits Settlements dollar was pegged at 2s 4d (a parity which remained until 1967) and was backed throughout this period by 100 per cent to 110 per cent sterling reserves.

hinterland marketing network to metropolitan centres. By providing such specialised function, Singapore established itself as the world's greatest market for rubber and tin.

The European banks mainly served the interests of the colonial government and foreign enterprises that engaged in trade with primary commodities and exports to and from Europe. Singapore was the main centre for finance in British Malaya, but had no local stock exchange. European banks in Singapore would not finance rubber estate development, either European or Chinese, or even advance to cultivators against crops. The banks considered the risk of locking up money for such a long time too high. Divergent property rights prevented Dutch banks in Singapore like the Netherlands Trading Society from lending long-term to rubber estates, as they did in the Dutch East Indies.

The local community, especially the overseas Chinese dealers and traders who also were actively engaged in the entrepot trade, found difficulties in accessing the credit facilities of these banks. In 1896, fewer than 50 Chinese traders had European bank accounts. The strict regulations of these European banks and their reluctance to accept small accounts further made them inaccessible to many locals. To overcome cash flow problems, especially for the trading of products between Malaya and China and other parts of Southeast Asia, the Chinese community began to form their own banks. These Chinese banks overcame their cash flow difficulties by transacting business on a close personal basis and mobilised Chinese savings for productive investment and integrated it with the region's credit and capital markets. The first two decades of the twentieth century saw the establishment of the Chinese Commercial Bank, the Ho Hong Bank and the Overseas-Chinese Bank.

A dual banking system consisting of foreign and local Chinese banks took shape. The local Chinese banks were smaller in scale in terms of employment, capital assets and operational network compared to their western counterparts. There were also differences among the local Chinese banks. Each bank catered specifically to the interests of its respective clients, who tended to come from the same ancestral place or dialect group as the founder of the bank. The OCBC, for example, looked after Chinese who came from the Fujian province of China, that is, the Hokkiens, while the Kwong Yik bank, catered to the Cantonese from the Guangdong province. Other ethnic-based banks such as the Indian banks came into existence much later. The Indian Overseas Bank was founded in 1937 and the Indian Bank in 1941.

These Chinese commercial banks performed the same functions as the foreign banks. They financed the distributive trade, construction and rubber industries, and granted personal loans. They were mainly interested in granting short-term advances to avoid runs on liquidity. They also provided remittance facilities to their clients. They adopted Western banking principles and practices from the foreign banks, such as accepting deposits, providing loans and advances, financing trade bills and transacting foreign exchange. They did not, however, have foreign networks to deal with foreign exchange.

The lack of adequate knowledge of local traders in the early days also prompted Chinese banks to adopt the *compradore* system initiated by Western banks. Under this system, prestigious wealthy middlemen helped introduce credit-worthy clients to the banks and acted as guarantors. In return, these *compradores* were given a commission. Banking transactions that were operated under such a patron-clientele system relied heavily on mutual trust and long-term banking relations. Hence they tended to be more liberal in their lending practices, which subsequently led to the failure of some.

These Chinese banks lacked the flexibility of Western banks in terms of cash flows, reserves and diversity of investments. The assets of Singapore's Chinese banks were without any doubt small compared to those of the European banks. Furthermore, the ability of Chinese banks to lend was seriously constrained by the ever-present danger of severe contractions in the money supply caused by sharp falls in rubber prices. At such times of liquidity crisis, the Chinese banks were highly exposed to risk due to the absence of either a central bank in Singapore to act as a lender of last resort or access to an external wholesale credit market, except possibly through the city's European banks. They thus failed to withstand the impact generated by the Great Depression of the 1930s.

The commercial opportunities of the city-state invited huge flows of immigrants who were later absorbed into the labour force to contribute further to the expansion of the local economy as well as that of the hinterland. The savings and remittances of this Chinese labour formed one of the major source of deposits and transactions of Chinese banks. The export-led growth, the need for social exchange and community leadership in the business sector and dealings with the government encouraged the formation of the Chinese Chamber of Commerce in 1906 and later the Chinese Produce Exchange.

Besides rubber and pineapple cultivation, the processing activities of the two commodities from hinterland Malaya and Indonesia became the most conspicuous source of Singapore Chinese wealth and provided a springboard for industrialisation and the growth of local Chinese deposit banking. After 1910, major administrative and economic changes affected Singapore's Chinese business community. The monopoly rights to collect taxes on the sale of opium, which had formerly yielded Chinese syndicates substantial profits, was abolished. New technological requirements and greater capital intensity in tin mining and local shipping had turned these once predominantly Chinese activities into European-dominated industries. Meanwhile the tropical produce and rice trades stagnated. The advent of rubber offset such unfavourable changes and fuelled an expansion of Chinese economic interest.

The subsequent plummeting rubber and tin prices in the 1930s brought with it the collapse of the commodities market. Local banks suffered heavy financial losses as their major borrowers - the tin and rubber merchants - were unable to repay their loans. The banks suffered more losses as Britain went off the Gold Standard and devalued the pound sterling in 1931. As a result, the three Hokkien banks (Chinese Commercial Bank, Ho Hong Bank and the Overseas Chinese

Bank) were merged to form a consortium known as the Overseas Chinese Banking Corporation (OCBC).

Although Singapore's European banks efficiently met the needs of export trade and helped to lay the basis for subsequent financial development, they did relatively little to push financial sector growth beyond its early stages. The European banks had not been responsible for the early development of trade, nor did they subsequently make Singapore an independent financial centre. Finance from the banks was mainly in response to demand in Singapore itself and dependent on primary commodity exports traded through the city. Singapore's pre-WW2 capital market remained weak and fragmented. There was as yet no local stock exchange, and the banking system was the only significant European institutional source of finance. The ability of larger European businesses to borrow on the London capital market and the comparative lack of European interest in the manufacturing industry in Singapore gave European banks no strong reason to depart from a policy of finance against stocks only and not for capital expenditure.

Besides the banks, a number of brokers, organised as the Singapore Exchange Brokers Association, also helped merchants to obtain competitive rates when dealing with the banks. For exports, finance from the European banks frequently involved a system of trust receipts against goods stored in warehouses or godowns. Small dealers without warehouse facilities lacked direct access to European bank finance. They depended on European merchants to store their produce and make advances against it. The produce was stored and held 'on trust' by the merchants for a certain period during which the dealers could close the sale at the price then current. European merchants were in turn financed by the European banks. The property in the stored produce was vested in the bank, and the term 'trust receipts' referred to the fact that the European merchants held the produce in trust for the bank.

In Singapore, Europeans responded to the problem of financing estate rubber development by transforming merchant houses into managing agency houses, and using the latter to float rubber companies on the London stock market. Through these activities, Singapore houses had a pioneering role in the development of Malayan estate rubber. The long experience of Singapore merchant houses in British Malaya, the existing business relationships in supplying and financing proprietary planters and the well known position of large merchant houses in Eastern trade gave prospectors confidence in their investment. Limited liability and public status for rubber companies meant that overseas shareholders bore most of the financial risk of rubber estate development.

The Singapore Chinese gradually built up their financial resources by accumulating profits from their estates in British Malaya. Constrained by size and capital outlay they developed rubber estates largely in conjunction with pineapples. When rubber was first established on the island, they often planted rubber seeds along with other crops, generally pineapples or gambier, and kept it in cultivation until the rubber trees had matured. The quick maturing fruits produce a cash flow to help meet outgoings, and could also be abandoned without much loss when the rubber trees were mature. There was no sacrifice in terms of rubber density, since the grown trees required the inter-row space occupied by the pineapples. With the cash crop and less elaborate fixed capital expenditure compared to the Europeans, it cost roughly $150 per acre to develop a Chinese estate, or about a quarter of the investment per acre necessary for European estates. Finance on this scale was within the reach of Singapore Chinese entrepreneurs. During the 1920s, this practice spread. By 1932 the bulk of Chinese rubber estate acreage in Johore was being developed with a catch crop of pineapples.

During the inter-war period a few of the most important Chinese businessmen could borrow from European banks to finance exports and stocks in progress, for example in rubber milling. However for all Asian entrepreneurs, the poor Singapore market for long-term institutional finance was a constraint. The development of a local Chinese banking system helped to overcome this, as did the planting of pineapples to finance rubber estate development. Nevertheless, Asian entrepreneurs could not borrow in London, and secure long-term finance was difficult to obtain. Singapore's capital market was thus segmented and Asian entrepreneurs borrowed mainly from Chinese banks or from a large informal sector including Chettiar moneylenders and rotating credit and savings associations (chit funds).

Under the currency board system and sterling exchange standard in the Straits Settlements, the adverse movements in the current account of the balance of payments, which almost certainly followed any substantial fall in rubber prices, did not have to lead to a large contraction in the money supply. European banks were protected from Singapore's liquidity crises consequent on sharp contractions in the money supply, since London, where their head offices were located, effectively acted as a central banker. But these liquidity crises had a severe effect on the Chinese business community and anyone dependent on the informal financial sector. Chinese banks, unlike European ones, did not have direct access to an external wholesale credit market, and Singapore had no central bank to serve as a lender of last resort. Hence unlike the European agency houses and banks, which survived the 1930s slump without undue difficulty, several of the largest Chinese businesses were forced into amalgamation or liquidation, while bankruptcies among smaller Chinese firms multiplied.

The assets of Singapore's Chinese banks were without doubt small compared to those of the European banks. Furthermore, the ability of Chinese banks to lend was seriously constrained by the ever-present danger of severe contractions in the money supply caused by sharp falls in rubber prices. At such times of liquidity crisis, the Chinese banks were highly exposed to risk due to the absence of either a central bank in Singapore to act as a lender of last resort or access to an external wholesale credit market, except possibly through the city's European banks. Between June 1920 and January 1921 total deposits in Singapore's five local Chinese banks fell by a third; similar contraction occurred in the six months from December 1931. These considerations probably encouraged Chinese bankers towards a conservative reserve ratio policy. As a rule, Chinese banks kept 50 per cent of their current account money liquid, and maintained large reserves as currency and deposits with the British banks. In the interwar period the Chinese and British banks remained complementary because of this deposit reserve practice and the fact that the Chinese banks developed a constituency which their British counterparts were largely unable to reach.

Hence, first tin, next rubber and pineapples, and then the associated industrial activities were fundamental to the development of Chinese banking, which depended on successful Chinese businessmen as founders and organisers. The first Chinese bank, Kwong Yik Bank (1903), had a large gambier planter as its principal promoter and managing director. The second, Sze Hai Tong Bank (1906), had a produce merchant as its leading shareholder. However, the Kwong Yik Bank failed in 1913, while the Sze Hai Tong Bank never became very significant. The Ban Hin Lee Bank founded in 1918 was closely connected with Chop Ban Hin Lee and its rice trade. The Lee Wah Bank specialised in remittance business.

Given the need for mercantile sponsorship, a major drawback of these local banks was that their promoters had a base in either a declining or a small sector of the economy. The sponsors associated with the rubber-pineapple complex and the Chinese industrial conglomerates later helped to overcome this limitation. They gave the Hokkien banks a status that instilled confidence in a wider Chinese public by providing substantial deposits or becoming directors of the banks. During the interwar years, Singapore's Chinese banks functioned like the financial syndicates of the Chinese community or the rich men's clubs like the Ee Ho Ean, which Chinese rubber magnates frequented. The Hokkien banks succeeded in spreading the banking habit by attracting a broadly-based clientele who learnt to make use of the banking facilities.

Without the crucial personal and economic links with those involved in the rubber-pineapple complex and the associated Chinese industrial conglomerates, it is doubtful that the local deposit banking could have achieved its strength. The Chinese Commercial Bank and the Ho Hong Bank depended heavily on rubber. Lim Nee Soon was vice-chairman of the Chinese Commercial Bank, and Lee Kong Chian served on its board of directors and later as vice-chairman. Both were rubber magnates. The first chairman of the bank's board of directors and one of its founders, Lee Choon Guan, had large rubber interests. So did another founder, Tan Chay Yan, who was a pioneer planter. Another man in the rubber business, See Boh Ih, was a director and managing director of the Chinese Commercial Bank. The Ho Hong Bank was largely an offshoot of the Ho Hong conglomerate, and like the Chinese Commercial Bank, it was supported by Lee Choon Guan. Important rubber estate owners closely involved with the bank included Tan Cheng Lock, Chee Swee Cheng and Lee Choon Seng. Tan and Chee were among its first directors, while they and Lee were later associated with the OCBC. Rubber also played a conspicuous role in the founding of the Overseas Chinese Bank. Its founders included two rubber magnates, Lim Nee Soon and Tan Ean Kiam. Lim became the bank's second chairman while also being the vice-chairman of the Chinese Commercial Bank. In the early 1930s, Tan was almost the idol of the Hokkien community and served as the Overseas-Chinese Bank's managing director.

During the 1930s slump, all three Hokkien banks encountered financial difficulties which led to the merger and forming of the OCBC in 1932. One of the four architects of the merger, Yap Twee, managing director of the Chinese

Commercial Bank and proprietor of a successful hardware business, was an ally of Tan Kah Kee. The other three had extensive rubber interests. They were Chee Swee Cheng, owner of rubber estates including 5,000 acres in Borneo and founder of the Atlas Ice Company, who was OCBC's first chairman; Tan Ean Kiam, who emerged as sole managing director soon after the bank's formation; and Lee Kong Chian, who became the OCBC's second chairman in 1938. At that time Lee Choon Seng was elected vice-chairman of the bank. Thus in 1939, rubber entrepreneurs filled the bank's top three positions. The United Chinese Bank (UCB), founded in 1935, boosted its strength by employing two former senior officers of the Overseas-Chinese Bank and the Ho Hong Bank. Like other banks, UCB depended on the backing of prominent merchants.

By giving hinterland producers access to new markets in the West, Singapore promoted a development that was based on export expansion in the region, namely rice, tin, rubber and timber from Malaya, Indonesia, Thailand, Myanmar and Vietnam. Simultaneously, Singapore stimulated the export expansion of the West by linking the West to the expanding regional market for food and manufactured goods and other materials necessary for economic specialisation. As an intermediary of trade flows and supply of services arising from the international and regional economies, Singapore helped to transform the surrounding countries and was itself transformed in the process.

Modernisation, Diversification and Expansion Years: 1965 – 1979

The recovery of the economy in the post-Second World War period brought about the start-up of four more local banks, namely, the Overseas Union Bank (OUB), the Chung Khiaw Bank, the Industrial and Commercial Bank and the Bank of Singapore. The opening of more local banks finally saturated the market for financing the trade of rubber and tin. Some began to carve their own niches. OUB, for example, established linkages with Thai and Burmese rice farmers and by the 1950s, it financed up to 80 per cent of the Thai rice trade in the region.

The nature of their operations began to change, though their traditional links with the region remained strong. After gaining independence in 1965, besides the chronic scarcity of natural resources, the Republic was plagued by political problems related to the British withdrawal, the break-up of the Malaysian Federation, confrontation with Indonesia and the imminent threat of communist infiltration. To deal with the more immediate socio-economic problems of a high unemployment rate and small domestic market, an active foreign investment-based, labour-intensive and export-oriented industrialisation programme was launched.

The industrialisation programme demanded large capital flows and investment intensity that largely came from abroad. Local banks, on the other hand, serviced the needs of local small and medium-sized industries but evidently lacked the capacity for the new industries. The Development Bank of Singapore (DBS) was set up in 1968 to provide long-term credits to industries. Generally, local commercial banks are supportive as they serve as directors of government-linked companies or statutory boards while former officials filled up senior positions in private banks. Local banks worked in partnership with the State in building the nation.

The external environment of the sixties benefited Singapore's economy. Regional oil exploration activities, the rapid economic growth of neighbouring countries, spill-over effects of the Vietnam War and Hong Kong's political developments boosted the demand for goods and services which Singapore quickly sought to provide. Singapore's outward-looking, export-oriented industrialisation policy was timely as world trade expanded and the relatively protectionist attitude of other developing countries helped it to capitalise on emerging opportunities.

To further stimulate the inflow of foreign investment to boost its manufacturing sector, the government introduced the Asian Currency Unit (ACU) in 1968. Banks engaged in ACU were allowed to accept deposits of non-residents in US dollars and other key currencies. This was attractive to investors, as tax was not charged for interest earned by foreign depositors in this market. The popularity of the ACUs fuelled the process of industrialisation and at the same time turned Singapore into a regional financial hub, providing funds not only to industries on the island but also to the development of industrial and agricultural projects in the ASEAN region. Most of the funds were inter-bank borrowings, thus the ACU acted as a source of funds for the region's commercial banks. The ACU assets were valued at US$30.5 million in 1968. Other new financial instruments, such as Asian dollar floating and fixed rate certificates of deposit, Asian dollar bonds and Asian commercial papers, were gradually introduced to attract broader off-shore financial activities to Singapore. The roles of the banks as credit suppliers to commodity merchants remain but the nature of their business has increasingly shifted towards dealing with manufactured and professional services like property development, foreign exchange, shares speculation, estate management and stock exchange.

By the seventies, Singapore reached the take-off stage (unlike most other developing economies which were badly affected by the oil crises). While the export-led manufacturing industry had helped cushion the impact of the oil crisis, petro-dollars flushed the markets with the cash that was needed to help fund the region's recovery activities and new developments. Singapore with its more developed financial sector again became the centre of such flows. To strengthen the financial sector, the Monetary Authority of Singapore (MAS) was set up in 1971. The objective was to attract a larger presence of international banks and to supervise and regulate the expansion and diversification of local banks. Four types of licenses were issued to banks, namely a full license and a restricted license for commercial banks, an ACU license and an offshore license. Fully licensed banks could operate fully in the domestic and international market. All banks that received banking licenses prior to 1971 were issued full licenses. Restricted-licensed banks were free to participate in the international market, but had limited access to the domestic market. The development of local banks in terms of services diversification at this stage was evidently state-led.

In the seventies, government-linked banks, namely DBS and the Keppel Bank and new local commercial ones such as the Tat Lee Bank joined the original group of local banks. The traditional British banks operated side by side with numerous other foreign banks from the USA, Japan and Western European countries. The increase in competition brought about better customer services, more competitive prices for financial services, stable difference in the deposit and lending rates and greater variety of securities. However, the original local giants, namely, OCBC, UOB, OUB and DBS continue to dominate. The smaller banks were the Industrial and Commercial Bank, the Asian Commercial Banking Corporation, the Far Eastern Bank and the Tat Lee Bank. As in the 1930s, the worldwide recession of the late seventies led to a series of acquisitions and mergers. OUB for instance, merged with the Chung Khiaw Bank in 1971. OCBC, after incorporating the Bank of Singapore, was merged with the Four Seas Communications Bank in 1972. In the same year UOB bought the Lee Wah Bank.

The seventies were a period of diversification as local banks went into joint ventures with other merchant banks, discount houses and leasing companies with very reputable international financial institutions. They also undertook regional expansion by opening overseas branches in neighbouring countries and other major financial centres of the world, notably New York, London and Tokyo. In 1975, to further bolster the financial strength of the Republic in overseas ventures, local banks partnered with the State by forming a consortium called the International Bank of Singapore. It is an equal partnership of OCBC, OUB, UOB and DBS. It established branches or representative offices overseas and actively opened branches in various cities including Seoul and Manila. At the same time local giants established branches and/or agencies in Malaysia, Hong Kong, Tokyo, London and New York. Nevertheless, the East Asian region remains the market stronghold of these banks. Physical proximity, market familiarity and cultural affinity developed during the colonial years have most likely influenced such preference, besides the advantage of a time zone which allows round-the-clock dealings.

Another external development in the 1970s triggered by the breakdown of the Bretton Woods System forced an inevitable change in the financial system of Singapore. Under the Bretton Woods system many developing countries tapped on the monetary stability supported by the U.S.A. as the value of the US dollar was pegged against gold reserves. During this period the US also exported large sums of money to the developing world. However, the increasingly huge volumes of trade finally outgrew the growth of the US gold reserves. The glut of US dollars in the world market was worsened by the defeat of the USA in the Vietnam War. There was an exodus of funds to Japan and Europe. The US government devalued the US dollar in 1973 and stopped backing the dollar by gold, triggering a massive wave of devaluation. The Singaporean economy could have been seriously affected by the dollar depreciation as it had strong trading and investment relationships with the USA, Japan and UK. To prevent imported inflation, the Singapore currency was floated in 1973. The Sterling was finally replaced by the US dollar as the intervention currency to 'manage' the floating of the Singapore dollar with a basket of ten currencies. The Singapore dollar was thus prevented from excessive fluctuation, thereby stabilising the economy. Positive growth returned, though its GDP growth rate fell to a low of 2.8 per cent in 1975. The spill-over effects of the oil boom in Indonesia and Malaysia and the availability of funds in the Asian Dollar market certainly also helped to solve the country's liquidity problems. The funds peaked at US$10,357.3 millions by 1974. The government also encouraged banks to maintain local rather than foreign liquidity. Huge inflow of funds poured into Singapore as local banks drew finance from abroad and foreign banks used head-office funds for financing domestic investments.

Unlike other developing countries, the Republic had low external debt, owing to the availability of a large pool of public funds. During times of economic crisis, the resources from the Post Office Savings Bank (POSB) and the Central Provident Fund (CPF) were unlocked. In 1972, the POSB was turned into a national savings bank by offering guarantees of up to S$100,000 of deposits and granting tax exemption on interest earned on deposits in the bank. This prompted the accumulation of large pools of savings. The restrictive regulations discourage the easy withdrawal of CPF and further buttress the country's financial resources. Such a ready supply of capital greatly reduced the government's need to borrow money from abroad and contributed to the strong Singapore dollar.

With a strong international payments position, the government eventually liberalised the foreign exchange market in 1978, enabling the free flow of trade-related transactions and capital-related receipts and payments. For instance it replaced the cartel system of foreign exchange rate quotation by commercial banks with competitive interest-rate setting, issued Singapore dollar-negotiable certificates of deposit and removed other exchange controls. These measures served to decontrol prices (exchange and interest rates) and open the country's current account, thereby boosting foreign exchange, money, and capital market transactions. The moves also paved the way for the development of the Asian Dollar Market and Asian Bond Market.

Consolidation and New Roles: 1980 Onwards

In the early eighties, the government attempted to enforce its restructuring plan, promoting high-tech industries, R&D and services. A number of government companies were privatised and became listed companies to add large amounts of assets to the main share market. Before the plan could take full force, another world recession occurred in the mid-eighties. Singapore had to confront problems of declining demand for its main export, electronic goods, high labour costs, strong Singapore dollar, rising protectionism and equally attractive incentive packages in its regional counterparts such as Malaysia, Indonesia and Thailand. The sluggish global demand for commodities and the declining prices affected the agriculture-based ASEAN economies, and in turn impacted on Singapore's entrepot, services, oil-related, shipbuilding and ship-repairing industries.

A series of remedial measures were taken to overcome the impasse. Wages and taxes were cut to reduce costs. Poor business resulted in many developers and property buyers defaulting on their loans. This culminated in the collapse of the stock market, and stricter regulations were consequently imposed in the aftermath of the Pan-Electric crisis. In 1985, the economy registered a negative GDP growth of 1.8 per cent for the first time since independence. To combat the recession and to sustain the country's competitiveness, capacity-enhancing measures were introduced with the aim of augmenting the supply and quality of labour and the level of technology and nurturing the growth of local businesses. An attempt was made to beef up the ailing financial sector by injecting an S$100 million Venture Capital Fund. Incentives were given for fund management and S$38 billion worth of bonds were to be issued over the next five years. It also planned to reform the Stock Exchange of Singapore and open it up to banks, merchant banks and foreign stockbrokers.

By 1986, the economy recovered. A high growth rate of over 8 per cent was maintained for the period from 1991 to 1997. By this time, the four local banks (OCBC, OUB, UOB and DBS), and major foreign banks such as Citibank, Standard Chartered Bank, HSBC, as well as local finance companies had benefited from the strong growth of the financial sector. To prevent massive speculation, the government retained tight control over the currency. Being highly dependent on world trade, the government remained cautious and disallowed the internationalisation of the Singapore dollar. Singapore residents and securities are excluded from the offshore international investment management market and the Singapore dollar is excluded as one of the financial futures contracts on a new international futures exchange. The stabilisation of the currency and strong economic fundamentals encouraged a large inflow of foreign direct investments in the early 1990s.

In the Strategic Economic Plan, the financial sector was one of the key clusters recommended by the Economic Planning Committee[2] to spearhead the growth and development of the city-state in the twenty-first century. The private sector was to play a greater and more active role in steering the economy. Before long, Singapore was engulfed in the 1997 crisis sparked off by its close partners in the region. For the second time, it registered a negative growth of 1.38 per cent between 1997 and 1998. The loss of investors' confidence arising from the crisis led to a massive outflow of funds. Singapore's cost competitiveness was further eroded as the Singaporean dollar appreciated against other currencies in the region which went through a period of devaluation.

In response to the crisis, besides enforcing short-term measures like a cost/tax cutting package through wage restraint and reducing employers' CPF contributions, other long-term measures were also put in place. To boost the financial sector, the government announced several liberalisation measures. Tax incentives were given for trading of swaps and financial derivatives, stamp duty on share transactions was abolished, and development of debt and swap markets was encouraged. A positive growth of 3.9 per cent was registered during the period of 1998-1999. The growth of the financial sector remains sluggish however, partly due to poor real estate earnings, weak commercial lending and poor regional demand for funds.

---

2 The Economic Planning Committee came up with the Strategic Economic Plan in 1991 to chart the economic course for the country in the next two to three decades. The plan's vision was for Singapore to catch up with the first league of developed nations by the decade of 2020-2030.

Singapore's position as one of the leading financial centres of Asia is increasingly threatened. Round-the-clock trading made possible by IT may soon weaken Singapore's strategic position in the Asian time zone. The use of a common currency in the EU will lead to more bank mergers and the formation of fewer but stronger banks. The competition for fluid capital flows has become more intense, and the tendency for money to flow into less regulated markets may get stronger. Although the government's cautious approach to the financial and banking system has enabled the economy to avert the BCCI collapse and tide over the 1995 Barings financial crisis and the 1997-1998 East Asian financial crisis, the government began to rethink its inward-looking and interventionist approach (which has helped insulate industry from external shocks and risks but has also resulted in an under-performing industry and low returns). Singapore was ranked as the most competitive economy in the world in the 1999 competitiveness report by the World Economic Forum (WEF), but scored badly in corporate disclosure and governance. In year 2000, it fell to second position. The lack of foreign access to the capital market and restricted entry into the domestic banking industry and hence the lack of competition may have stiffened the growth and expansion of the banking sector.

In contrast to early liberalisation of the banking and finance sector in debt-ridden Thailand and Indonesia, the two most affected by the 1997 crisis, Singapore has been rather conservative. In 1999 MAS announced a programme to liberalise commercial banking and to strengthen corporate governance of local banks. Competition, not protection, is the key to foster the development of strong and large local banks, capable of holding their own against the major international banks domestically and in Asian markets. To pre-empt the move of foreign banks using electronic delivery channels to reach out extensively to domestic consumers, the government set to neutralise the advantages that local banks enjoy from having an extensive branch network and government protection. Hence the 40 per cent foreign shareholding limit on local banks has been removed. However, since about two-thirds of the deposits of Singapore residents are in Singapore-owned banks, MAS wanted to maintain this share at no less than 50 per cent of the market. The political economy factors continue to sway the logic of economics, efficiency and competition.

To promote competition, a Qualifying Full Banking (QFB) license is now issued to foreign banks which can open up branches and install automated-teller machines in up to ten locations in the country. They may also share a common network or tap the existing ATM ones. However, to ensure that these banks would not leave under adverse conditions, careful selection was made based on the banks' commitment to developing Singapore into a global financial centre. Four foreign banks, namely, ABN Amro, Citibank, Standard Chartered and the *Banque Nationale de Paris*, have received a QFB license. Additionally, MAS has issued more restricted bank licenses. Offshore banks approved by MAS will have their lending limit raised from S$300 million to S$1 billion; all other offshore banks will have their limit raised to S$500 million. Created in 1973, offshore banks cannot accept interest-bearing S$ deposits from persons other than approved institutions.

Offshore banks are now given greater flexibility in Singapore dollar wholesale business. To encourage greater capital flow, MAS has released billions of dollars in the CPF to lure fund managers to Singapore. It has also merged the stock and futures exchange and relaxed the banks' capital adequacy ratio. Although the government is sticking to its non-internationalisation policy on the Singapore dollar, it has relaxed its lending rules for currency and the transaction in options to non-residents.

Local banks are urged to merge into large groups in order to bolster asset size and reap economies of scale. The government initiated the move by merging DBS with POSB in 1998. In 2000, local banks were told about the dim prospects of competing with global players even by merging or forming alliances. They were called on to make cross-border acquisitions or form partnerships with foreign players. Some local banks responded. UOB, for instance, has bought Westmont Bank in the Philippines and is currently building up its network in stock broking in Hong Kong, Indonesia, Malaysia, Singapore and the Philippines. It has also entered into a joint venture with a German corporation to create a China-based asset management company. It openly declared its aims to reap 50 per cent of its annual profit from overseas operations in ten years' time.

The Development Bank of Singapore (DBS), the government-linked development bank and Southeast Asia's biggest bank by assets, has also formed alliances with financial institutions with a specific service or talent. For instance, it formed alliances with fund managers like the Frank Russell Company and Boston-based State Street Bank and Trust Company. It also bought the PT Bank DBS Buana (Indonesia) in 1997. Its acquisition drives included Singapore's POSB, Bank Philippines, DBS Thai Dhanu and Hong Kong's Kwong On Bank, the Bank of South East Asia, Far East Bank and Dao Heng Bank in Hong Kong. It also formed an alliance with the Bank of the Philippine Islands. To expedite the integration of interests in Hong Kong, Indonesia, Thailand and the Philippines, DBS has set up a Regional Integration Centre (RIC).

The island's banking and finance sector quickly captured other emerging opportunities by capitalising on IT-related banking services. In 1999, local banks teamed up with eight other local broking firms to start Internet trading. Kay Hian, Singapore's oldest brokerage, has also gone into fund management and sells other financial products. The MAS is planning to issue an all-in-one finance license for fund managers and stockbrokers to spur greater innovation and competition. Internet banks or e-banks are promoted, as such IT-based services allow customers to transfer funds and trade forex, equity and unit trust online, thereby lowering costs as transactions and information can be processed and transmitted electronically. Settling transactions in virtual space reduces the needs for physical space for offices and buildings. The four local banks are leaders in the regional start-up of Internet banking. OUB set up the first internet bank by teaming up with foreign players. The paid-up capital requirement for setting up such services has been lowered from S$1.5 billion (for normal banks) to S$100 million (for Internet banks) as a means to ensure lower barriers of entry.

Other than DBS, Singapore's local banks remain mainly Chinese family businesses. The chairman of OCBC and UOB, for instance, are sons of their founders. For the UOB, the current deputy chairman is the first grandson of the founder. Doubt is now cast on the viability of traditional practices in the management and business operations of these banks in challenging the global players. The government has strongly criticised the local banks' lack of initiative to modernise and globalise their banking practices. The DPM has commented that it is no longer possible for local MNCs to do business based on family relationship and *'guanxi'*. The failures of some of the major banks in neighbouring countries during the Asian financial crisis are linked to poor business ethics and moral hazards that are closely associated with such traditional practices and hence easily open to the abuse of the legal system. To ensure greater transparency and minimise opportunities for malpractice, a 3-year deadline was set for banks to streamline their financial with non-financial entities. Banks are to concentrate on their core capabilities. Deadlines are also set for local banks to form nominating committees to choose top personnel, especially foreign talents, in their management boards.

Local Banks' Current Position

In 2001 OCBC acquired the Keppel Tat Lee bank and in 2002 UOB merged with OUB, closing another chapter in the history of local banks. Both banks have employed high calibre foreign expertise to lead the management team. With the liberalisation of the housing loans market and the relaxation of the CPF management scheme, local banks are now competing with foreign banks in the consumer market. Local consumers stand to benefit to a certain extent in having better deals and more options. As the 40 per cent limit on foreign ownership has been lifted, local banks now move extensively into other non-interest financial activities such as insurance markets, asset fund management, and venture capital and securities investment rather than relying on traditional sources of income from interest loans. They chart out their income earnings from three different business relations – transactional, loan services and purchase of equities (Figure 11.8). They continue to be committed players in the SME market. While retaining relationships with old clients, relationships with customers in selective new and growth industries are fostered. Currently they are looking into other areas to provide them support in overseas ventures.

Setting the goal to be the premier banks in the Asia Pacific region, these banks intensify their efforts to set up branches and make more acquisitions abroad. They intend to leverage on their existing cultural heritage and linkages as well as extensive regional knowledge and network to explore markets beyond Malaysia. In 1999, UOB for example, besides acquiring Westmont Bank of the Philippines and its network of 97 branches, also acquired a majority interest in Radanasin Bank in Thailand and its network of 68 branches. It has placed Taiwan, Hong Kong and China in its Greater China market expansion plan. It now has branches in 6 Chinese cities namely, Xiamen, Shanghai, Beijing, Chengdu, Shenzhen and

Guangzhou. A new agenda is in the making for the overseas Chinese network since these banks are some of the largest overseas Chinese Banks in East Asia. Has *guanxi* embedded in the established network really lost its lustre as an engine of business pursuit?

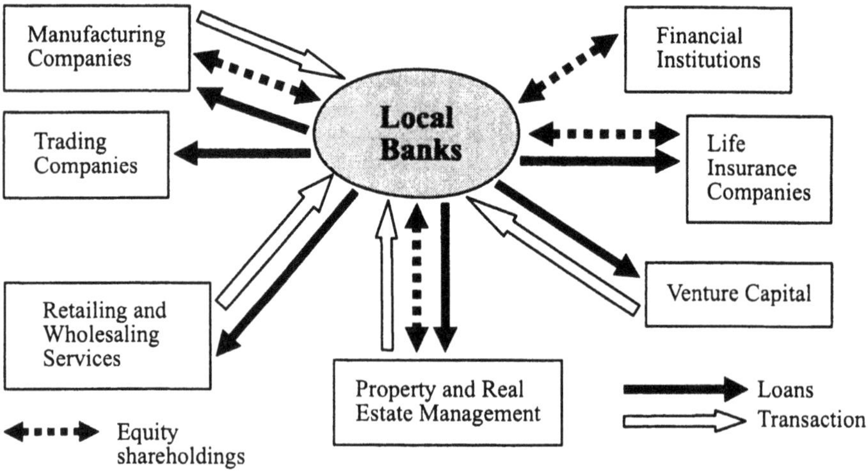

Figure 11.8 Local banks' business network

Though the remaining three local banks are actively expanding their geographical spread, the limits are evident. The dependence on the domestic market remains strong. Even after many mergers, the geographical ratio of asset in Singapore of the three banks still takes up about three-quarters of their totals. The concentration in Malaysia has changed little. The geographical distribution of DBS is the most widespread, with about 37 per cent of its asset overseas. There are still scopes for UOB and OCBC to extend their territorial influence provided more shareholdings are subscribed to by heavyweight partners from abroad, whose networks local banks have yet to weave into. This raises another issue of ownership and control to be discussed later (Table 11.3).

Table 11.3 Exposure to regional economies, Hong Kong and China as of December 31, 2003 (% of group assets)

Market	UOB	DBS	OCBC
Malaysia	8.8	0.6	11.5
Indonesia	0.5	0.3	0.9
Thailand	2.6	0.2*	0.3
South Korea	0.9	1.1	0.4
Philippines	0.5	0.7	0.1
Greater China	4.0	21.6	3.6

*Excludes DBS Thai Danu Bank

Source: Banks' 2001 Results.

Chinese banks in Singapore are still small players. They have yet to come to grips with domestic business needs, market segmentation and innovative banking. Though they have captured the local SME markets, there are rising complaints from local SMEs about their diffident loan services and the lack of empathy towards them, especially during the crisis of the late nineties. It is not unlikely that they may turn to foreign banks when business needs and interests override the importance of established ties or communal bonds.

The liberalisation measure to force banks to separate financial and non-financial activities in three years' time may enable local banks to concentrate on their core business and develop competencies and at the same time reduce the risk of resource depletion especially in unregulated activities such as land development. Banks can hold a stake of less than 10 per cent in non-financial companies as equity portfolio investment as long as they do not exert control. Stakes above 10 per cent can be retained by principal shareholders by restructuring current banking groups to place banking, insurance, securities, fund management and other financial activities under a financial arm. Banks, like many others, have invested in real estate and land development to improve their asset base and profit margins. For example UOB has United Overseas Land, OCBC has Fraser & Neave and DBS has NatSteel. DBS, being a GLC, has moved out of these non-core areas upon instruction from the authorities. It will be hard on local Chinese banks which have traditional interests and faith in land and real estate. Under this new measure they have to sell properties despite the glut of funds.

While DBS may have tapped the Singapore Inc label more to market its services abroad, its GLC status gives it more leeway to increase resources and may even allow it to use state-to-state relations to promote investments in overseas ventures. Its links with other GLCs present further opportunities. For Chinese banks, the Chinese networks will remain useful. They may also ride on the

State-led opportunities such as providing services to companies and people in the Suzhou Industrial Park project (SIP), a Singapore joint venture in China. Ethnic Chinese business structures and networks are advocated to forge greater links and business development with China. The fact that the project was not a resounding success has to a certain extent affected the enthusiasm for launching into the China market. It is now realised that similar ancestral origin, cultural links and language proficiency alone are not sufficient to deal with the new generation of actors in the post-Mao or post-Deng economy. Neither can state-to-state relations safeguard any promising results (Wong, 2001).

Technical efficiency and macro-economic perspective have to be supported by a pragmatic approach to engaging in cross-border activities. It is not just in the Chinese market that the banks have to learn how to adapt and change. Strategies that carry a broader perspective and take into account the people, organisational structure, management and practices of the place have to be construed. The Chinese network has played a role in laying the foundation for these Chinese banks which managed to co-exist and subsequently collaborated with foreign banks in pre-war Singapore. The Overseas Union Bank, which now becomes part of the UOB, was for example born and registered in China and not Singapore. The current situation offers the best chance for these banks to revitalise their network for more cross-border activities. The financial liberalisation measures imposed by the authorities signal to the banks the need for a survival strategy. In Indonesia, Lippo Bank, a bank founded by a Chinese Indonesian (Mochtar Riady or Lee Mo Tie), has made a strident move. His sons have invited AIG from the USA and ING from the Netherlands to be the bank's partners, at the same time introducing Western trained business management practices and e-banking. There are very few foreign holdings and local banks are predominantly locally owned. This is in sharp contrast to local banks in Hong Kong where many shares have been acquired by foreign banks. The biggest shareholders of banks are private limited companies although there are numerous individual shareholders, indicating the desire to remain anonymous or to avoid estate duties.

Although Singaporean banks have been employing more professionals to do routine work, policy making remains in the hands of major shareholders and directors who tend to form a nucleus group of networks. The shares of the banks also tend to concentrate in the hands of the directors. In theory, directors are supposed to be elected by shareholders at general meetings. Often they are appointed or invited by the nucleus group who organises and controls the bank. Not all directors are shareholders, as some are invited to join the board on account of their professionalism, expertise, social standing and business connections, all of which contribute to the growth of the bank. Yong Pung How, for example, is a director and vice-chairman of OCBC because of his enormous experience as a lawyer and because he has been serving as the chairman of the former Malaysian Singapore Airlines and has been director of the Malayan Banking and managing director of Singapore International Merchant Bank.

There is also a strong correlation between ownership and control. This is apparent in the board composition, such as the Tan and Lee families at OCBC and the Wee family at UOB. The same is found in other banks such as the Lien family at OUB, the Goh family at Tat Lee and Ng family at FEB before they were merged with either OCBC or UOB. These families form the nucleus group in their respective banks. The nucleus group comprises members who are either the pioneers or descendants of the pioneers. Lee Seng Wee, the chairman of OCBC, for example, is the son of Lee Kong Chian, the rubber tycoon who contributed greatly to the 1932 merger that formed the OCBC. They are dedicated, giving advantages to small shareholders who are more interested in capital appreciation and dividends and less concerned with the long-term affairs of the bank. Managers are recruited on the basis of professionalism with little reference to family ties. This stands in sharp contrast to the past when the banks were established on a communal basis and bank officers were preferably drawn from the same communities. DBS being a GLC, its managers and officers are recruited in open competition, where personal or family relationships do not play a part. Its ownership and control are linked together in the hands of the government.

Cross directorate among local banks is rare. DBS again differs and has invited local bankers as directors, such as UOB's Wee Cho Yaw. Within the same banking group, however, such as the OCBC group or UOB group, there are directors sitting on the board of their associate banks. Vertical integration within the group has become more common than before. Within the UOB group, Chung Khiaw Bank and Lee Wah Bank have the same chairman, namely Wee Cho Yaw (Figure 11.9). Within the OCBC group, Lee Seng Wee is the chairman. It was reorganised in late 2002 displacing the management of its component banks from the hands of different persons.[3] The nucleus group is firmly in control. In addition, the organisation has been streamlined into four groups covering customers, products, support functions and geography, which report directly to a newly appointed expatriate Chief Executive Officer from the USA who replaced the former CEO from Hong Kong. The other banks have similarly undergone restructuring corresponding to the government's emphasis on effective corporate governance as a means of internationalisation.

3 For example, Four Seas Communications Bank was managed by Lee Hiok Siang and Bank of Singapore by Runme Shaw.

Figure 11.9 Local banks' business structure

The UOB Group – Chairman (Wee Cho Yaw)	
Chung Khiaw Bank Far Eastern Bank Lee Wah Bank Industrial & Commercial Bank United Overseas Bank	United Overseas Finance Industrial & Commercial Insurance United IBV United International Securities UOB Life Assurance United Overseas Insurance United Overseas Land Group Haw Par Brothers International Group
The OCBC Group – Chairman (Lee Seng Wee)*	
Bank of Singapore BOS Ventures OCBC Asset Management OCBC Capital Management OCBC Finance OCBC Property OCBC Securities OCBC Securities (Hong Kong) OCBC Trustee (*to be succeeded by Cheong Choong Kong from 1 July 2003)	
The DBS Group – Chairman (S Dhanabalan)	
DBS Group Holdings Ltd DBS Ltd DBS Diamond Holdings Ltd DBS Group Holdings (Hong Kong) Ltd DBS Thai Danu Bank Public Co Ltd PT Bank DBS Indonesia Bank of the Philippine Islands DBS Vickers Securities Holdings Pte Ltd Dao Heng Bank Group Lted DBS Kwong On Bank Ltd	DBS Asia Capital Ltd DBS Asset Management Ltd DBS Capital Investments Ltd DBS Nominees Pte Ltd DBS Trustee Ltd

Source: Annual Reports of UOB, OCBC, DBS.

The merger will generate reciprocal benefits as the large ones offer resources while the small ones give their special strengths, such as their connections. Keppel Tat Lee, for example, will help boost the connection of OCBC with Indonesia. The founder of Tat Lee, Goh Tjoei Kok, came from Indonesia, was engaged in trade in Indonesia and Singapore, before investing in National Iron and Steel Mills of Singapore. Similarly, OUB's strong links with China will pave the way for

UOB's Greater China plan. In the future, local banks will be further institutionalised and include more professionals in their management to meet the challenge of international competition and to ensure perpetual growth.

Future Prospects: Globalisation and Dilution of Ethnic Networks in Banking?

The development of Singapore's local banks and its financial sector has shown how changing power relations and social, cultural and institutional factors have shaped its growth, organisational structure and economic space. Local banks are the outgrowth of the staple economy based on rubber and tin, the compelling forces of change will bring them to a greater height, diversifying into more cross-border activities as they integrate deeper into the global economy. The question remains, to what extent can existing networks continue to generate the expediency and efficiency in these banks? Will the amalgamation reinforce the relations and the banks? How do the established traditional networks based on trust, cultural affinity, family control and ownership and regional familiarity compromise market efficiency and the moral standards upheld by the new generation of bankers?

As banks globalise in the era of ICT and KBE, the distinction between local and foreign may be blurred as resources become more mobile and accessible. Common interests, competent performance, and accountability are going to be the foundations of mutual trust and bargaining tools and the key to advance the global standing of Singapore banks. Nevertheless, it is doubtful that the relations embedded in the network would decay into obsolescence. There are already visible attempts at rejuvenating the ties. The real tasks ahead are identifying what kind of new recipes for success can be designed to reinvent the wheel of connections. The networks of individuals have been galvanised into networks of sizable collective entities as a consequence of the changing rules of the game in the financial system. With the declining role of the State, will the laissez-faire model of the pre-war era be resuscitated? This seems unlikely as actors in the network of today bear little semblance to those of yesterday. The current generation of young Chinese in today's Singapore and post-Deng China are quite a different race.

References

Brown, R.A. (1994) *Capital and Entrepreneurship in Southeast Asia*, St. Martin's Press, Great Britain.

Cheng Yuk-Sheng, Chia Siow Yue, Christopher Findlay, 'Governance in the city-states: Hong Kong and Singapore 'in Drysdale Peter. (2000). *Reform and recovery in East Asia: the role of the state and economic enterprise.* New York: Routledge, 2000.

Chia Siow Yue. (1993) 'Foreign Direct Investment in Singapore', in *Evolution of Asia - pacific economies : international trade and direct investment* edited by Ippei Yamazawa, Fu-Chen Lo. Kuala Lumpur, Malaysia: Asian and Pacific Development Centre.

Dodd, N. (1994) *The Sociology of Money: Economics, Reason and Contemporary Society.* Cambridge: Polity Press.
Haggard Stephen and Linda Low. (2002) 'State, politics, and business in Singapore', in *Political business in East Asia*, edited by Edmund Terence Gomez. London; New York: Routledge.
Hamilton, G. (1991) *Business Networks and Economic Development in East and Southeast Asia.* Hong Kong: Center of Asian Studies. University of Hong Kong.
Lee, S.Y. (1981) 'Ownership and Control of Local Banks', *Singapore Banking and Finance 1980/81*, pp. 109-117.
Lim C. Y. et. al. (1988) *Policy Options*, McGraw-Hill Book Co. Singapore.
Loh, G., Goh, C.B. and Tan, T.L. (20009. *Building Bridges, Carving Niches: An Enduring Legacy.* Oxford University. Singapore.
Low Linda and Douglas M. Johnston. (2001) *Singapore Inc.: public policy options in the Third millennium.* Singapore: Asia Pacific Press.
Low Linda. (1999) *Singapore: towards a developed status.* Oxford: Oxford University.
Masuyama Seiichi, Donna Vandenbrink, Chia Siow Yue. 1999. *East Asia's financial systems: evolution & crisis.* Singapore: Institute of Southeast Asian Studies.
Overseas Union Bank. (1974) *Overseas Union Bank Limited: 25th anniversary, 1949-1974.* OUB, Singapore.
Perry Martin, Lily Kong and Brenda Yeoh. (eds).(1997) *Singapore: a developmental city state.* New York: Wiley.
Rodan Garry. (1989)*The political economy of Singapore's industrialization: national state and international capital.* Basingstoke: Macmillan. Singapore Banking and Finance (1977) "Domestic Banks' Problems in a Growing Financial Centre", *Singapore Banking and Finance* 77, pp. 20-22.
Tan, C.H. (1990) *Financial Markets and Institutions in Singapore.* Tenth Edition. Singapore University Press. Singapore.
Thrift, N. and Kris Olds. (1996) 'Refiguring the economic in economic geography', *Progress in Human Geography.* 20 (3). pp. 331-337.
Toh Mun Heng, and Tan Kong Yam. (1998) *Competitiveness of the Singapore economy: a Strategic perspective.* Singapore: Singapore University Press: World Scientific.
Wilson, D. (1972) *Solid as a Rock,* OCBC, Singapore.
Wong Shuang Yann. (2001) 'Competition and Collaboration: problems of overseas investment: the experience of a city-state'. Paper presented at the 2001 residential conference organized by the IGU commission on the dynamics of economic spaces in Turin, Italy, July 10-14, 2001.
Yap, P.G. (1982) *Scholar, Banker, Gentleman, Soldier*, Times Books International, Singapore.

Chapter 12

'The Region is Local': Restructuring the Clothing Industry of Southern Africa

Christian M. Rogerson

Introduction

In his classic work *Global Shift*, Peter Dicken (2003) aptly observes that the clothing industry was perhaps the first manufacturing sector to take on a truly global character. Moreover, it was a form of industrial activity that during the 1990s was changing very rapidly in terms of its global geography, organisation and technology. As a consequence of such shifts and of the importance of garments production in both newly industrialising as well as less industrialised countries, the clothing industry was at once a source of political friction and even the basis for 'global political football' (Dicken, 2003, p. 317). Within what might be called the 'World Cup' of clothing manufacture, several observers discern the emergence of a series of 'regional production networks' in the global garments industry that integrate together consumers in the developed world with producers in the developing world (Gereffi, 1994, 1999; Gibbon, 2002). Within these networks and the international league tables of clothing production, Africa is only a minor player, particularly as compared to the dominance of East Asia.

The aim in this chapter is to examine restructuring trends taking place in the landscape of clothing production in Southern Africa, a peripheral part of the global clothing economy. Despite its peripheral situation in global clothing production, significant local changes are taking place in the competitive organisation as well as the spatial structure of garments manufacture within the overarching framework of the fourteen-nation Southern African Development Community (SADC). Post-apartheid South Africa is the economic heart of SADC and has since 1994 moved out of a status of relative international isolation, confronting challenges of re-integration and re-engagement in the global economy (Lemon and Rogerson, 2002). The analysis here focuses geographically upon South Africa, which is the SADC's and Africa's largest manufacturer of clothing (Figure 12.1). Special attention centres on analysing certain 'local responses' by enterprises and other stakeholders towards changes occurring in the global economy of clothing production. The term 'local response' refers here (1) to sets of changes occurring at the enterprise level towards heightened competition in global garments manufacture and (2) to emerging sets of larger-scale regional or Southern African responses which are beginning to take place in order to manage global change. Overall, it is argued that in order to enhance the competitiveness of garment producers across the region of Southern Africa, local network formation and planned cooperation are essential for success in an increasingly competitive global environment.

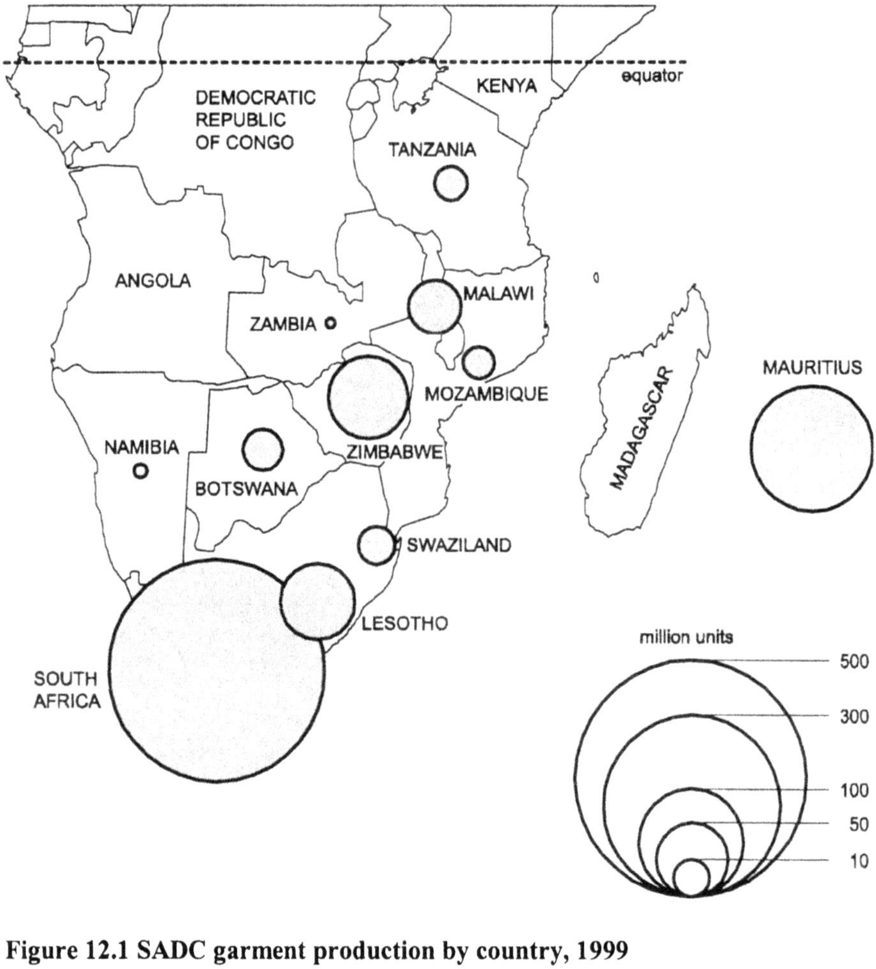

Figure 12.1 SADC garment production by country, 1999

Structurally, this chapter unfolds through three uneven sections of discussion. First, several critical elements in the changing global production environment of the clothing industry are identified, particularly concerning changes in retail practices and international trading regimes. Second, the local responses to global change in the context of the South African clothing industry are interrogated. Issues of concern relate to particularly the growing *segmentation* of clothing enterprises most especially in relation to end-markets. Finally, in the last section, attention broadens to the Southern African context in order to examine regional level responses or initiatives that have been instituted to strengthen the clothing value chain across the region. In conclusion, it is argued that the adoption of a strategy and credo that recognises 'the region as local' offers the best prospects for regional competitiveness, prosperity and future economic health for the clothing economy of Southern Africa.

Global Changes in the Production Environment for Clothing

The global patterns and dramatic geographical shifts taking place in the clothing industry have been a response to diverse sets of pressures. As Dicken (2003) observed, the general trend has been of a relative decline in the older established centres of production in Europe, especially, and a corresponding rapid growth of production in a number of developing countries. Between 1965 and 1997 the share of world garment production accounted for by developing countries expanded from 14.8 per cent to 65.2 per cent with the share for developed countries falling from 69.7 to 34.8 per cent (Coughlin et al, 2001). Technological levels of development in production and corresponding levels of labour intensity involved have been important factors underpinning this geographical shift of production. Important changes have occurred also in the type and range of fibres used in clothing with a significant growth in the use of man-made fibres, the production of which is still mainly the domain of developed countries. Further, the dynamics of the clothing industry have been shaped by consumer demands for greater product diversity. Although basic items of clothing still remain the major part of markets, even in Western Europe, USA or Japan, the rise in demand for fashion and fashion-basic items has implications for clothing producers in terms of their need to respond to the fluidity and pace of change of consumer demands in shapes, colours and fashion items (Coughlin et al, 2001). In understanding the overall dynamics of global change in clothing manufacture, however, the two most important themes relate to changing retailing practices and changing international trade agreements and regulation.

Shifting Retail Practices and New Global Geographies

Of greatest significance in terms of global changes in the production environment for clothing is the strengthened power of retailers. Within a commodity-chain analytical framework the heightened significance of retailers is recognised with the clothing industry viewed as a classic example of a buyer-driven chain (Gereffi, 1994, 1999). In all the major developed markets, the United States, the European Union and Japan, the trend is observed for clothing demands to be increasingly dominated by the purchasing policies of large retail chains such as Sears-Roebuck, Marks and Spencer or Daiei. The highly concentrated purchasing power of these retail chains has resulted in considerable leverage and power exerted over clothing producers. Most importantly, with the shift from standardised to more differentiated clothing goods with more frequent fashion changes manufacturers must respond rapidly to retailer demands and specifications (Gibbon, 2004).

Changing retail management practices are imposing new conditions on production planning, sourcing strategies and delivery times. The new paradigm of 'lean retailing' demands a reduction in the costs of maintaining large stocks of a wide variety of items. Moreover, the application of information technology with bar coding and scanning, allowing for real time point of sale information on replenishment needs, further imposes new terms of competitiveness on producers. Although a recent study of global sourcing patterns by retail chains in the United Kingdom disclosed that sourcing patterns remained heavily price-based, it observed escalated qualifying criteria for suppliers in line with the imperatives of lean retailing (Gibbon, 2002). This implies that only 'the most resourceful' developing country producers can secure a place in the global sourcing networks of major UK clothing retailers (Gibbon, 2002). Indeed, retail buyers for the European Union now demand that factories in Mauritius go so far as to attach retail price tags and put garments on hangers in the containers ready to go to the retail outlet (Gibbon, 2000; Coughlin et al, 2001).

Taken together, the strengthening of retail buyers, the exigencies of short-cycle production and fast time to market are re-moulding the global geography of clothing. In Gereffi's (1994) work the role of specific producers and producing countries in global clothing commodity chains was viewed initially in terms of a series of 'concentric rings' of production, ranging from developed countries to 'the fringes of the international production frontier', a designation that would refer to much of Africa. In more recent studies, this conceptualisation is now re-configured in terms of the crystallisation of new regional production zones (Gibbon, 2002, 2003). More especially, the major regional production zones – United States-Mexico-Caribbean Basin, Western Europe-Eastern Europe-Mediterranean-North Africa and Japan-Southeast Asia – combine developed markets with developing areas of production (Dicken, 2003). Overall, therefore in the contemporary global environment confronting clothing producers, competition is:

> much more than a simple matter of price and cost: it involves the ability to respond efficiently to frequent shifts in the supply chain and to stringent demands imposed by customers often arising from new retailing practices and the consumers' changed preferences (Coughlin et al, 2001, p. 5).

Typically, in the United Kingdom, global sourcing for clothing involves the balancing or trading-off of cost considerations by issues of lead times and production flexibility (Gibbon, 2002, p. 11).

Changing International Trade Regimes Concerning Clothing

A second critical set of issues impacting upon sourcing strategies and global production patterns of clothing relate to changes in international trade agreements. Alongside textiles, the garment sector is viewed as the only industry in the world economy to which special international trade restrictions apply through the Multi-Fibre Arrangement (MFA) (Dicken, 2003). The MFA has existed in various forms since 1974 and under this international regulatory agreement individual quotas were negotiated which set precise limits on the quantities of textiles and clothing products which can be exported from one country to another. Once the quota amount for any individual item or product is reached, no further imports were permitted. The MFA agreement has been subject of much controversy for it contravenes GATT agreements concerning international trade. Although certain developing world countries, such as Bangladesh and Vietnam have taken advantage of the restrictions imposed on other countries' exports by the MFA (Nadvi and Thorburn, 2003), the major impact of this agreement has been to greatly restrict the growth of exports from most developing countries (Dicken, 2003, p. 338). As a result of the so-called Agreement on Textiles and Clothing the textiles and clothing sectors were to be integrated into GATT with full integration targeted for 2005. The major implication of the impending phase out of the MFA is that 'markets will be free from quotas, but tariffs will remain' (Coughlin et al, 2001). Correspondingly, the significance is enhanced of other newer trade agreements in shaping new global geographies of clothing. In terms of Africa, the two most significant sets of new trade agreements are those which relate to securing preferential entry for clothing goods into the developed markets of the United States and the European Union.

The United States African Growth and Opportunity Act (AGOA), which came into force in October 2000, is designed to facilitate economic growth in Africa through trade rather than aid (Otiso, 2004). AGOA represents a non-reciprocal unilaterally determined trade agreement offered by the USA to sub-Saharan Africa. The act opens up access to the world's largest clothing market for African preferential exports for a period of eight years ending 2008. The opportunity for African producers is for a 3 per cent share of the US consumer market for clothing, an amount worth roughly US $3 billion. As a unilateral declaration of preference AGOA's renewal beyond 2008 depends upon circumstances at that time. In order to benefit from the quota free and duty-free preferential access to the US market, African countries have to fulfil two conditions under AGOA. The first is a political condition that countries must meet certain political and human rights criteria, most importantly that they cannot engage in activities that undermine US national security or foreign policy (Coughlin et al, 2001). The second concerns legal and bureaucratic requirements that countries put in place a customs visa system that is officially approved by US customs authorities to guarantee compliance with rules of origin. Under the stringent AGOA rules of origin it is stated that, with very specific exceptions, garments to be imported to the US and assembled in sub-Saharan Africa must be from fabric wholly formed in sub-Saharan Africa from yarn made either in the USA or in Africa. The specific exemption is for a group of designated African Less Developed Countries (LDCs) which until September 2004 may produce clothing from fabrics imported from anywhere in the world. It should be noted that South Africa, along with Botswana and Mauritius, do not qualify under the African LDC designation (Coughlin et al, 2001; Stern and Netshitomboni, 2002; Otiso, 2004).

Access to the clothing markets of the European Union for African producers has been secured through the Lome-Cotonou agreements which provide for duty free and quota free access for most African producers (with the exception of South Africa) to the European Union. Other countries unaffected by Lome-Cotonou agreement secure access through the so-termed 'Everything-But-Arms' Amendment to the European Union's Generalised Scheme of Preferences under which duties and quotas were eliminated for essentially all products (excepting sugar, rice and bananas) from the world's 48 poorest countries as of March 2001. Improved access for South African clothing producers to the European Union was secured through the 2000 EU-South Africa Trade, Development and Cooperation Agreement which requires removal of customs duties on 'substantially' all trade over a maximum 12 year period. In terms of clothing, South Africa is seen as benefiting from this agreement as South African tariff cuts will take place only in the second half of the 12 year agreement thus allowing a period for restructuring and strengthening of the sector in order to meet international competition.

Local Response and Restructuring in South African Clothing

With the transition to democracy, post-apartheid South Africa resumed old international linkages, such as with the International Monetary Fund and the International Labour Organisation, as well as cemented new relationships, most importantly its involvement in the Uruguay Round of the GATT and membership of the World Trade Organisation in 1995 (Lemon and Rogerson, 2002). Fresh regional relations were forged in Africa, not least with South Africa's accession to the Southern African Development Community. This shifting pattern of international relations brought with it both the dangers of tightening competitive forces as well as new opportunities (Coughlin et al, 2001). The most important opportunities are those that arise through the Africa Growth and Opportunity Act and the South Africa-European Union Free Trade Agreement, which together offer easier access for South African producers to developed country markets (Jafta and Jeetah, 2001; Coughlin et al, 2001; Rubin and Coughlin, 2002; Stern and Netshitomboni, 2002).

Of greatest policy significance since 1994 has been the abandonment of South Africa's former programmes of heavy industrial protectionism and of import-substitution towards new emphases on greater trade liberalisation and a more outward-oriented industrial strategy (Kaplinsky and Morris, 1998). The protection formerly provided to South African producers has been lifted with the government's adoption of an accelerated phase down of import tariffs (Kamaha, 2004). Moreover, as Gibbon (2004) argues, real levels of protection were further reduced by the public service's apparent inability to enforce customs regulations which resulted in a rapid escalation of clothing imports. Overall, this re-orientation of post-1994 trade and industrial policy in South Africa inevitably spawned the challenge of enhancing the capacity of local manufacturing enterprises to compete more effectively in both global and domestic markets, an imperative which is of immediate and pressing concern to the country's clothing producers (Rogerson, 2004). In this section the key features of South Africa's clothing industry are sketched and local enterprises responses to heightened levels of competition are reviewed (Figure 12.2).

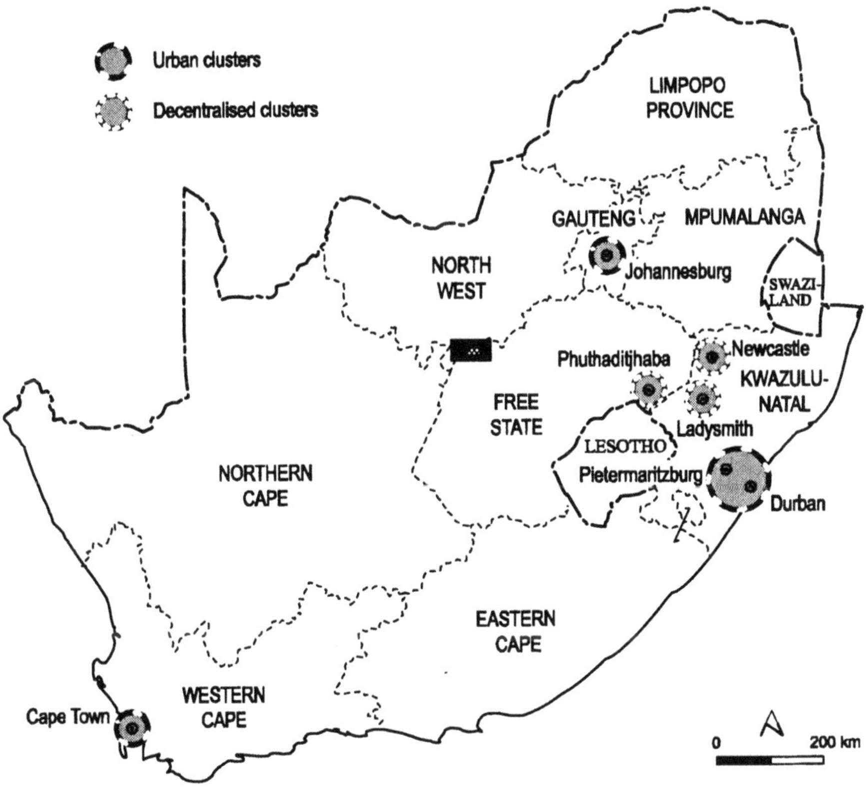

Figure 12.2 South Africa: Location of major clothing production

Key Features of the South African Clothing Industry

The contemporary South African clothing economy, Dunne (2000, p. 1) observes, 'is dominated by a small number of large retail groups, each comprising a number of chains focussing on different consumer groups' and operating in various sub-sectors of clothing, textiles and footwear markets. Like much of South African industry, the manufacture of garments is oriented largely towards the domestic market which operates as a buyer-driven commodity chain with the major clothing retailers assuming a critical role in terms of the governance of the apparel chain (Dunne, 2000; Gibbon, 2004). Overall, the clothing industry in South Africa caters to a wide range of retail customer needs and products. At the high income end of the South African market, production occurs of quality styled garments, fashion products, boutique clothing and of products for high fashion chains. The middle-market production is geared more towards semi-styled garments, often with frequent style changes. The low income end of the market is represented by mass volume production which has experienced, post-trade liberalisation, highly intense competition from imported clothing, both of a legal and illegal nature (Harrison and Dunne, 1998, p. 14). Market conditions in South Africa are shifting rapidly particularly in the context of political and economic changes which have given impetus 'to a massive black aspirant market that is increasingly demanding and sophisticated' (Harrison and Dunne, 1998, p. 13).

Organisationally, clothing manufacture in South Africa evidences a dualistic structure with both a formal and an 'informal' element of production (Altman, 1994; House and Williams, 2000). Although clothing is often perceived as an industry with many small firms, the reality is that the sector in South Africa is dominated by a small group of large firms (Kamaha, 2004). Ownership of the large enterprises in the clothing industry is overwhelmingly in South African hands. There is a small but significant element of foreign owned clothing manufacturers which operate in South Africa. This group of foreign owned producing establishments includes large export-oriented investors originating from Singapore, Hong Kong, Malaysia and, most recently, from mainland China (Gibbon, 2004). In addition, there are a larger number of small and medium sized firms, mainly from Taiwan, who produce primarily for the domestic market. The chequered development of the cluster of Taiwanese-owned clothing establishments must be traced back to the apartheid sanctions period of the 1980s when developing country multinationals were the leading source of foreign investment in South Africa and attracted to decentralised areas by a combination of low labour costs, generous locational incentives and aggressive South African government marketing (Rogerson, 1987; Todes, 1999; Hart, 2002).

Spatially, the formal component of the industry governed by institutionalised bargaining council arrangements remains largely concentrated in South Africa's three largest metropolitan areas with the Western Cape and Durban the two leading centres (Altman, 1994, 1996; Rogerson, 2004). The Gauteng area, with Johannesburg at its hub, contains the third largest cluster of producers (Rogerson, 2000, 2001). The remainder of the industry is spread widely, albeit focussed in a number of peripheral decentralised locations, with Newcastle, Ladysmith and Phuthaditjhaba, the most notable production clusters (Todes, 1999; Nattrass, 2000; Hart, 2002). Important spatial wage differentials exist between particularly the three metropolitan clothing clusters and of producers in peripheral areas. According to the most recent estimates those factories operating in decentralised spaces with large pools of surplus labour enjoy wage levels of between a quarter and half of urban levels (Hart, 2002; Gibbon, 2004).

Alongside the formal clothing industry, there is emerging a very significant segment of informal production of clothing (Kamaha, 2004). This type of production activity is manifest in the form of home workers and of new clusters of small garment producers which operate in surplus office space of inner-city areas of South Africa's major cities (Manning, 1993; Rogerson, 2001; Cachalia et al, 2004). In addition, within Johannesburg, at least, many immigrant entrepreneurs from other parts of sub-Saharan Africa also have established informal clothing businesses (Peberdy and Rogerson, 2000; Cachalia et al, 2004). Although informal producers are engaged in the production of a range of different types of garments their domination is strongest in the manufacture of traditional garments such as pinafores, low-cost garments, which represent both comfortable and affordable workwear for lower-income people (Manning, 1993). Nevertheless, the spectrum of informal sector producers ranges 'from 'backyard' machinists producing low quality tracksuits and school uniforms for neighbours to relatively unsophisticated 'design houses' producing for the major retailers' (Williams and Hanival, 1999, p. 74).

The clothing industry of South Africa clearly encompasses a complex mix of formal and informal producers as well as specialist firms which link into clothing production (McCormick, 1999; Kamaha, 2004). The group of clothing manufacturers comprises broadly three types of enterprises: full-range manufacturers, cut-make-and-trim (CMT) operators and informal sector producers (McCormick, 1999). South Africa's largest firms generally pursue the full spectrum of pre-assembly (such as design or pattern making), assembly (sewing) and post-assembly operations (such as inspecting, pressing). By contrast, many medium-sized or small enterprises do not carry out all the functions of design, production and finishing. Rather, many of these establishments operate as CMT operators which undertake only cutting and assembly work on a contract basis (McCormick, 1999). Much of the group of informal clothing enterprises function as CMT operations with the birth of such enterprises, in many instances, following the closure and retrenchment of workers from formal factories (Fakude, 2000).

These categories of producing enterprises are frequently, however, blurred for, in addition to having their own lines of production, many larger enterprises occasionally take on CMT work. Additionally, whilst the majority of informal sector enterprises work on a subcontracting basis or a CMT capacity some informal producers have been found to be full-range manufacturers (Kamaha, 2004). Alongside this complexity of enterprises and fragmentation of production there has occurred the emergence of so-termed 'design houses' in the clothing industry. Such design houses specialise in cutting, finishing, design, merchandising and marketing. Indeed, Fakude (2001, p. 5) observes that the emergence and development of these design houses 'reflects a strategic positioning of new firms and a downsizing trend in existing manufacturing firms to maximise gains through focusing on the core function of design whilst outsourcing the less lucrative manufacturing and assembly functions'.

Globalisation and Local Response

Since the early 1990s the South African clothing industry has been strongly affected by the impacts of globalisation and of trade liberalisation (Williams and Hanival, 1999; Kamaha, 2004). Of greatest significance is the emergence and consolidation of new global production regimes for clothing goods in which transnational production networks are coordinated spatially to allow specialised functions to be carried out in different countries (cf Gereffi, 1994, 1999; Gibbon, 2002). Overlapping the impress of the making of these global production networks has been the consequences of the era of trade liberalisation in post-1994 South Africa which was heralded by the democratic government's offer to GATT (Kaplinsky and Morris, 1998; McCormick, 1999; House and Williams, 2001). Under the GATT, the duty phase down on South African clothing should have been spread over 12 years, albeit in 1995 the government opted to accelerate the removal of protective tariff walls with the intention of more rapidly completing the process by September 2002, when the duty on clothing imports will be 40 per cent (Dunne, 2000; Jafta and Jeetah, 2001). Taken together, trade liberalisation and globalisation have precipitated a 'steep learning curve' for South African clothing enterprises, traditionally protected by high tariff and non-tariff barriers from overseas competition, which are compelled now to compete with international suppliers in the domestic market as well as in neighbouring countries and abroad (Fakude, 2001, p. 5). Indeed, during the past decade it is evident that the clothing sector in South Africa has undergone a metamorphosis as a 'protected domestic market dominated by domestic manufacturers has given way to a market which is increasingly exposed to international competition' (Harrison and Dunne, 1998, p. 45).

Considerable and rapid changes therefore have been taking place in the size, organisation and complexion of South African clothing manufacture since 1994. At one level, it is observed that in terms of output indicators, 'the sector has undergone a substantial real decline' between 1995-2000, whilst output escalated in nominal terms by 17 per cent, in real terms a decline of 19.6 per cent is recorded (Gibbon, 2004). Far less clear is the employment situation in clothing manufacture with different estimates of total employment in the South African clothing sector varying considerably and even fluctuating dramatically from between 130-200 000 workers; the higher figure often includes estimates for the magnitude of informal clothing employment in South Africa (McCormick, 1999). Despite the murky statistical situation concerning employment numbers, it is clear that the clothing industry in South Africa remains overwhelmingly a women-dominated sector with the best estimates suggesting that female labour represents three-quarters of the total labour force (Pape, 2000, p. 6).

The general consensus is that by 2000 the size of the formal South African clothing industry was approximately 130-140 000 jobs, a level which suggests that employment numbers are little different from the time of political transition (House and Williams, 2000; Jafta and Jeetah, 2001). For some observers this poor performance of the clothing sector is interpreted that the industry has been 'crippled by the advent of imports in the trade liberalisation era' with the influx of lower cost imported garments primarily from low-cost producers in Asia (see Fakude, 2001, p. 5). More especially, the disappointing performance of the industry is expressed in terms of concerns about job losses in many formal sector enterprises as a result of restructuring in defence of international competition (Williams and Hanival, 1999, p. 73). Nevertheless, the overall picture is complex and it is important that conclusions are not drawn based upon superficial inspection of employment data alone (House and Williams, 2000). Firm closures, downsizing and retrenchment by many urban-based South African enterprises is accompanied by a flow of export-led investment, in particular by large transnational investors locating in decentralised areas. Sometimes this demise of many local producers is in line with global trends; in other cases it is a reflection of enterprise's inability to grow or move into export activities to counteract the loss of local market share (House and Williams, 2000, p. 26).

The South African clothing industry is clearly facing serious problems of readjustment from the formerly highly protected environment of the previous import substitution era (Rogerson, 2004). The stagnant picture of total clothing employment, however, masks a number of very significant, complex and sometimes seemingly contradictory changes that are taking place in terms of the re-organisation of production and the nature of employment in response to this increasingly competitive market conditions. Enterprises in the South African clothing industry are pursuing a variety of restructuring strategies singly and in combination to respond to the competitive pressures that they face (Fakude, 2001). This rearrangement of firm-level strategies includes new methods of organising production, spatial shifts and the articulation of new regional production systems, casualisation, outsourcing, subcontracting and informalisation of production.

The works of Altman (1994, 1996) stress the importance of clothing producers adjusting to new market conditions by dynamic organisational changes in which enterprises implement practices that permanently address their capacity to respond to changing market circumstances. Further emphasis on the imperative for introducing world class manufacturing standards and organisation in the industry was expressed at the Clothing and Textiles Summit in 2000, a gathering which brought together representatives of government, labour and business with the goal of 'transforming the industry into a strong internationally competitive and growing industry capable of sustaining and attracting jobs and investment' (Department of Trade and Industry 2001). Amongst the kinds of strategies highlighted are process innovations, organisational changes, the introduction of new technologies or programmes for skills development that enhance human resource development. Often such changes are interdependent with process innovations. The application of new technologies goes hand-in-hand with new organisational changes and the encouragement of multi-skilling practices. In particular, new organisational forms are seen as consistent with shorter production cycles, modular production, total quality control, reduced stock inventories and more skilled workforces. Moreover, new technologies, such as computer aided design, mechanised laying-up or cutting, facilitates the adoption of unit production systems, which greatly enhance efficiency.

In terms of the spatial organisation of production, it is evident that major changes currently are reshaping the geographical landscape of production in both South African clothing and the wider regional clothing pipeline in the Southern African Development Community as a whole (cf Nkoko, 2000; Coughlin *et al*, 2001). Locationally, the South African clothing industry remains highly competitive and footloose with cost considerations driving location preferences. According to Nattrass (2000) this 'tendency appears to be getting worse' with many clothing enterprises responding to wage pressures by relocation strategies and the articulation and development of *regional* production networks that maximise the production possibilities inherent in the Southern African Development Community region as a whole. Horton (1999) and House and Williams (2001) record the relocation of South African clothing manufacturers to Botswana, Malawi and Mozambique either completely or by putting their South African workforce on short-time and subcontracting a large proportion of their work to low-cost labour in CMT firms in these countries. For the old production spaces in South Africa the loss of these clothing jobs through regionalisation is particularly disastrous for rural women as these low-paid jobs are an important source of income and independence for them (Phalatse, 2000).

Another local response to competition has been that of restructuring which involves production subcontracting and informalisation (Mosoetsa, 2000). The growth of survivalist informal clothing producers in South African urban areas has been observed for almost a decade. Nevertheless, the dramatic advance of informalisation over recent years is now confirmed as part of the wider restructuring processes of existing medium and small enterprises, many of which have informalised or gone underground (Williams and Hanival, 1999; Fakude, 2000). Closely related to informalisation there has been a shift toward more flexible work arrangements, including short-time, contract work and a host of other arrangements which 'ultimately mean less income for workers already living on the edge of poverty' (Pape, 2000, p. 7). Flexibilisation is associated with the expansion of labour broking and subcontracting. In Kwazulu-Natal the Confederation of Employers of South Africa (COFESA) is an organisation which extensively advises employers on restructuring production in such a way that it converts employees contracts into service contracts and employees into independent contractors (Skinner and Valodia, 2004). For former employees the terms of such contracts simply state production targets and time frames with no paid holidays, no sick pay, pension or unemployment insurance (Pape, 2000). The activities of COFESA in brokering and subcontracting allows companies to by-pass labour legislation on the basis that their workers are no longer classed as employees in labour contracts thus radically weakening the position of workers in terms of employment security (Skinner and Valodia, 2004).

The complex of emerging restructuring strategies of South African clothing producers are interpreted as shifting the trajectory of South African clothing from being insular and geared only to domestic markets to being export-oriented and aimed at maximising the potential for producers to access new markets in Europe and the USA through preferential market access arrangements (House and Williams, 2001). The findings from a recent analysis of the export performance of domestic versus foreign-owned South African clothing manufacturers suggests that domestic South African enterprises need to borrow certain elements of the business and manufacturing models of these successful foreign exporting enterprises (Gibbon, 2004).The net balance and long term impacts of the radical changes taking place will determine ultimately whether South African clothing is a sunset industry in decline or an industry that is restructured, re-energised and looking to a new sunrise (Rogerson, 2004).Using the theoretical lens of commodity chains, Gibbon's (2004) research highlights the *segmentation* and *differentiation* that is taking place in the landscape of clothing production of South Africa. For example, typically, producers engaged in export clothing chains are based geographically in decentralised cheap labour areas and source the majority of their yarn and fabric from non-South African sources. Many of these manufacturers are vertically-integrated operations within a global production network. The organisation of production in these primarily Asian-owned factories is targeted at serving the needs of United States (and to a lesser extent European Union) consumers. By contrast to these highly profitable export factories, the group of garment producers engaged in manufacture for the domestic market are confronting the competitive challenge of cost pressures from growing imports, particularly at the low-income range of clothing. The majority of these producers are found in South African ownership and are situated in the relatively high wage areas of the three metropolitan areas of Cape Town, Durban and Johannesburg. Moreover, they draw a large proportion of their yarn and fabric from the local South African textiles industry. For these South African enterprises engaged in domestic commodity producing chains, growing competitive pressures from imported clothing are forcing major restructuring and reorganisation of production through such means as outsourcing, casualisation and informalisation.

The Local Response is Regional

Notwithstanding these local enterprise responses to global change, it is argued that the long-term ability of South African producers to survive the competitive challenges of export markets is crucially dependent upon a number of necessary improvements in the wider regional clothing value chain, which includes the textiles sector of South Africa and the SADC region as a whole (Coughlin et al, 2001). In Southern Africa there is a sense of optimism that the prospects and opportunities may be improving for Africa's clothing industry, in particular given a context of rapid changes taking place in international trade environments. None the less, producers and other stakeholders, in particular governments, must acknowledge the need for a regional level response in order to enhance the competitiveness (if not the survival) of the garments industry in Southern Africa.

The strengthening of regional industrial linkages and of initiatives for regional restructuring is a central platform of post-apartheid trade and industrial planning which is supported by other programmes such as those surrounding Spatial Development Initiatives (SDIs) in Southern Africa (Rogerson, 2002). Among their goals, the planned regional SDI initiatives in Southern Africa are to enhance the competitiveness of regional enterprises in the global production economy (Lemon and Rogerson, 2002). For the clothing industry the imperative for developing a specific regional initiative designed to strengthen the competitive base of local producers is given great impetus by the introduction of AGOA and of the recent changed trade agreements with the European Union (Coughlin *et al*, 2001; Otiso, 2003). Nevertheless, the evidence from the first two years of AGOA's operation is that, as yet, the response by producers has been slow (Rubin and Coughlin, 2002).

During this period, total shipments of clothing goods under AGOA were only 36 per cent of the allocated quota for the whole of sub-Saharan Africa. Although South Africa is by far the most significant continental clothing manufacturer, the leading sources of clothing exports from Africa to both the United States and the European Union are Mauritius followed by Madagascar (Gibbon, 2000; Jeetah and Coughlin, 2001; Rubin and Coughlin, 2002). In terms of exports, South Africa occupies only a poor third place and in the case of exports to the USA was surpassed in 2001 even by Lesotho (Rubin, 2001). The key findings of Rubin and Coughlin (2002), as shown on Figure 12.3, underscore that the response so far by South African producers to the export opportunities offered by AGOA and the EU has been limited (if not laggard). Indeed, as is evident from Figure 12.3, the impact of AGOA in terms of export expansion has been greatest in terms of South Africa's neighbour, Lesotho, which has experienced a major boom in employment in garments and textiles manufacture. Several large vertically integrated foreign-owned manufacturers have located in Lesotho and benefit (until 2004) from the short-term exemptions accorded to these countries as African LDCs (Rubin, 2001). In the case of South Africa, whilst exports to the USA have shown some signs of growth, by 2001 only an estimated 9 per cent of total clothing exports from South Africa to the USA were classed as AGOA compliant; the remaining 91 per cent were duty paid and thus did not benefit from AGOA.

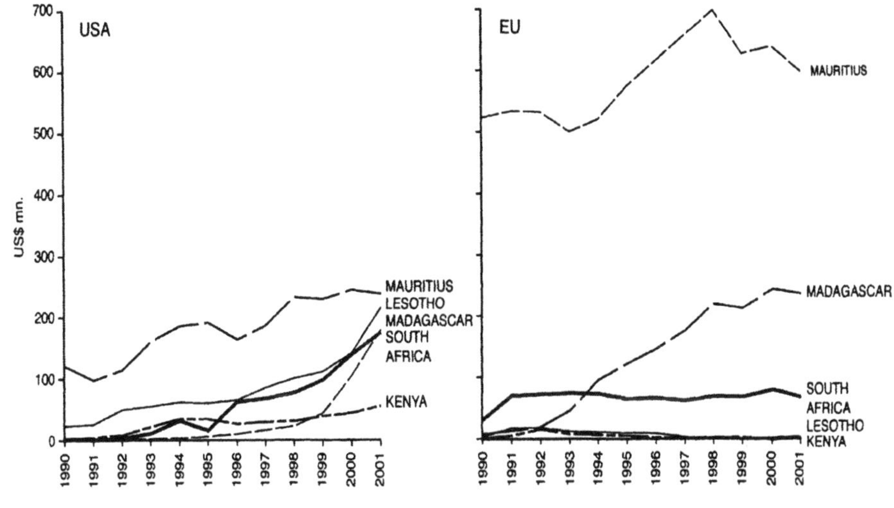

Figure 12.3 African clothing exports to the USA and the EU, 1990-2001

Overall, these findings point to an urgent need for evolving a more coordinated long-term regional response by governments and producers towards 'growing' the clothing economy of Southern African as a whole. The outlines of this regional response are becoming clearer (Coughlin et al, 2001). Recent research for the SADC Secretariat highlights the need to strengthen the clothing value chain across the region through capitalising on the comparative advantages of the different countries in the SADC for different stages in the clothing value chain (Coughlin et al, 2001). Essentially, the basis for proposed restructuring is that as no country in the SADC region has an absolute advantage across all stages of production, 'the most competitive pipeline for garment production will be a regional one with each country contributing what it does best' (Coughlin et al, 2001, p. xii). It is apparent that across the Southern African region whilst there are many world class producers of textiles and garments that can support the entire clothing pipeline, at the same time there are several important constraints that urgently need to be addressed. Among the most significant constraints are *inter alia*, the presently insufficient range and supply of man-made fibre and yarn as well as fabric for regional clothing producers to be AGOA compliant; shortages of experienced staff and adequate training; an inadequate intra-regional transport infrastructure; long and unpredictable customs procedures across the region; and, perhaps most critical of all, the absence of a clear institutional structure for coordinating and leading a process of regional restructuring.

Under the umbrella of the Southern African Development Community, the initial steps towards seeking to implement a suite of programmes designed to strengthen the region's clothing industry are based on the premise that 'the region is local' (Coughlin et al, 2001). This important project involves all stakeholders in the clothing and textiles value chain and in particular concerns the large private sector producers and governments in the region.

Conclusion

This chapter sought to investigate local responses to global changes taking place in the production environment of the clothing industry. Internationally, the case study is situated in a peripheral part of the clothing economy which, Dicken (2003, p. 353) stresses, is increasingly a sector of manufacturing that is 'globally regionalised'. Currently, in global terms, the region of Southern Africa is a marginal producer of garments and lies beyond the major established global production regions. Nevertheless, it was evidenced that this peripheral region has been greatly impacted by globalisation. In particular, the clothing industry of South Africa has experienced restructuring pressures due to the country's re-insertion in the global economy. Enterprise responses to heightened competition have taken a number of different forms from organisational changes, spatial restructuring, expanded subcontracting and informalisation. Increasing recognition is given to the fact that the long-term future of the clothing industry in South Africa is bound up with that of the SADC region as a whole. Accordingly, stakeholders within the Southern African Development Community are beginning to take the first steps in an economic project which is targeted at strengthening the region's clothing industry on a collective basis (Coughlin et al, 2001).

In final analysis, the logic is clear for regional networking and cooperation in order to build competitiveness in the face of the window of new opportunities afforded by a changed international trade environment, most importantly of the introduction of AGOA. Nevertheless, it must be cautioned that there are many significant political and logistical hurdles that need to be overcome for the successful operationalisation of this regional project in Southern Africa designed to manage globalisation.

Acknowledgements

Thanks are due to Mrs Wendy Job for the preparation of the diagrams that accompany this chapter and to the editors for useful comments on an earlier draft of this chapter.

References

Altman, M., (1994) *An Industrial Strategy for the Clothing Sector*, University of Cape Town Press, Cape Town.

Altman, M., (1996) Labour regulation and enterprise strategies in the South African clothing industry, *Regional Studies*, 30, pp. 387-399.

Cachalia, F., Jocum, M. and Rogerson, C.M., (2004) 'The urban edge of African fashion': the evolution of Johannesburg's planned fashion district, in D. McCormick and C.M. Rogerson (eds), *Clothing and Footwear in African Industrialisation*, Africa Institute of South Africa, Pretoria, pp. 527-546.

Coughlin, P., Rubin, M. and Darga, L.A., (2001) *The SADC Textile and garment Industries: constraints and opportunities – myopia or global vision*, Unpublished report prepared for the Southern African Development Community.

Department of Trade and Industry, (2001) *Clothing and textiles sector summit 2000*, Unpublished report, South African Department of Trade and Industry, Pretoria.

Dicken, P., (2003) *Global Shift: Reshaping the Global Economic Map in the 21st Century*, Sage, London (4th Edition).

Dunne, N., (2000) *Understanding the South African clothing manufacturing sector from the perspective of leading South African clothing retailers*, Paper prepared for the Department of Trade and Industry Policy Support Programme, International Competitiveness and Value Chains in Selected Manufacturing Sectors Study, available at www.tips.org.za.

Fakude, G., (2000) *Informalisation in Kwazulu-Natal's clothing sector*, Paper prepared for the Department of Trade and Industry Policy Support Programme, International Competitiveness and Value Chains in Selected Manufacturing Sectors Study, available at www.tips.org.za.

Fakude, G., (2001) *Final report on design houses in clothing and retail value chain*, Paper prepared for the Department of Trade and Industry Policy Support Programme, International Competitiveness and Value Chains in Selected Manufacturing Sectors Study, available at www.tips.org.za.

Gereffi, G., (1994) The organisation of buyer-driven global commodity chains: how US retailers shape overseas production networks, in G. Gereffi and M. Korzeniewicz (eds), *Commodity Chains and Global Capitalism,* Greenwood Press, Westport, pp. 95-122.

Gereffi, G., (1999) International trade and industrial upgrading in the apparel commodity chain, *Journal of International Economics*, 48, pp. 37-70.

Gibbon, P., (2000) *'Back to basics' through delocalisation: the Mauritian garment industry at the end of the twentieth century*, Working Paper, Centre for Development Research, Copenhagen.

Gibbon, P., (2002) *At the cutting edge?: financialisation and UK clothing retailers' global sourcing patterns and practices*, Unpublished paper presented at the International conference on Clothing and Footwear in African industrialization, Mombasa, Kenya, 5-6 July.

Gibbon, P., (2004) South Africa and the global commodity chain for clothing: export performance and constraints, in D. McCormick and C.M. Rogerson (eds), *Clothing and Footwear in African Industrialisation*, Africa Institute of South Africa, Pretoria, pp. 148-192.

Harrison, K. and Dunne, N., (1998) *KwaZulu-Natal's Clothing Industry: The Importance of Firm-Level and Pipeline Competitiveness Issues*, Research Report No. 14, Centre for Social and Development Studies, University of Natal, Durban.

Hart, G., (2002) *Disabling Globalization: Places of Power in Post-Apartheid South Africa*, University of Natal Press, Pietermaritzburg.

Horton, C., (1999) Made in SADC: sold in South Africa, *South African Labour Bulletin*, 23 (1), pp. 47-50.

House, K. and Williams, A.J., (2000) *The status of employment in the South African clothing sector – diverting a 'race to the bottom'*, Unpublished paper presented at the 2000 Annual Forum of the Trade and Industrial Policy Secretariat, Muldersdrift, 18-20 September.

Jafta, R. and Jeetah, R., (2001) *South Africa*, Unpublished paper prepared for the Southern African Development Community SADC Study of the Textile and Garment Industries, SADC, Gaborone.

Jeetah, R. and Coughlin, P., (2001) *Mauritius*, Unpublished paper prepared for the Southern African Development Community SADC Study of the Textile and Garment Industries, SADC, Gaborone.

Kamaha, A.P., (2004) South Africa's clothing industry: the case of successful SMMEs in Johannesburg, in D. McCormick and C.M. Rogerson (eds), *Clothing and Footwear in African Industrialisation*, Africa Institute of South Africa, Pretoria, in press.

Kaplinsky, R. and Morris, M., (1998) Trade policy reform and the competitive response in KwaZulu-Natal province, South Africa, *World Development* 26, pp. 717-737.

Lemon, A. and Rogerson, C.M. (eds), (2002) *Geography and Economy in South Africa and Its Neighbours*, Ashgate, London.

Manning, C., (1993) *Informal Manufacturing in the South African Economy*, Working Paper No. 11, Centre for Social and Development Studies, University of Natal, Durban.

McCormick, D., (1999) *Institutions and industrial development: a case study of garments and textiles in South Africa*, Unpublished paper presented at the Workshop on African Business Systems, Nairobi, 2 December.

Mosoetsa, S., (2000) South Africa: moving into the fourth world?, *South African Labour Bulletin*, 24 (1), pp. 10-15.

Nadvi, K. and Thorburn, J., (2003) Vietnam in the global and textile value chain: impacts on firms and workers, *Journal of International Development*, in press.

Nattrass, N., (2000) Wage strategies and minimum wages in decentralized regions: the case of the clothing industry in Phuthaditjhaba, South Africa, *International Journal of Urban and Regional Research*, 24, pp. 873-888.

Nkoko, M. A., (2000) 'We must forget our borders': regionalisation and regional trade union structures, *South African Labour Bulletin*, 24 (5), pp. 30-33.

October, L., (1996) *A Study of the Cape Clothing Industry*, Development Policy Research Unit Industrial Strategy Project: Phase 2, Working Paper 2, University of Cape Town.

Otiso, K.M., (2004) The African Growth and Opportunity Act and the textile and footwear industry in Africa, in D. McCormick and C.M. Rogerson (eds), *Clothing and Footwear in African Industrialisation*, Africa Institute of South Africa, Pretoria, pp. 78-107.

Pape, J., (2000) *Gender and Globalisation in South Africa: Some Preliminary Reflections on Working Women and Poverty*, Occasional Paper, ILRIG, Cape Town.

Peberdy S. and Rogerson, C.M., (2000): Transnationalism and non-South African entrepreneurs in South Africa's small, medium and micro-enterprise (SMME) economy, *Canadian Journal of African Studies*, 34, pp. 20-40.

Phalatse, M.R., (2000) From industrialization to de-industrialization in the former South African homelands: the case of Mogwase, North-West Province, *Urban Forum*, 11, pp. 149-161.

Rogerson, C.M., (1987) Decentralization and the location of Third World multinationals in South Africa, in R. Tomlinson and M. Addleson (eds), *Regional Restructuring Under Apartheid: Urban and Regional Policies in Contemporary South Africa*, Ravan, Johannesburg, pp. 294-308.

Rogerson, C.M., (2000) Successful SMEs in South Africa: the case of clothing producers in the Witwatersrand, *Development Southern Africa,* 17, pp. 712-740.

Rogerson, C.M., (2001) Inner-city economic revitalization through cluster support: the Johannesburg clothing industry, *Urban Forum*, 12, pp. 49-70.

Rogerson, C.M., (2002) Spatial Development Initiatives in South Africa: elements, evolution and evaluation, *Geography*, 87, pp. 38-48.

Rogerson, C.M., (2004) Sunrise or sunset industries?: South Africa's clothing and footwear sectors, in D. McCormick and C.M. Rogerson (eds), *Clothing and Footwear in African Industrialisation*, Africa Institute of South Africa, Pretoria, pp. 110-147.

Rubin, M., (2001) *Lesotho*, Unpublished paper prepared for the Southern African Development Community SADC Study of the Textile and Garment Industries, SADC, Gaborone.

Rubin, M. and Coughlin, P., (2002) *AGOA Update*, Unpublished paper EconPolicy Research Group, Maputo.

Skinner, C. and Valodia, I., (2004) Informalisation and the reconfiguration of labour relations in the Durban clothing industry, in D. McCormick and C.M. Rogerson (eds), *Clothing and Footwear in African Industrialisation*, Africa Institute of South Africa, Pretoria, pp. 508-526.

Stern, M. and Neshitomboni, N., (2002) *Africa Growth and Opportunity Act: hot air or hot stuff*, Available at www.tips.org.za.

Todes, A., (1999) Industrial restructuring in South Africa: the case of Newcastle, *Tijdschrift voor Economische en Sociale Geografie*, 90, pp. 379-390.

Williams, A.J. and Hanival, S., (1999) Clothing and textiles: ensuring survival of the South African industry, *South African Labour Bulletin*, 23 (6), pp. 71-76.

Index

For Product Safety Concerns and Information please contact our EU
representative GPSR@taylorandfrancis.com
Taylor & Francis Verlag GmbH, Kaufingerstraße 24, 80331 München, Germany

www.ingramcontent.com/pod-product-compliance
Ingram Content Group UK Ltd.
Pitfield, Milton Keynes, MK11 3LW, UK
UKHW022216190926
475702UK00003B/392
9780815390244